Optical Hall Effect in the Sharp Focus of Laser Light

Victor V. Kotlyar · Alexey A. Kovalev · Anton G. Nalimov

Optical Hall Effect in the Sharp Focus of Laser Light

Springer

Victor V. Kotlyar
Image Processing Systems Institute
National Research Center
Kurchatov Institute
Samara, Russia

Alexey A. Kovalev
Department of Technical Cybernetics
Samara National Research University
Samara, Russia

Anton G. Nalimov
Image Processing Systems Institute
National Research Center
Kurchatov Institute
Samara, Russia

ISBN 978-3-031-64682-9 ISBN 978-3-031-64683-6 (eBook)
https://doi.org/10.1007/978-3-031-64683-6

This Springer imprint is published by the registered company Springer Nature Switzerland AG
The registered company address is: Gewerbestrasse 11, 6330 Cham, Switzerland

Preface

Since ancient times, mankind has been able to focus light with the help of lenses and mirrors. At the focus, not only is the energy of light concentrated, but all six projections of the electromagnetic field vectors are added at the focus, forming complex three-dimensional distributions of amplitude, phase, and polarization states. Recently, many interesting optical effects have been discovered in a sharp focus of coherent laser radiation: rotation of polarization vectors only in the longitudinal plane—photon wheels; the presence of points and lines of polarization singularity, in which the direction of the linear polarization vector is not determined; the reverse flow of light energy, when the longitudinal projection of the Poynting vector is directed in the opposite direction with respect to the direction of propagation of the focused light; spin-orbit conversion, when a transverse energy flux is formed at the focus of a Gaussian beam with circular polarization, which can rotate a microparticle along a circular trajectory. Another interesting effect that has recently been discovered in the focus of laser light is the optical Hall effect. The usual Hall effect consists in a transverse displacement, in different directions, in a magnetic field of charges of different signs or different spins, which carry an electric current. In optics, particles with different spins correspond to light beams with left and right circular polarizations. Therefore, the optical spin Hall effect in focus consists in the formation of regions separated in space, in which light has a different direction of elliptical or circular polarization.

The book will be of interest to a wide range of scientists, engineers working in the field of optics, photonics, laser physics, optoinformation technologies, and optical instrumentation. It can also be useful for bachelors and masters in the specialties applied mathematics and physics, applied mathematics and informatics, optics and graduate students specializing in these areas.

Samara, Russia

Victor V. Kotlyar
Alexey A. Kovalev
Anton G. Nalimov

Introduction

The Hall effect consists in the occurrence of a transverse voltage (Hall voltage) in a metal conductor at the edges of a sample placed in a transverse magnetic field when a current flows perpendicular to the field. Hall voltage was discovered by Edwin Hall in 1879, and the effect is named after him. Due to the many types of Hall effects, for clarity, the original effect is sometimes referred to as the ordinary Hall effect to distinguish it from other types that may have additional physical mechanisms. In semiconductors, the Hall effect leads to separation of electrons and holes in space. Note that electrons and holes have different spins. In the absence of a magnetic field in non-magnetic conductors, current carriers with opposite directions of spins can be deflected in different directions perpendicular to the electric field. This phenomenon, called the spin Hall effect, was theoretically predicted by Dyakonov and Perel in 1971. There are external and internal spin effects. The first of them is associated with spin-dependent scattering, and the second, with spin-orbit interaction. The spin Hall effect is closely related to another interesting effect, the Magnus effect. The Magnus effect was discovered by Heinrich Magnus in 1853 and occurs when a liquid or gas flows around a rotating body. In this case, a force acting perpendicular to the flow acts on the body. This phenomenon is often used in sports, for example, a dry leaf football kick, as well as in a twisted serve in table tennis (a spinning tennis ball deviates from a straight line). In optics, an effect similar to the Magnus effect was discovered in multimode fibers in 1990 by Zel'dovich B. Ya. It was shown that the vortex modes of a fiber with left and right circular polarization propagate at different angles to the optical axis of the fiber. In 1992, A. V. Volyar discovered a similar Magnus effect in uniaxial crystals. The spin Hall effect in optics was discovered later. In 2004, Onoda M. et al. theoretically showed that when reflected from the interface between two media, linearly polarized light is divided into two beams with left and right circular polarization, propagating at different angles to the surface. In 2005, Kavokin A. experimentally observed the optical spin Hall effect during the passage of light with linear polarization through a multilayer structure. In the transmitted light, the regions with left and right circular polarizations were separated in space. The theory of optical effects of Magnus and Hall based on the geometric phase of Berry and spin-orbit interaction was developed in 2004 by Bliokh K. Y. The geometric phase of Berry

(1984) as applied to light with circular polarization consists in the fact that light with circular polarization, passing the same distance in a gradient medium with light with linear polarization, acquires an additional phase incursion. Also, between the light with left circular and right circular polarizations, an additional phase delay can occur. This leads to the fact that in a gradient (or other inhomogeneous) medium, light with left and right circular polarizations can propagate along different paths and be separated in space. The question arises: Can the optical Hall effect manifest itself not due to the interaction of light with the medium, but when propagating in free space, for example, in a sharp focus? In 1992, Allen L. et al showed that an individual photon has an orbital angular momentum. And in 2011, Bliokh K. et al. showed that spin-orbit conversion takes place in a sharp focus of light. That is, even at the focus, one can find the conditions under which the Hall effect occurs.

This book is devoted to the optical Hall effect in the sharp focus of laser radiation. On the basis of the theory of Richards-Wolf (1959), which adequately describes the behavior of light at the focus, many specific examples of light fields show that both spin and orbital Hall effects take place near the focus. In this case, there can be many regions with left and right circular polarization at the focus. Their number depends on some parameter of the focused beam. The spin Hall effect is when local regions are formed in the plane of focus (or near it), in some of which the light has a right elliptical polarization, and in others it has a left one. In this case, in the initial plane, the light had linear polarization at each point. The orbital Hall effect appears when there are local areas in the plane of focus, in some of which the transverse energy flow rotates clockwise, and in others—counterclockwise. In the simplest case, the spin Hall effect occurs when a linearly polarized Gaussian beam is focused. In this case, four local regions are formed near the focus (before and after, but not in the focus itself), in which the light has a left-hand elliptical polarization along one diagonal, and a right-hand polarization along the other diagonal. The spin and orbital Hall effects also arise at the focus of light fields with non-uniform linear polarization, when at each point in the beam cross section the polarization is linear, but changes its direction from point to point. For such cylindrical vector fields, under certain conditions, the Hall effect occurs at the focus itself, and for other beam parameters, near the focus. For such beams, the presence or absence of the Hall effect at the focus is associated with the absence or presence of polarization singularity points in the initial light field.

The authors are grateful for the numerical simulation by Ph.D. Stafeev S. S. and Ph.D. Kozlova E. S.

The results included in the monograph were supported by the Russian Science Foundation grant 23-12-00236.

Contents

About the Authors

Victor V. Kotlyar is a head of Laboratory at the Image Processing Systems Institute of National Research Center, Kurchatov Institute, Samara, Russia and professor of Computer Science department at Samara National Research University. He received his MS, Ph.D., and Dr.Sc. degrees in Physics and Mathematics from Samara State University (1979), Saratov State University (1988) and Moscow Central Design Institute of Unique Instrumentation, the Russian Academy of Sciences (1992). He is co-author of 400 scientific papers, five books and seven inventions. e-mail: kotlyar@ipsiras.ru

Alexey A. Kovalev (b. 1979) graduated (2002) from Samara National Research University, majoring in Applied Mathematics. He received his Doctor in Physics and Maths degree in 2012. He is a senior researcher of Laser Measurements laboratory at Image Processing Systems Institute National Research Center, Kurchatov Institute, Samara, Russia and professor of Computer Science department at Samara National Research University. He is a co-author of more than 200 scientific papers. e-mail: alanko@ipsiras.ru

Anton G. Nalimov (b. 1980) graduated from Samara State Aerospace University in February, 2003. Entered in postgraduate study in 2003, finished it in 2006 with speciality 01.04.05 "Optics". A. G. Nalimov works on Technical Cybernetics department in Samara National Research University as an associate professor, works as a scientist in the Image Processing Systems Institute of National Research Center, Kurchatov Institute, Samara, Russia. Candidate in Physics and Mathematics, co-author of 100 papers and three inventions. e-mail: anton@ipsiras.ru

Chapter 1
Spin Hall Effect at the Focus for Light with Linear Polarization

1.1 Circular Polarization Near the Tight Focus of Linearly Polarized Light

Sharp focusing of laser radiation is understood as the focusing of light by lenses with a high numerical aperture, and it is no longer possible to neglect the vector nature of the light wave. In this case, to calculate the light field at the focus, it is necessary to take into account all the components of the strength of the electric and magnetic field of the light wave. The classical formulas for calculating the light field in a sharp focus were obtained by Richards and Wolf in [1].

At present, a large number of works are devoted to the sharp focusing of light. However, most of the works are devoted to studying the behavior of the intensity at the focus, for example, obtaining focal spots of various shapes [2–7]. Much less work is presented on the study of other characteristics of the light field, such as the energy flux (Poynting vector) [8–10], spin or orbital angular momentum [11–14]. We also note that the main attention of researchers is focused on the study of the behavior of light directly in focus; less attention is paid to the behavior of light at some distance from the plane of sharp focus.

In this section, the sharp focusing of linearly polarized light is considered. It was shown that, with distance from the focal plane, regions arise in which the polarization ceases to be linear. In this case, when passing through the plane of focus, the direction of polarization in these regions changes to the opposite—in regions with right circular polarization, the direction changes to left circular and vice versa.

1.1.1 Theoretical Background

In [1], expressions were obtained for the projections of the electric field strength vector at the focus of the aplanatic system. The Jones vector for an initial field with

V. V. Kotlyar et al., *Optical Hall Effect in the Sharp Focus of Laser Light*,
https://doi.org/10.1007/978-3-031-64683-6_1

linear polarization directed along the y-axis has the form:

$$\boldsymbol{E}_{\text{lin}} = A(\theta)\begin{pmatrix} 0 \\ 1 \end{pmatrix} \tag{1.1}$$

and the projections of the vector of the electric field strength and magnetic field strength near the focus for the initial field (1.1) have the form:

$$\begin{aligned} E_x &= -iI_{2,2}\sin 2\varphi, \\ E_y &= -i\big(I_{0,0} - I_{2,2}\cos 2\varphi\big), \\ E_z &= -2I_{1,1}\sin\varphi, \\ H_x &= i\big(I_{0,0} + I_{2,2}\cos 2\varphi\big), \\ H_y &= iI_{2,2}\sin 2\varphi, \\ H_z &= 2I_{1,1}\cos\varphi, \end{aligned} \tag{1.2}$$

where

$$\begin{aligned} I_{\nu,\mu} = \left(\frac{4\pi f}{\lambda}\right)\int\limits_0^{\theta_0} \sin^{\nu+1}\left(\frac{\theta}{2}\right)\cos^{3-\nu}\left(\frac{\theta}{2}\right) \\ \cos^{1/2}(\theta)A(\theta)\mathrm{e}^{ikz\cos\theta}J_\mu(x)\mathrm{d}\theta \end{aligned} \tag{1.3}$$

where λ is the wavelength of light, f is the focal length of the aplanatic system, $x = kr\sin\theta$, $J\mu(x)$ is the Bessel function of the first kind, and $\text{NA} = \sin\theta_0$ is the numerical aperture. The angle φ in Eq. (1.2) is the conventional polar (or azimuthal) angle in the transverse planes, including the focal plane. A positive angle value increases counterclockwise from the horizontal x-axis. In the initial plane, the light field has only linear polarization directed along the vertical y-axis, and the Jones vector (1) does not depend on the polar angle φ. In Eqs. (1.2) and (1.3), angle θ is the tilt angle of the rays to the optical axis, θ_0 is the maximal tilt angle, determining the numerical aperture NA, z is the direction of the optical axis, $z = 0$ is the focal plane, k is the wavenumber of light, (x, y) are the Cartesian coordinates in the cross sections of the light beam converging into the focus (x is the horizontal axis, y is the vertical axis). The initial amplitude function $A(\theta)$ (suppose it is a real function) can be constant (plane wave) or in the form of a Gaussian beam. From (1.2), one can obtained the intensity distributions of each component of the electric vector

$$\begin{aligned} I_x &= I_{2,2}^2\sin^2(2\varphi), \\ I_y &= I_{0,0}^2 + I_{2,2}^2\cos^2(2\varphi) - 2I_{0,0}I_{2,2}\cos(2\varphi), \\ I_z &= 4I_{1,1}^2\sin^2(\varphi). \end{aligned} \tag{1.4}$$

We note that formulae (1.1)–(1.4) differ from the formulae obtained in [1], since the initial field (1.1) is polarized along the *y*-axis, whereas in [1] the initial field was polarized along the *x*-axis. Despite the initial light field (1.1) has only one component E_y, Maxwell's equations indicate that, upon light propagation, all three components of the E-field appear. If the light field propagates at a small angle to the optical axis, then the other two field components (E_x and E_z) are small and can be neglected. At tight focusing, light propagates at large angles to the optical axis, so that all three components of the E-field (1.2) have a comparable value [15, 16]. It can be seen from (1.1) that the intensity distribution I_x of the horizontal projection of the electric vector in the plane of focus will have the form of four local maxima (light spots), the centers of which are located on a circle centered on the optical axis and lying on the rays emanating from the center at angles $\varphi = \pi/4$, $3\pi/4$, $5\pi/4$, $7\pi/4$.

The intensity distribution I_y will have the form of an almost circular spot with a maximum on the optical axis $I_y = I_{0,0}^2$. The difference from the round shape of the spot arises from the fact that the distribution of intensity I_y along the vertical axis ($\varphi = \pi/2$) will be greater ($I_y = \left(I_{0,0} + I_{2,2}\right)^2$) than along the horizontal axis ($\varphi = 0$, $I_y = \left(I_{0,0} - I_{2,2}\right)^2$). Intensity distribution (1.4) at the focus of the longitudinal component of the electric vector I_z will have the form of two light spots, the centers of which lie on the vertical axis. This type of intensity distribution of electric vector individual components leads to the fact that the distribution of the total intensity at the focus has the form of an ellipse elongated along the vertical axis:

$$I = I_x + I_y + I_z = I_{0,0}^2 + I_{2,2}^2 + 2I_{1,1}^2 - 2\left(I_{1,1}^2 + I_{0,0}I_{2,2}\right)\cos(2\varphi). \tag{1.5}$$

Let us find the longitudinal component of the spin angular momentum (SAM) vector near the field focus (1.1) using the formula [17]:

$$\boldsymbol{S} = \left(\frac{c^2\varepsilon_0}{2\omega}\right)\mathrm{Im}\left(\boldsymbol{E}^* \times \boldsymbol{E}\right) \tag{1.6}$$

where *c* is the speed of light in vacuum, ω is the angular frequency of the monochromatic light, ε_0 is the vacuum permittivity, Im is the imaginary part of the number, × is the sign of vector multiplication, * is the sign of complex conjugation. Below, we omit the constant $[(c^2\varepsilon_0)/(2\omega)]$ for brevity. We note that sometimes, due to the electric–magnetic democracy, Eq. (1.6) is written with two terms rather than one: $[c^2/(2\omega)]\mathrm{Im}\left[\varepsilon_0(\boldsymbol{E}^* \times \boldsymbol{E}) + \mu_0(\boldsymbol{H}^* \times \boldsymbol{H})\right]$, with μ_0 being the vacuum permeability ($c^2\varepsilon_0 = \mu_0^{-1}$). However, immediately from the expression for the Poynting vector, only one term is obtained either for the E-vector or for the H-vector [17]. In addition, due to different constants, both terms will give different contribution to the components of the SAM vector. Thus, expression (1.6) is correct. Substituting from (1.2) into (1.6), we will assume that integrals (1.3) are complex, since *z* is different from zero. We get:

$$S_z = 2\mathrm{Im}\left(E_x^* E_y\right) = 2\sin(2\varphi)\mathrm{Im}\left(I_{2,2}^*\left(I_{0,0} - \cos(2\varphi)I_{2,2}\right)\right) \tag{1.7}$$

Certainly, near the tight focus, all 6 components of the E- and H-vectors (1.2) are significant, and none of these components can be neglected. Therefore, similarly to Eq. (1.7), we can write expressions for the components S_x and S_y:

$$S_x = 2\mathrm{Im}\left(E_y^* E_z\right) = 4\sin(\varphi)\mathrm{Re}\left(I_{1,1}\left(\cos(2\varphi)I_{2,2}^* - I_{0,0}^*\right)\right),$$
$$S_y = 2\mathrm{Im}\left(E_z^* E_x\right) = 4\sin(\varphi)\sin(2\varphi)\mathrm{Re}\left(I_{1,1}^* I_{2,2}\right). \quad (1.8)$$

Let us single out the real and imaginary parts of the integrals included in (1.7) $I_{0,0} = R_0 + iI_0$, $I_{2,2} = R_2 + iI_2$. Then, instead of (1.7), we write:

$$S_z = 2\sin(2\varphi)(I_0 R_2 - I_2 R_0) \quad (1.9)$$

The integrals R_0, R_2 in (1.9) include the co-multiplier $\cos(kz\cos\theta) \approx 1$ at $kz \ll 1$, and the integrals I_0 , I_2 include the co-multiplier $\sin(kz\cos\theta) \approx kz\cos\theta$ at $kz \ll 1$. With this in mind, instead of (1.9), we write:

$$S_z \approx 2kz\sin(2\varphi)\left(\overline{I}_0\overline{R}_2 - \overline{I}_2\overline{R}_0\right). \quad (1.10)$$

In (1.10), the following notations are used:

$$\overline{R}_0 = I_{0,0}(z=0),\ \overline{R}_2 = I_{2,2}(z=0),$$
$$\overline{I}_0 = \overline{I}_{0,0}(z=0),\ \overline{I}_2 = \overline{I}_{2,2}(z=0),$$
$$\overline{I}_{\nu,\mu} = \left(\frac{4\pi f}{\lambda}\right)\int_0^{\theta_0} \sin^{\nu+1}\left(\frac{\theta}{2}\right)\cos^{3-\nu}\left(\frac{\theta}{2}\right)\cos^{3/2}(\theta)A(\theta)\mathrm{e}^{ikz\cos\theta}J_\mu(x)\mathrm{d}\theta. \quad (1.11)$$

Let on a circle of some radius the expression in parentheses in (1.10) be greater than zero $\overline{I}_0\overline{R}_2 - \overline{I}_2\overline{R}_0 > 0$, and since $\sin(2\varphi)$ in (1.10) is positive in 1 and 3 quadrants, and negative in 2 and 4, then before the focus ($z < 0$) the longitudinal component SAM S_z in (1.10) will be positive in 2 and 4 quadrants, and negative in 1 and 3. And since the sign of the entire expression after focus ($z > 0$) will change to the opposite, the longitudinal component of SAM S_z in (1.9) will be positive in 1 and 3 quadrants, and negative in 2 and 4. This means that before the focus in the 2 and 4 quadrants the polarization vector will rotate counterclockwise (right circular or elliptical polarization), and after focus in these quadrants, the polarization vector will rotate clockwise (left circular or elliptical polarization). Recall that in the plane of focus, the light at each point has only linear polarization, since at $z = 0$ the longitudinal component of the SAM S_z in (1.10) is equal to zero. The defocusing magnitude z in Eq. (1.10) affects the size of the areas in the transverse plane, where polarization is not linear. At a distance z nearly equal to λ, the size of the circular polarization area is maximal (for NA = 0.95 it is approximately $\lambda/2$). As z tends to zero (i.e., in the focus), the size of the area with circular polarization decreases to zero.

Note also that the longitudinal component of the SAM is exactly equal to the third component of the Stokes vector:

$$S_z = 2\mathrm{Im}\big(E_x^* E_y\big) = s_3, \tag{1.12}$$

which shows the presence of circular and elliptical polarization in the light field. In the next section, the presented theoretical predictions will be confirmed by simulation. We note that the change in the rotation direction of the polarization vector to the opposite beyond the focal plane, as follows from Eq. (1.10), can be explained by the angular momentum (AM) conservation law. Since polarization in the initial plane and in the focal plane is locally linear, $S_z = 0$. Therefore, if there are areas with left-handed circular polarization before the focus, then beyond the focus, circular polarization in these areas should become right-handed. However, the presence of such areas near the focus does not follow from the AM conservation.

1.1.2 Simulation by Richards-Wolf Formula

In this work, using the Richards-Wolf formulas, focusing of a linearly polarized plane wave (wavelength 633 nm) was simulated by choosing a lens with NA = 0.95. The field near the tight focus was calculated using the integrals [1]:

$$\begin{aligned} \boldsymbol{U}(\rho, \psi, z) = -\frac{if}{\lambda} \int_0^{\theta_0} \int_0^{2\pi} & B(\theta, \varphi) T(\theta) \boldsymbol{P}(\theta, \varphi) \\ & \times \exp\{ik[\rho \sin\theta \cos(\varphi - \psi) + z \cos\vartheta]\} \\ & \sin\theta d\theta d\varphi, \end{aligned} \tag{1.13}$$

where $\mathbf{U}(\rho, \psi, z)$ is the strength of the electric or magnetic field, $B\,(\vartheta, \varphi)$ is the electric or magnetic field at the input of the wide-aperture system in coordinates of the exit pupil (θ is the polar angle, φ is the azimuthal angle), $T\,(\theta)$ is the lens apodization function, f is the focal length, $k = 2\pi/\lambda$ is the wavenumber, λ is the wavelength (in the simulation it was considered equal to 633 nm), θ_0 is the maximum polar angle determined by the numerical aperture of the lens (NA = $\sin\theta_0$), $\mathbf{P}\,(\theta, \varphi)$ is the polarization vector, for the strength of the electric and magnetic fields has the form:

$$\begin{aligned} \boldsymbol{P}(\theta, \varphi) = & \begin{bmatrix} 1 + \cos^2\varphi(\cos\theta - 1) \\ \sin\varphi\cos\varphi(\cos\theta - 1) \\ -\sin\theta\cos\varphi \end{bmatrix} a(\theta, \varphi) \\ & + \begin{bmatrix} \sin\varphi\cos\varphi(\cos\theta - 1) \\ 1 + \sin^2\varphi(\cos\theta - 1) \\ -\sin\theta\sin\varphi \end{bmatrix} b(\theta, \varphi), \end{aligned} \tag{1.14}$$

where a (θ, φ) and b (θ, φ) are functions describing the polarization state of the x- and y-components intensities of the focused beam. In contrast to formulae (1.2) and (1.3), we gave Eqs. (1.13) and (1.14) in a general form to show that further modeling is carried out by the general formulae (1.13), (1.14) and that the simulation results confirm the theoretical conclusions, following from the expressions (1.11), (1.12). After calculating the components of the electric field, the behavior of the components of the Stokes vector near the sharp focus was calculated. The Stokes vector components are calculated using the formulas:

$$\begin{aligned} s_0 &= E_x E_x^* + E_y E_y^*, \\ s_1 &= E_x E_x^* - E_y E_y^*, \\ s_2 &= 2\mathrm{Re}\big(E_x^* E_y\big), \\ s_3 &= 2\mathrm{Im}\big(E_x^* E_y\big). \end{aligned} \tag{1.15}$$

Similarly to the expressions (1.7)–(1.9), substitution of Eq. (1.2) into Eq. (1.15) allows obtaining explicit expressions for the Stokes components s_1 and s_2 near the focus. For instance, more simple expression is derived for s_2 at $kz \ll 1$:

$$s_2 \approx 2\sin(2\varphi)\overline{R}_2\big(\overline{R}_0 - \overline{R}_2\cos(2\varphi)\big). \tag{1.16}$$

At small $kz \ll 1$, the second Stokes component (1.16) does not depend on z and therefore does not change sign when passing through the focus ($z = 0$). Below, this is confirmed by simulation. Similarly, the first Stokes component s_1 in Eq. (1.15), expressed via the components of the E-vector (1.2), is also independent of z near the focus.

To estimate the relative contribution of individual polarization components, it is convenient to use the Stokes vector components normalized to the transverse intensity: $(S_1, S_2, S_3) = \big(s_1/s_0, s_2/s_0, s_3/s_0\big)$. It is known that when focusing light of linear polarization at the focus, all three components of the electric field strength are observed [18]. Figure 1.1 illustrates the distribution of the total intensity and its individual components in the focus of an aplanatic lens with NA = 0.95 when focusing a plane wave with a wavelength of 633 nm and polarization along the y-axis. To estimate the effect of defocusing, Fig. 1.2 shows the same distributions of the total intensity and of the individual intensity components as in Fig. 1.1, but at a distance λ from the focal plane. The intensity distributions have the same shape at the same distance before and after the focus.

Figure 1.1 shows that the initial component makes the main contribution to the focal spot formation, but the longitudinal component of the intensity also begins to make a significant contribution. The component perpendicular to the input polarization is rather small but present, while the light at the focus is still linearly polarized. Note that the distributions of the total intensity at the focus and the intensity of individual components in Fig. 1.1 confirm the theoretical predictions that follow from

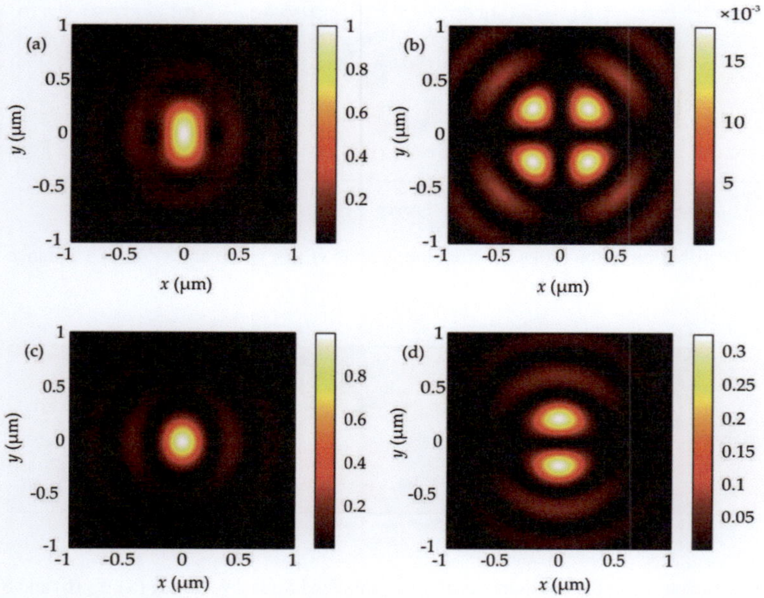

Fig. 1.1 Distribution of the total intensity $I_x + I_y + I_z$ (**a**) and individual components of the intensity I_x (**b**), I_y (**c**), I_z (**d**) in the plane of focus

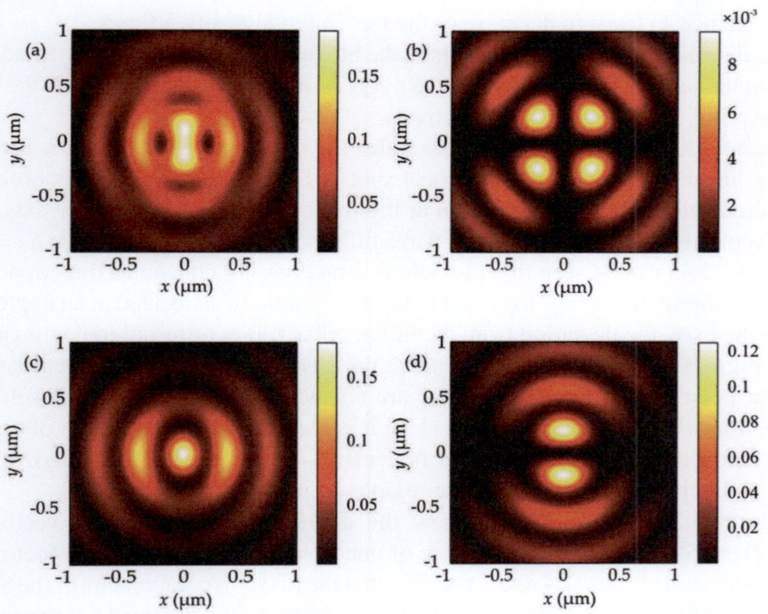

Fig. 1.2 Distribution of the total intensity $I_x + I_y + I_z$ (**a**) and individual components of the intensity I_x (**b**), I_y (**c**), I_z (**d**) at a distance λ after the focus

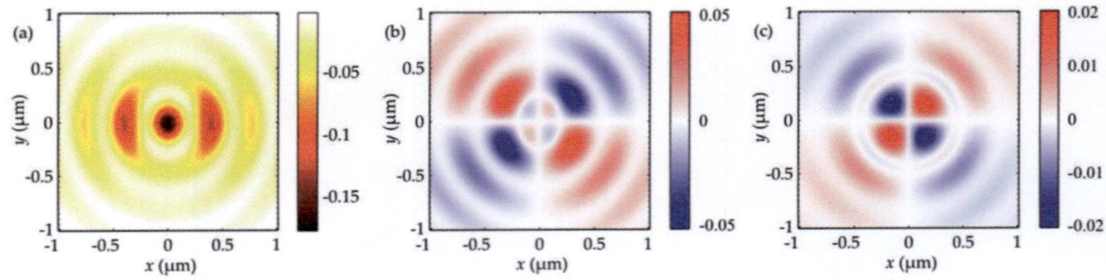

Fig. 1.3 Distribution of the Stokes vector components s_1 (**a**), s_2 (**b**) and s_3 (**c**) at a distance λ after the focus

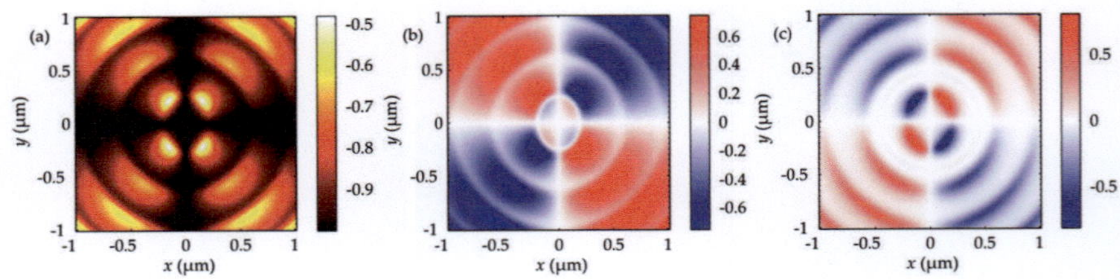

Fig. 1.4 Distribution of the components of the normalized Stokes vector S_1 (**a**), S_2 (**b**) and S_3 (**c**) at a distance λ after the focus

expressions (1.4) and (1.5). Figure 1.2 indicates that a small shift from the focal plane (by a distance λ) leads to decrease of the maximum intensity 5 times.

The distribution of the components of the Stokes vector (s_1, s_2, s_3) and the normalized components of the Stokes vector (S_1, S_2, S_3) at the distance $z = \lambda$ after the focus is shown in Figs. 1.3 and 1.4, respectively.

From Figs. 1.3 and 1.4, it can be seen that the polarization after focus is predominantly linear. In the center of the focal spot in Fig. 1.3a, a minimum is observed, which indicates that the polarization at the focus is directed along the y-axis. This is also confirmed by Fig. 1.4a: for a wave fully polarized along the y-axis $S_1 = -1$. From Fig. 1.4a can be seen that the polarization does not change its direction at the focus and along the x and y axes, but along the straight lines located at an angle of ± 45° to the axes, the deviation from the initial polarization turns out to be maximum. From Figs. 1.3 and 1.4, it is also seen that the diverging beam contains regions with circular polarization. Recall that there are no such regions at the focus itself—the light is linearly polarized. From Fig. 1.4c it is seen that the contribution of circular polarization in such regions is quite noticeable—for $S_3 = \pm 1$ the polarization is completely circular, but here in some regions S_3 reaches values of ± 0.8.

Figures 1.5and 1.6 similarly show the distribution of the Stokes vector and normalized Stokes vector at a distance of one wavelength in front of the focus.

Comparison of Figs. 1.4 and 1.6 shows that the first two components of the Stokes vector describing linear polarization have not changed, and the third has changed its sign to the opposite. After passing the plane of the focus, the direction of circular (elliptical) polarization is reversed—for example, in the first quarter, the light in front

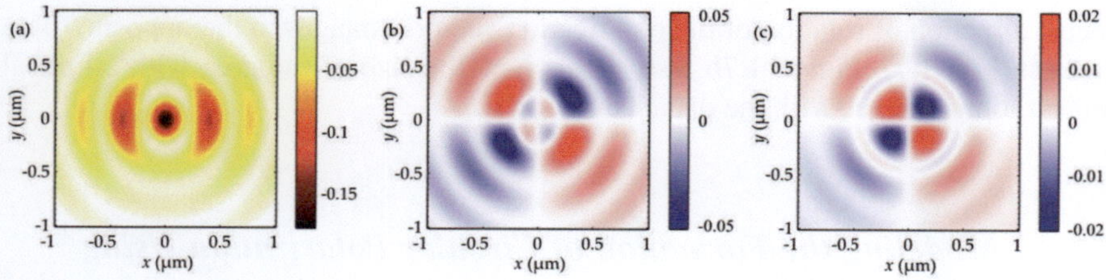

Fig. 1.5 Stokes vector components s_1 (**a**), s_2 (**b**), and s_3 (**c**) at a distance λ before the focal plane

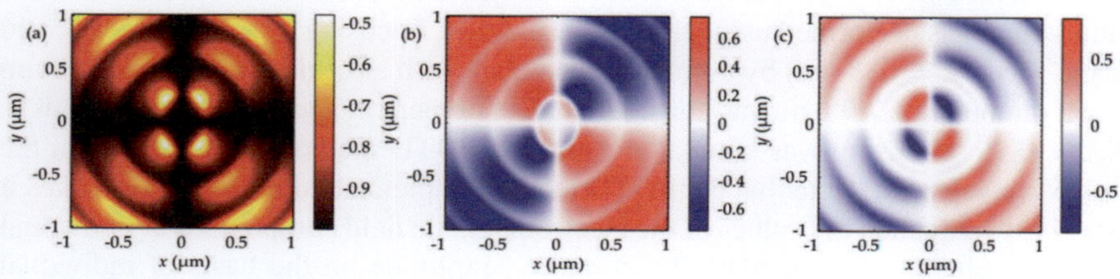

Fig. 1.6 Distribution of the components of the normalized Stokes vector S_1 (**a**), S_2 (**b**) and S_3 (**c**) at a distance λ before the focal plane

of the focus plane was with left circular polarization, and after focus—with right polarization. Before the focus, the right circular (elliptical) polarization appears in the 2nd and 4th quadrants and the left circular polarization appears in the 1st and 3rd quadrants (Fig. 1.6c). It agrees with the theoretical prediction based on expression (1.10). And the change in the direction of rotation of the polarization vector in these quadrants after passing through the focus also follows from (1.10).

Below we show how the distribution of S_3 changes with the distance from the focal plane. Figure 1.7 shows the intensity distribution (Fig. 1.7a) and the longitudinal Stokes component S_3 (Fig. 1.7b) in the longitudinal plane *yz* along the *z*-axis, rotated by an angle $\varphi = 45°$ (i.e., passing through the S_3 maximum in Fig. 1.6).

Figure 1.7 demonstrates that in the focal plane, the light field is linearly polarized. However, directly beyond the focal plane, areas with elliptical polarization are generated (red areas in Fig. 1.7). It is also interesting that as we move away from the

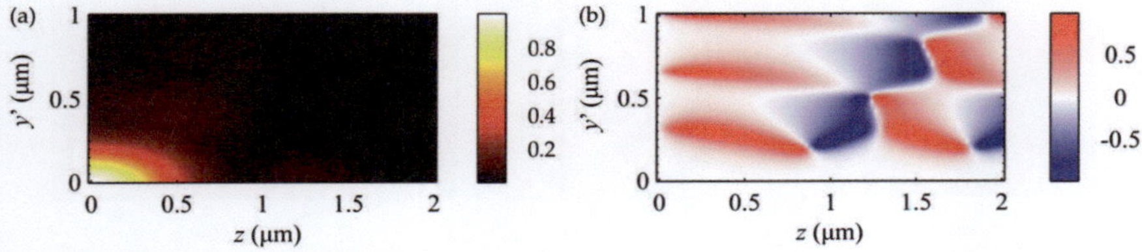

Fig. 1.7 Distributions of the intensity and of the third Stokes component in the longitudinal plane *yz* along the *z*-axis (by an angle 45°)

focus, direction of rotation of the polarization vectors changes to the opposite (blue areas in Fig. 1.7). Figure 1.7b also shows how the size of the area with elliptical polarization changes with the distance z.

1.1.3 Modeling the Formation of Circular Polarization Using the FDTD Method

To check the correctness of calculations by the Richards-Wolf formulas, an additional simulation was performed using the FDTD method. Focusing of a linearly polarized plane wave ($\lambda = 633$ nm) by a Fresnel zone plate with a focal length of $f = 500$ nm and a diameter of 7.9 μm was considered. The numerical aperture of such a lens is NA = 0.99. Focusing was simulated using the FDTD method implemented in the FullWave software. Note that the FDTD method implemented in FullWave makes it possible to calculate the values of the electromagnetic field components at individual moments of time. To calculate the complex amplitude on the basis of individual instantaneous values of the field amplitudes, the method proposed in [19] was used. Figure 1.8 shows the distribution of the components of the normalized Stokes vector at a distance of one wavelength after the focus.

From Fig. 1.8, it can be seen that simulating using the FDTD method confirms the results obtained using the Richards-Wolf formulas. In particular, Fig. 1.8a shows that light is predominantly linearly polarized along the y-axis, and Fig. 1.8c shows that quadrants 1 and 3 contain right-handed circular polarization, and quadrants 2 and 4—left.

Comparison of Figs. 1.4 and 1.8 indicates that although the structures of both patterns are similar, there are also significant differences. This is because the simulations by the Richards-Wolf method [1] and by the FDTD method [19] were carried out under different conditions. In the latter case, tight focusing of light was simulated by passing the light field through a real Fresnel zone plate with a focal length equal to the wavelength ($f = \lambda$) and with a numerical aperture NA = 0.99. At the same time, the Richards-Wolf formalism adequately describes the light field at the focus of an ideal spherical lens if $f \gg \lambda$. Thus, the Richards-Wolf formalism approximately

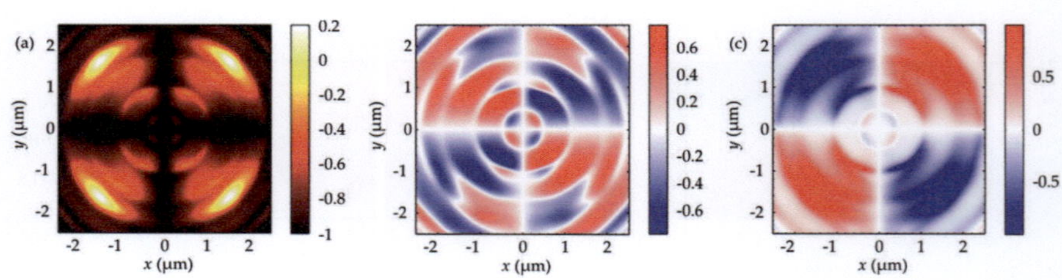

Fig. 1.8 Components of the Stokes vector S_1 (**a**), S_2 (**b**) and S_3 (**c**) when calculating using the FullWAVE software at a distance of $z = 0.65$ μm after the actual focus

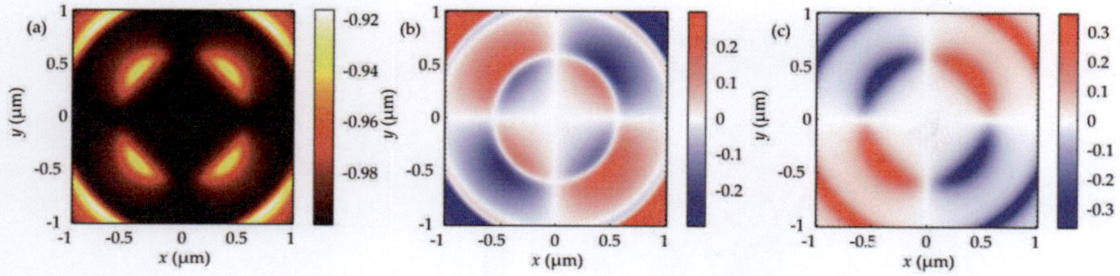

Fig. 1.9 Distribution of the components of the normalized Stokes vector S_1 (**a**), S_2 (**b**) and S_3 (**c**) for a lens with a numerical aperture NA = 0.6

describes the behavior of light near the focus, whereas the FDTD method, based on a rigorous solution of the Maxwell equations, adequately describes the behavior of light at the focus near the surface of the focusing zone plate. Therefore, modeling by the FDTD method expands the boundaries of the discovered optical phenomenon: generation of local areas with circular (elliptical) polarization near the tight focus of light with initially linear polarization.

1.1.4 Reducing the Contribution of Circular Polarization with Decreasing Numerical Aperture of the Lens

Let us now consider the contribution of reducing the numerical aperture of the lens to NA = 0.6 (corresponding to a standard 40 × aplanatic lens). The result is shown in Fig. 1.9. Figure 1.9 shows that the maximum S_3 has decreased by 2 times. And from Fig. 1.9a, it can be seen that the relative contribution of linear polarization (along the y-axis) increased significantly: the maximum in Fig. 1.4 was equal to − 0.5, and in Fig. 1.9a to − 0.92. Recall that for $S_1 = \pm 1$, the polarization is completely linear.

1.1.5 Calculation of the Moment of Forces Acting on a Dielectric Microparticle Near the Focus

Let us calculate a force and a torque, acting onto a microbed from the light field. The force **F** and the torque **M** relative to an arbitrary point A, are equal to [20, 21].

$$\boldsymbol{F} = -\oint_S (\sigma \cdot \boldsymbol{n})\mathrm{d}S \tag{1.17}$$

$$\boldsymbol{M} = \oint_S [\boldsymbol{r} \times (\sigma \cdot \boldsymbol{n})]\mathrm{d}S \tag{1.18}$$

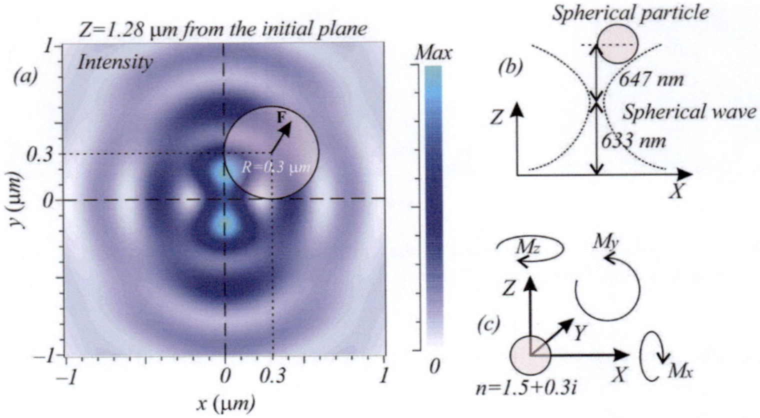

Fig. 1.10 Intensity pattern (**a**) and a spherical bed with radius $R = 0.3$ μm. The position of the bed is $x_p = 0.3$ μm, $y_p = 0.3$ μm. Shown at the right are a schematic position of the bed (**b**) and directions of the positive torques values along *x, y* and *z* axis (**c**)

where **r** is the radius-vector from the point $A(x, y, z)$ to the point of integration on the surface *S*, **n** is an external normal vector to the surface *S*, *A* is the point relative to which the torque **M** is calculated, and σ is the Maxwell stress tensor, the components of which in the CGS system can be written [22]

$$\sigma_{ik} = \frac{1}{4\pi}\left(\frac{|\boldsymbol{E}|^2 + |\boldsymbol{H}|^2}{2}\delta_{ik} - E_i E_k - H_i H_k\right) \tag{1.19}$$

where E_i, H_i are the electric and magnetic field components and δ_{ik} is the Kronecker symbol ($\delta_{i=k} = 1$, $\delta_{i\neq k} = 0$).

Shown in Fig. 1.10 is a simulation result of the torque and force calculation acting on the spherical microbed.

Calculations show that for the position of the particle $x_p = 0.3$ μm, $y_p = 0.3$ μm the force projections are $F_x = 2.79$ pN, $F_y = 3.7$ pN, $F_z = 8.78$ pN. The torque projections are $M_x = 2.81 \cdot 10^{-19}$ Nm, $M_y = -5.55 \times 10^{-19}$, $M_z = 1.73 \times 10^{-19}$ Nm. If shift the bed at the position $x_p = 0.3$ μm, $y_p = -0.3$ μm, then the result force projections will be $F_x = 2.66$ pN, $F_y = -3.58$ pN, $F_z = 8.9$ pN, and the torque projections will be $M_x = -3.0 \times 10^{-19}$ Nm, $M_y = -5.9 \times 10^{-19}$, $M_z = -1.5 \times 10^{-19}$ Nm. Figure 1.10 shows that in the first quadrant the axial moment of forces is positive ($M_z = 1.73 \times 10^{-19}$), and in the fourth quadrant the moment of forces is negative ($M_z = -1.5 \times 10^{-19}$). This proves that the longitudinal projection of the SAM is positive in the first quadrant and negative in the fourth (Figs. 1.8 and 1.9).

In this section, theoretically, using the Richards-Wolf formalism and using two different modeling methods, it was shown that with sharp focusing of light with linear polarization in the planes before and after the focus, there are regions that arise in pairs in even and odd quadrants, and in which light is circularly or elliptically polarized (e.g., even to the right and to the odd to the left) [22]. Moreover, after passing through the focus in these areas, the direction of rotation of the polarization vector changes to the opposite (in even quadrants, it is now left-handed, and in odd quadrants, it is right-handed circular or elliptical polarization). This result allows the use of linearly polarized light to rotate microparticles (the size of the circularly polarized region is about 0.6 μm by 0.6 μm) around its center of mass. We note that a similar result has been obtained in [23]. It has been shown that certain structures allow generating before the focus and beyond the focus two conjugate optical vortices with opposite-sign topological charges and with longitudinal axial polarization. In our work, we have not used any additional structures.

1.2 Focusing a Vortex Laser Beam with Polarization Conversion

When strongly focusing a circularly polarized Gaussian beam, a near-focus orbital energy flow has been generated thanks to spin-orbital conversion [24–30]. In the original plane, such a beam has no orbital angular momentum (OAM), only having a non-zero on-axis projection of the spin angular momentum (SAM) vector thanks to circular polarization. However, a non-zero longitudinal component of the electric vector that occurs in the strong focus leads to the generation of a transverse energy flow, which produces the non-zero longitudinal OAM component. Behavioral patterns of SAM and OAM in the tight focus of optical vortices were studied in [31–35]. On the other hand, there have been publications concerned with a reverse energy flow in the tight focus of optical vortices [1, 10, 36, 37] and some laser beams, like vector X-waves [38], non-paraxial Airy beams [39], Weber beams [40], vector Bessel beams [41], and fractional Bessel vortex beams [42].

In this section, using Richards-Wolf formulae, we derive analytical relationships to describe projections of the Poynting vector (the energy flow) and the SAM vector when tightly focusing a linearly polarized optical vortex with the topological charge 2. In the original plane, all SAM vector components of such a beam are zero, but they all become non-zero near the strong focus. This can be explained by the effect inverse to the spin-orbital conversion. Thus, in the case under study, thanks to the orbital-spin conversion, the original linearly polarized vortex beam generates a circularly polarized vortex beam in the tight focus. It is important to mention that a vortex beam with the topological charge (TC) $m = 2$ has a specific feature—that of generating an on-axis reverse energy flow in the tight focus (characterized by the negative longitudinal projection of the Poynting vector). Besides, there will be non-zero on-axis intensity. At any $m > 2$, except for $m = 1$ and $m = 2$, both the on-axis intensity of light and energy flow are zero.

We note that for spin-orbital coupling to occur the beam needs to propagate in a medium and because of this throughout the text below, we use the notion of spin-orbital conversion. Thanks to the beam rays converging to the focus, there appears a non-zero longitudinal projection of the electric field vector that, combined with the transverse components, produces a transverse energy flow (although the original energy flow has only a longitudinal component), which, in turn, produces a longitudinal projection of the OAM vector. In the focus, two transverse projections of the electric field vector have a relative phase shift of $\pi/2$, generating a circularly polarized beam, which, in turn, generates the longitudinal component of the SAM vector.

1.2.1 Energy Flow and SAM in the Strong Focus

Previously, relationships to describe projections of the electric and magnetic fields in the vicinity of the tight focus of an original linearly polarized optical vortex with an arbitrary integer TC m have been derived [29]. In this case, a near-axis reverse energy flow in the focus was shown to occur at any $m \geq 2$. However, with the reverse energy flow being maximal on the optical axis only at $m = 2$, below, we look into focusing a linearly polarized optical vortex with TC $m = 2$. Based on the Richards-Wolf theory [1], it is possible to derive projections of the electric field vector in the tight focus of an aplanatic optical system. If the original light field is given by

$$\boldsymbol{E} = A(\theta)\mathrm{e}^{i2\varphi}\begin{pmatrix}1\\0\end{pmatrix}, \boldsymbol{H} = A(\theta)\mathrm{e}^{i2\varphi}\begin{pmatrix}0\\1\end{pmatrix} \quad (1.20)$$

where **E** and **H** are the electric and magnetic field of Jones vectors, projections of the electric field vector in the focal plane will be given by [29]

$$\begin{aligned} E_x &= i\mathrm{e}^{2i\varphi}\left(I_{0,2} + \frac{1}{\sqrt{2}}\mathrm{e}^{2i\varphi}I_{2,4} + \frac{1}{\sqrt{2}}\mathrm{e}^{-2i\varphi}I_{2,0}\right), \\ E_y &= -\mathrm{e}^{2i\varphi}\left(-\frac{1}{\sqrt{2}}\mathrm{e}^{2i\varphi}I_{2,4} + \frac{1}{\sqrt{2}}\mathrm{e}^{-2i\varphi}I_{2,0}\right), \\ E_z &= 2\mathrm{e}^{2i\varphi}\left(\frac{1}{\sqrt{2}}\mathrm{e}^{i\varphi}I_{1,3} - \frac{1}{\sqrt{2}}\mathrm{e}^{-i\varphi}I_{1,1}\right), \end{aligned} \quad (1.21)$$

where

$$I_{\nu,\mu} = \left(\frac{4\pi f}{\lambda}\right)\int_0^{\theta_0} \sin^{\nu+1}\left(\frac{\theta}{2}\right)\cos^{3-\nu}\left(\frac{\theta}{2}\right)\cos^{1/2}(\theta)A(\theta)\mathrm{e}^{ikz\cos\theta}J_\mu(x)\mathrm{d}\theta, \quad (1.22)$$

and λ is the incident wavelength, f is the focal length of the aplanatic system, $x = kr\sin\theta$, $J_\mu(x)$ is the first-kind Bessel function, and $NA = \sin\theta_0$ is the numerical aperture. Assuming the initial amplitude $A(\theta)$ to be a real function, it can be given by a constant (plane wave) or a Gaussian beam

$$A(\theta) = \exp\left(\frac{-\gamma^2 \sin^2\theta}{\sin^2\theta_0}\right) \quad (1.23)$$

where γ is constant. We seek to derive projections of the SAM vector

$$\boldsymbol{S} = \frac{1}{2}\mathrm{Im}\left[\boldsymbol{E}^* \times \boldsymbol{E}\right] \quad (1.24)$$

where Im is the imaginary part of the number, $\mathbf{E}^*$ denotes complex conjugation of the electric field vector, ×—vector multiplication sign. Substituting (1.21) into (1.24) yields expressions for projections of the SAM vector in the focal plane ($z = 0$) for an initially linearly polarized optical vortex ($m = 2$), Eq. (1.20):

$$\begin{aligned} S_x &= \left(I_{1,1}I_{2,0} - I_{1,3}I_{2,4}\right)\sin\varphi + \left(I_{1,1}I_{2,4} - I_{1,3}I_{2,0}\right)\sin 3\varphi, \\ S_y &= -\left(I_{1,1}I_{2,0} - I_{1,3}I_{2,4}\right)\cos\varphi - \\ &\quad - \left(I_{1,1}I_{2,4} - I_{1,3}I_{2,0}\right)\cos 3\varphi - \sqrt{2}I_{2,0}\left(I_{1,1} - I_{1,3}\right)\cos\varphi, \\ S_z &= \frac{1}{2}\left(I_{2,0} - I_{2,4}\right)\left(I_{2,0} + I_{2,4} + \sqrt{2}I_{0,2}\cos 2\varphi\right). \end{aligned} \quad (1.25)$$

From Eq. (1.25), the on-axis longitudinal projection of the SAM vector is seen to be non-zero and positive:

$$S_z(r = z = 0) = \frac{1}{2}I_{2,0}^2. \quad (1.26)$$

This means that the light wave in the focus near the optical axis has right-handed circular polarization (electric vector rotates anticlockwise). From the last equation in (1.25) the light is also seen to be inhomogeneously polarized in the focal plane. For instance, light will be linearly polarized on the radii where the inequality $I_{2,0} = I_{2,4}$ holds, because $S_z = 0$. Meanwhile, in the regions where $S_z < 0$, the light wave will be left-handed circularly polarized. Along the rays in the focal plane outgoing from the center at angles φ: $\pi/4$, $3\pi/4$, $5\pi/4$, and $7\pi/4$, alternating polarization states will occur: being right-handed circular at $I_{2,0}^2 > I_{2,4}^2$, linear at $I_{2,0}^2 = I_{2,4}^2$, and left-handed circular at $I_{2,0}^2 < I_{2,4}^2$. From the first two equations of Eq. (1.25), it is seen that at $\varphi = \pi n$, $n = 1, 2, \ldots$ $S_x = 0$ and at $\varphi = \pi/2 + \pi n$, $n = 1, 2, \ldots$ $S_y = 0$. This means that in the longitudinal planes *yz* and *xz*, light is circularly (or elliptically) polarized near the strong focus.

Next, let us consider expressions for projections of the Poynting vector (energy flow) $\boldsymbol{P} = \frac{1}{2}\mathrm{Re}\left[\boldsymbol{E}^* \times \boldsymbol{H}\right]$ in the focal plane when focusing an optical vortex ($m = 2$) with linear original polarization (1.20):

$$\begin{aligned} P_x &= -Q(r)\sin\varphi, \\ P_x &= Q(r)\cos\varphi, \end{aligned}$$

$$P_z = \frac{1}{2}\left(I_{0,2}^2 - I_{2,4}^2 - I_{2,0}^2\right),$$
$$Q(r) = \frac{1}{2}\left[I_{1,3}\left(I_{0,2} + I_{2,4}\right) + I_{1,1}\left(I_{0,2} + I_{2,0}\right)\right]. \quad (1.27)$$

From the comparison of (1.25) and (1.27), the focal-plane distribution of the SAM vector is seen to be radially asymmetric because the original light is linearly polarized, whereas the distribution of the longitudinal projection of the energy flow is circularly symmetric. We note that, as is the case with the SAM distribution, the intensity distribution in the focal plane is also radially asymmetric:

$$\begin{aligned} I &= |E_x|^2 + \left|E_y\right|^2 + |E_z|^2 = \\ &= I_{2,0}^2 + I_{0,2}^2 + I_{2,4}^2 + 2I_{1,1}^2 + 2I_{1,3}^2 - \\ &\quad - 2I_{1,1}I_{1,3}\cos\varphi + \sqrt{2}I_{0,2}\left(I_{2,0} + I_{2,4}\right)\cos 2\varphi \end{aligned} \quad (1.28)$$

being non-zero on the optical axis: $(I(r = 0) = I_{2,0}^2)$. From Eq. (1.27), it follows that the on-axis energy flow equals in magnitude the longitudinal projection of the SAM vector in Eq. (1.26), but opposite in sign:

$$P_z(r = z = 0) = -S_z(r = z = 0) = -\frac{1}{2}I_{2,0}^2. \quad (1.29)$$

Hence, we infer that the on-axis energy flow is negative in the focus (being directed oppositely to the incident light beam). Equation (1.27) also suggests that the transverse energy flow rotates anticlockwise around the optical axis:

$$P_r = 0, P_\varphi = Q(r). \quad (1.30)$$

Directly on the optical axis, the transverse energy flow is zero $(P_\varphi(r = 0) = Q(0) = 0)$. Hence, we can infer that in the focal plane near the optical axis both the transverse energy flow and the polarization vector rotate anticlockwise. However, as distinct from the zero on-axis value of the transverse energy flow of (1.30), the longitudinal component of the SAM of (1.26) takes the maximum positive value on the axis. If an absorbing microsphere centered on the optical axis is placed in the focus, the longitudinal SAM projection may be expected to make the microsphere rotate about the optical axis anticlockwise [43].

Below, we demonstrate that if TC of the linearly polarized optical axis changes sign $(m = -2)$, the on-axis reverse energy flow in the focus will still occur, whereas the longitudinal SAM component will become negative. Actually, instead of Eq. (1.21), projections of the electric field in the focus for the initial linearly polarized optical vortex $(m = -2)$ will be given by

$$E_x = i\mathrm{e}^{-2i\varphi}\left(I_{0,2} + \frac{1}{\sqrt{2}}\mathrm{e}^{2i\varphi}I_{2,4} + \frac{1}{\sqrt{2}}\mathrm{e}^{-2i\varphi}I_{2,0}\right),$$

$$E_y = -e^{-2i\varphi}\left(-\frac{1}{\sqrt{2}}e^{2i\varphi}I_{2,4} + \frac{1}{\sqrt{2}}e^{-2i\varphi}I_{2,0}\right),$$
$$E_z = 2e^{-2i\varphi}\left(\frac{1}{\sqrt{2}}e^{i\varphi}I_{1,3} - \frac{1}{\sqrt{2}}e^{-i\varphi}I_{1,1}\right). \quad (1.31)$$

Making use of Eq. (1.31), the longitudinal components of the Poynting vector and the SAM vector can be defined as

$$P_z = \frac{1}{2}\left(I_{0,2}^2 - I_{2,4}^2 - I_{2,0}^2\right),$$
$$S_z = -\frac{1}{2}\left(I_{2,0} - I_{2,4}\right)\left(I_{2,0} + I_{2,4} + \sqrt{2}I_{0,2}\cos 2\varphi\right). \quad (1.32)$$

The comparison of (1.25), (1.27), and (1.31) suggests that when the sign of the optical vortex is reversed, the longitudinal component of the energy flow vector remains unchanged and the longitudinal SAM component only changes its sign. This means that a reverse on-axis energy flow and a negative on-axis projection of the SAM vector (left-handed circular polarization) occurs in the focus:

$$P_z(r = z = 0) = S_z(r = z = 0) = -\frac{1}{2}I_{2,0}^2. \quad (1.33)$$

Thus, thanks to the effect of orbital-spin conversion, in the focal plane of a strongly focused linearly polarized optical vortex is observed an on-axis reverse energy flow and the right-handed ($m = 2$) or left-handed ($m = -2$) circular polarization of light. Notably, the on-axis polarization vector and the transverse energy flow have the same handedness: anticlockwise ($m = 2$) or clockwise ($m = -2$).

The above-described reasoning can be shortly summarized as follows. In the initial plane, there is a linearly polarized optical vortex, which has only a longitudinal component of OAM (with the electric field vector having only components in the original plane (Eq. 1.20), with all SAM vector projections being equal to zero. The angular momentum vector is known to be composed of the sum of OAM and SAM vectors and conserve upon free-space propagation [27, 28, 33]. Therefore, thanks to all projections of the SAM vector becoming non-zero in the tight focus, all projections of the OAM vector in the focus may also be expected to become non-zero.

1.2.2 Numerical Simulation

In the numerical simulation, an incident beam of wavelength $\lambda = 633$ *nm* is passed through an 8-μm aperture. The field is described by a linearly polarized plane wave of unit intensity and a phase vortex with $m = 2$ (Fig. 1.11), multiplied by a spherical wave converging at a distance of $f = 1.31$ μm (NA = 0.95):

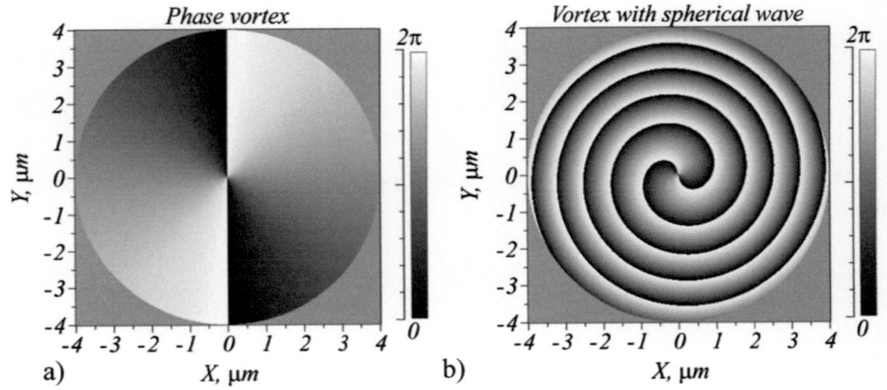

Fig. 1.11 **a** A phase vortex with $m = 2$ embedded into an incident field and, **b** the resulting phase of the initial field containing a spherical wave

$$\begin{aligned} E_x &= \exp(i(kf - m\varphi - \omega t)), \\ E_y &= 0. \end{aligned} \tag{1.34}$$

Assuming the above-defined original field, the intensity pattern depicted in Fig. 1.12 is observed in the focal plane. The FDTD-aided numerical simulation was conducted using the FullWave software. The calculation was conducted for λ/30 mesh along all three axes, with the initial field containing 601 × 601 pixels. By taking an odd number of pixels we ensured symmetry of the field relative to the origin.

Figure 1.12 suggests that in the strong focus of a linearly polarized wave, there is a 0.3 μm × 0.27 μm region (on the x- and y-axis, respectively) where the projection of the Poynting vector on the optical axis is negative. From Fig. 1.12e, it can also be seen that in this region a gradual linear-to-circular on-axis polarization conversion occurs, with the **E**-vector rotating anticlockwise on and near the axis. On the edges of this central region, there are two regions where the **E**-vector rotates clockwise: $x = 0$, $-0.2\ \mu\text{m} < y < -0.1\ \mu\text{m}$ and $0.1\ \mu\text{m} < y < 0.2\ \mu\text{m}$. Besides, in the distribution pattern of the Poynting vector, there is a ring of approximate radius 0.66 μm (e.g., $0.622\ \mu\text{m} < x < 0.693\ \mu\text{m}$ at $y = 0$) in which the Poynting vector projection is also negative (Fig. 1.12d). Within the ring, linear-to-circular polarization conversion also occurs, with the **E**-vector rotating clockwise, as seen from Fig. 1.12e.

Fig. 1.13 depicts a two-dimensional distribution of the SAM vector projection S_z (Eq. (1.24)) alongside its x- and y-axis profiles on the optical axis. The z-axis projection of the SAM vector is seen to be non-zero and positive on the optical axis within the region $(0.267\ \mu\text{m} < x < 0.267\ \mu\text{m})$, $(-0.13\ \mu\text{m} < y < 0.13\ \mu\text{m})$.

An absorbing nanoparticle placed in the field depicted in Fig. 1.12 may be expected to experience a torque making it rotate around its center of mass thanks to circular polarization. Shown in Fig. 1.14 is the result of calculating the projection of a torque M_z exerted on a 0.3-μm nanoparticle (with refractive index $n = 1.5 + 0.3i$) relative

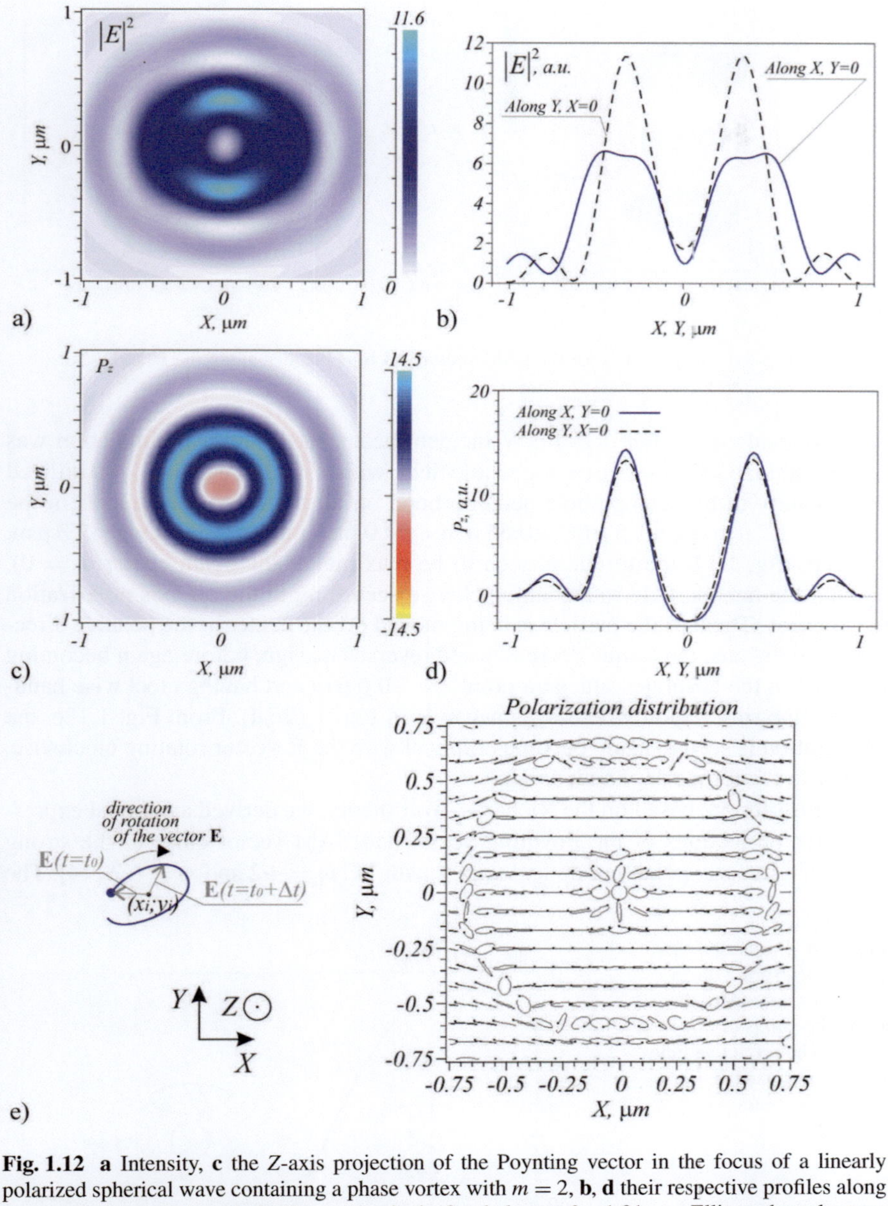

Fig. 1.12 **a** Intensity, **c** the Z-axis projection of the Poynting vector in the focus of a linearly polarized spherical wave containing a phase vortex with $m = 2$, **b**, **d** their respective profiles along the x- and y-axes, and **e** polarization pattern in the focal plane at $f = 1.31$ μm. Ellipse-shaped curves in Fig. 1.12e depict trajectories of the ends of the rotating polarization vectors, with their origins found at the 'ellipse center'. Bold dots mark begins of each single rotation periods. The z-axis is directed towards the reader

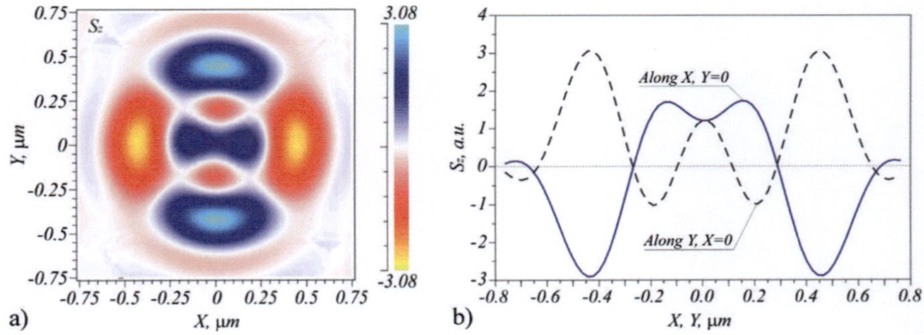

Fig. 1.13 **a** Z-axis projection S_z of the SAM vector and **b** its profiles along the *x*- and *y*-axes

to the optical *z*-axis. For a 100-mW incident beam, the numerical simulation was conducted on a λ/30 mesh over the whole field, with a smaller mesh of λ/80 utilized immediately in the nanoparticle neighborhood on all three coordinate axes in the region $-1.5\ \mu m < x < 1.5\ \mu m$, $-0.65\ \mu m < y < 0.65\ \mu m$, and $0.3\ \mu m < z < 2.3\ \mu m$.

From Fig. 1.14, the torque is seen to be maximal at the beam center ($x = 0$), making the nanoparticle rotate anticlockwise according to the on-axis polarization handedness. Then, as the particle is being moved off the center in the positive direction of the *x*-axis, the torque decreases and reverses its sign, before again becoming maximal in the absolute value at a point $x = \sim 0.6\ \mu m$ and having clockwise handedness (the ring region where S_z is negative in Fig. 1.12c,d). From Fig. 1.12e, the polarization is seen to again become elliptical with the **E**-vector rotating clockwise, similar to the torque M_z handedness.

In this section, based on the Richards-Wolf theory, we derived analytical expressions for projections of the Poynting vector and SAM vector close to the strong focus of a linearly polarized optical vortex with TC $m = +2$ and $m = -2$ [44]. The

Fig. 1.14 Projection Mz of the torque exerted on an off-center microparticle along the *x*-axis ($x = y = 0$, $z = f$). The refractive index of the microparticle—$n = 1.5 + 0.3i$, beam power is 100 mW

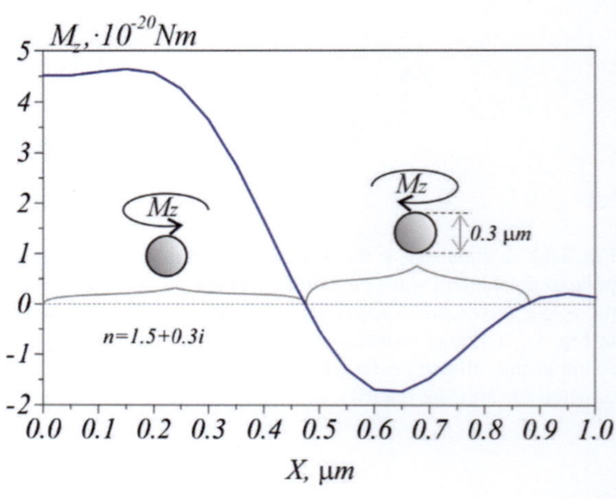

relations derived suggest that near the optical axis in the focal plane there occurs a reverse energy flow. In the reverse energy flow region, the right- (for $m = 2$) or left-handed ($m = -2$) polarization of light is observed. In the meantime, inhomogeneous polarization is found at the periphery of the focal plane, where polarization is alternating from right-handed circular (or elliptical) to left-handed circular (or elliptical) and to linear. At the same time, close to the optical axis, the handedness of the polarization vector rotation coincides with that of the transverse energy flow rotation: with the both rotating anticlockwise at $m = 2$ and clockwise at $m = -2$. The generation of circular polarization in the reverse energy flow region enables the latter to be detected. Actually, if an absorbing nanoparticle is placed near the focus in the reverse energy flow region, the SAM will make it rotate around its center of mass. With the on-axis transverse energy flow being zero, it will hardly contribute to the nanoparticle rotation. The handedness of the nanoparticle rotation will be defined by the TC sign of vortex beam.

1.3 Hall Effect at the Focus of an Optical Vortex with Linear Polarization

In 1909, Poynting [45] predicted that left-handed circularly polarized light has a spin angular momentum (SAM), or in short, a spin of -1, and right-handed circularly polarized light has a spin of $+1$. More precisely, he predicted that each photon could have a spin equal to Planck's constant: either $-\hbar$ or $\hbar$. In 1936, Beth [46] proved this experimentally by showing that when linearly polarized light passes through a quarter-wave plate, the plate acquires a torque. In 1992, Allen showed [47] that light, including each photon with a vortex phase described by the angular harmonic $\exp(in\varphi)$, has an orbital angular momentum (OAM) $\hbar n$, where n is the topological charge (TC). In the paraxial case, SAM and OAM are independent and are preserved separately during light propagation in free space. However, spin–orbit conversion (SOC) can occur when light is sharply focused near the focus [33]. At present, a lot of papers are devoted to the study of SAM, OAM, and SOC [48]. Paper [48] is a small review on SOC in a tight focus of structured light. In [49], the tight focusing of radially polarized light was studied. In this work, it is shown that the intensity of the longitudinal light component at the focus increases with increasing numerical aperture, and becomes equal to the intensity of the transverse component at a unit numerical aperture. The Hall effect [50] was investigated at the focus of an optical vortex with radially polarization [51]. It was shown in [50] that for tight focusing of an optical vortex with radial polarization the SAM is positive at the focus near the optical axis if the TC of the vortex is $+1$, and the SAM is negative if the TC of the vortex is -1. This is the so-called catalyst-like effect. In [51], the authors showed that when focusing an optical vortex with radial polarization the longitudinal projection of the SAM vector has different signs at different distances from the optical axis in the focal plane. This is the radial spin Hall effect. In [52], the 3D SAM was studied in the tight focus of an optical vortex with linear polarization. In this study, the force vector was calculated that will act on an ellipsoidal particle that is in focus. The tight focusing of an optical vortex with azimuthal polarization was observed in [13]. It was shown in [13] that when the optical vortex TC sign changes near the optical axis in the focus, the sign of the SAM longitudinal projection also changes, and therefore the particle placed in the focus changes the rotation direction around its own axis and around the optical axis. In [53], the angular momentum (AM) in a sharp focus of hybrid cylindrical vector beams was studied. It was shown in [53] that, for such light fields the longitudinal component of SAM is equal to zero at the focus. The orbital motion of microparticles in a tight focus of optical vortices with circular and radially polarization was investigated in [54]. In [55], SOC was considered in non-paraxial beams with hybrid polarization. It was shown in [55] that when light with a high-order hybrid polarization is tightly focused, regions are formed in the focus in which the longitudinal components of the OAM and the SAM change signs. That is, the spin and orbital Hall effects take place. Beams with hybrid polarization in a tight focus were observed in [56]. It was shown in [56] that when focusing light whose polarization changes only along the radius, the polarization in the focal plane will also change along the radius. Linear polarization and elliptical polarization will alternate, but of the same sign. In [57], the paraxial focusing of Bessel beams with circular polarization was studied. It was shown in [57] that the sign of the angular momentum vector will be different on different sides of the light intensity ring in the beam. The tight focusing of high-order Poincare beams was considered in [58]. In our recent works, we investigated the Hall effect in the tight focus of high-order cylindrical vector beams [59], beams with hybrid inhomogeneous polarization [60], Poincaré beams [61], and optical vortices with circular polarization [62]. Another version of the Hall effect in a sharp focus appears when the center of a vortex laser beam gravity is shifted at the case of its limitation by a diaphragm [63]. The Hall effect in the tight focus of an optical vortex with linear polarization has not been considered before.

We note that the spin Hall effect arises not only in a tight focus, but also when light is scattered by inhomogeneous structures. Thus, it was shown theoretically in [64] and experimentally in [65], that when a laser beam with linear polarization is reflected from a microresonator with Bragg mirrors, four regions with circular polarization of different signs are formed in the beam. And in [66], it was experimentally shown that the scattering of a Hermite-Gaussian ($HG_{0,1}$) beam with linear polarization on a silver nanowire (AgNW) also gives the spin Hall effect. It was shown in [14] that, due to the SOC gold particles placed in the tight focus of the Laguerre-Gaussian vortex beam ($LG_{0,1}$) rotate at different speeds for light with left and right circular polarization.

In this section, we consider the tight focusing of an optical vortex with an integer TC and linear polarization. Using the Richards-Wolf theory [1], which accurately describes light in the vicinity of a tight focus of coherent light, exact analytical expressions are obtained for the longitudinal components of the SAM, OAM, and AM vectors in the focal plane for an optical vortex with linear polarization. It is shown that the longitudinal SAM and OAM components, averaged over the beam cross section, are preserved in the initial and the focal planes. It is also demonstrated that, there is a separation of regions with different signs of the SAM longitudinal component and regions with different signs of the AM longitudinal component at the focus. It is found that the AM and the SAM values are independent and sufficient to describe the light at the focus, while the meaning of the OAM value at the focus is not clear since the AM is not the sum of the SAM and OAM. However, it is easy to prove the conservation of the OAM value, while it was not possible to prove the conservation of the AM.

1.3.1 Components of the Electric and the Magnetic Fields and the Energy Flux at the Focus

Consider the initial Jones vector for an optical vortex with linear polarization:

$$\mathbf{E_n}(\varphi) = \exp(\mathbf{i}n\varphi)\begin{pmatrix}1\\0\end{pmatrix} \tag{1.35}$$

where (r, φ) are polar coordinates in the beam cross section, n is the TC and is integer, the linear polarization vector is directed along the horizontal x-axis. In [64], the electric and the magnetic fields components in the plane of tight focus for the initial field (1) were obtained:

$$\begin{aligned}
E_x &= \frac{i^{n-1}}{2}\mathrm{e}^{in\varphi}\left(2I_{0,n} + \mathrm{e}^{2i\varphi}I_{2,n+2} + \mathrm{e}^{-2i\varphi}I_{2,n-2}\right),\\
E_y &= \frac{i^{n}}{2}\mathrm{e}^{in\varphi}\left(\mathrm{e}^{-2i\varphi}I_{2,n-2} - \mathrm{e}^{2i\varphi}I_{2,n+2}\right),\\
E_z &= i^{n}\mathrm{e}^{in\varphi}\left(\mathrm{e}^{-i\varphi}I_{1,n-1} - \mathrm{e}^{i\varphi}I_{1,n+1}\right),\\
H_x &= \frac{i^{n}}{2}\mathrm{e}^{in\varphi}\left(\mathrm{e}^{-2i\varphi}I_{2,n-2} - \mathrm{e}^{2i\varphi}I_{2,n+2}\right),\\
H_y &= \frac{i^{n-1}}{2}\mathrm{e}^{in\varphi}\left(2I_{0,n} - \mathrm{e}^{2i\varphi}I_{2,n+2} - \mathrm{e}^{-2i\varphi}I_{2,n-2}\right),\\
H_z &= i^{n+1}\mathrm{e}^{in\varphi}\left(\mathrm{e}^{-i\varphi}I_{1,n-1} + \mathrm{e}^{i\varphi}I_{1,n+1}\right).
\end{aligned} \tag{1.36}$$

Formulas (1.36) include functions $I_{\nu,\mu}$ depending only on the radial variable r:

$$
\begin{aligned}
I_{\nu,\mu} = 2kf \int\limits_0^\alpha \sin^{\nu+1}\left(\frac{\theta}{2}\right) \cos^{3-\nu}\left(\frac{\theta}{2}\right) \cos^{1/2}(\theta) \times \\
\times A(\theta) \mathrm{e}^{ikz\cos\theta} J_\mu(kr\sin\theta)\mathrm{d}\theta, \qquad (1.37)
\end{aligned}
$$

where $k = 2\pi/\lambda$ is the wave number of monochromatic light with a wavelength of λ, f is the focal length of the lens, α is the maximum angle of the rays inclination to the optical axis, which determines the numerical aperture (NA) of the aplanatic lens NA $= \sin\alpha$, $J_\mu(kr\sin\theta)$ is the μ-th order Bessel function of the first kind. In Eq. (1.36) and everywhere below, the indices ν and μ will take the following values: $\nu = 0, 1, 2$; $\mu = n-2, n-1, n, n+1, n+2$. $A(\theta)$ is a real function that determines the radially symmetric initial field amplitude which depends on the inclination angle θ of the beam emanating from a point on the initial spherical front and converging to the center of the focus plane. The description of the light field at the focus using functions (1.37) was obtained for the first time in the classic study by Richards and Wolf [1]. Next, we find the components of the Poynting vector:

$$
\boldsymbol{P} = \frac{c}{2\pi}\mathrm{Re}\left(\boldsymbol{E}^* \times \boldsymbol{H}\right) \qquad (1.38)
$$

where **E** and **H** are vectors of the electric and the magnetic fields, signs "*" and "× " mean complex conjugation and vector product, Re is the real part of a complex number, and c is the speed of light in vacuum. Next, we will omit the constant $c/(2\pi)$. Substituting (1.36) into (1.38), we obtain in polar coordinates at the focus of field (1.35):

$$
\begin{aligned}
P_r &= 0, \\
P_\varphi &= Q(r), \\
P_z &= \frac{1}{2}\left(2I_{0,n}^2 - I_{2,n+2}^2 - I_{2,n-2}^2\right), \\
Q(r) &= I_{1,n+1}\left(I_{0,n} + I_{2,n+2}\right) + I_{1,n-1}\left(I_{0,n} + I_{2,n-2}\right). \qquad (1.39)
\end{aligned}
$$

It follows from (1.39) that the transverse energy flux at the focus of the field (1.35) rotates counterclockwise if $Q(r) > 0$ and clockwise if $Q(r) < 0$. The longitudinal component of the energy flow at different radii r can be positive or negative. It can be shown that the total energy of each term in P_z at the focus is equal to the expression:

$$W_{\nu,\mu} = 2\pi \int\limits_0^\infty \left|I_{\nu,\mu}(r)\right|^2 r\mathrm{d}r$$
$$= 4\pi f^2 \int\limits_0^\alpha \sin^{2\nu+1}\left(\frac{\theta}{2}\right)\cos^{5-2\nu}\left(\frac{\theta}{2}\right)|A(\theta)|^2\mathrm{d}\theta = W_\nu \tag{1.40}$$

Equation (1.40) was obtained using Eqs. (1.37) and the orthogonality of the Bessel functions:

$$\int\limits_0^\infty J_\mu(k\sin\theta r)J_\mu(k\sin\theta' r)r\mathrm{d}r = \frac{1}{k^2}\frac{\delta(\sin\theta - \sin\theta')}{\sin\theta}.$$

It can be seen from (1.40) that the energy (or power) does not depend on the order of the Bessel function μ. Applying formula (1.40) to the axial energy flow crossing the focus (1.39), we obtain:

$$\hat{P}_z = \int\limits_0^\infty r\mathrm{d}r \int\limits_0^{2\pi} d\varphi P_z(r) = W_0 - W_2$$
$$= W - 2W_1 - 2W_2. \tag{1.41}$$

In (1.41) W is the total power of the laser beam. It can be shown that the power W_0 is approximately 7 times greater than the power W_2 (it is exactly greater 7 times for $\alpha = \pi/2$ and for $|A(\theta)| \equiv 1$). Therefore, the total flow (1.41) is always positive, although the energy flux density (1.39) at different radii r can be both positive and negative (reverse energy flux [65]). Equation (1.41) shows that not all of the power W crosses the focus plane from left to right (in the positive direction of the z-axis). The part of the power $2W_1$ propagates in the direction perpendicular to the optical axis and does not cross the focus plane. The part of the power W_2 crosses the focus plane in the opposite direction. And only the part of the power W_0 flows along the positive direction of the z-axis. It is interesting that the power ratio (1.41) does not depend on the TC of the beam (1.35).

1.3.2 The Longitudinal Component of the SAM Vector at the Focus

Next, we find the axial SAM component which shows the presence of light with elliptical and circular polarization in the focus. The longitudinal SAM component is defined as follows [66]:

$$S_z = 2\mathrm{Im}\{E_x^* E_y\} \tag{1.42}$$

where Im is the imaginary part of a complex number. Substituting (1.36) into (1.42), we obtain the axial SAM component at the focus for field (1.35):

$$S_z = \frac{1}{2}(I_{2,n+2} - I_{2,n-2})(I_{2,n+2} + I_{2,n-2} + 2\cos(2\varphi)I_{0,n}) \tag{1.43}$$

It can be seen from (1.43) that if the first factor is not equal to zero then there are 4 regions in the focus plane, where SAM sign is different. Centers of these areas lie on the Cartesian axes: two regions centered on the vertical axis and two regions centered on the horizontal axis. If $I_{2,n+2} - I_{2,n-2} > 0$, then at $\varphi = 0$ and $\varphi = \pi$ the second factor is positive and $S_z > 0$, while at $\varphi = \pi/2$ and $\varphi = 3\pi/4$ the second factor in (1.43) is negative and $S_z < 0$. If, conversely, $I_{2,n+2} - I_{2,n-2} < 0$, then SAM is positive on the vertical axis and negative on the horizontal axis. The first factor is equal to zero only in the absence of an optical vortex ($n = 0$). Thus, it follows from (1.43) that the spin Hall effect takes place for the field (1.35) at the focus at $n \neq 0$. It leads to the separation of the vectors with left and right elliptical polarizations (with different spins) from each other and their localization in four regions along in pairs on the vertical and horizontal axes. Since the axial SAM component in the initial plane (1.35) is equal to zero (due to linear polarization), then the total spin at the focus must be equal to zero. Indeed, if we integrate the SAM in (1.43), we get:

$$\begin{aligned}
\hat{S}_z &= \int_0^\infty r\mathrm{d}r \int_0^{2\pi} \mathrm{d}\varphi S_z(r,\varphi) \\
&= \frac{1}{2}\int_0^\infty r\mathrm{d}r \int_0^{2\pi} \mathrm{d}\varphi\left(I_{2,n+2}^2 - I_{2,n+2}^2 - 2\cos 2\varphi I_{0,n}(I_{2,n-2} - I_{2,n+2})\right) \\
&= \frac{1}{2}(W_2 - W_2 + 0) = 0.
\end{aligned} \tag{1.44}$$

Integration over the entire focus plane of the first and second terms gives the difference between two identical energies (1.40). The third term depending on $\cos(2\varphi)$ is zero when integrated over an integer number of the angle φ periods. Since the total spin at the focus is zero, regions with different spins must appear in pairs to cancel each other out.

1.3.3 *The Intensity and the Longitudinal OAM Component at the Focus*

Next, we find the longitudinal component of the OAM vector at the focus of the field (1.35). The longitudinal OAM component is equal to the expression [62]:

$$L_z = \mathrm{Im}\left(E_x^* \frac{\partial}{\partial \varphi} E_x + E_y^* \frac{\partial}{\partial \varphi} E_y + E_z^* \frac{\partial}{\partial \varphi} E_z \right), \tag{1.45}$$

Substituting (1.36) into (1.45), we get:

$$\begin{aligned} L_z = \frac{1}{2} \{ & 2nI_{0,n}^2 + 2(n+1)I_{1,n+1}^2 + 2(n-1)I_{1,n-1}^2 \\ & + (n+2)I_{2,n+2}^2 + (n-2)I_{2,n-2}^2 + 2\cos(2\varphi) \\ & \times \left[2(n+1)I_{0,n}I_{2,n+2} + 2(n-1)I_{0,n}I_{2,n-2} - nI_{1,n-1}I_{1,n+1} \right] \}. \end{aligned} \tag{1.46}$$

Since expression (1.46) has the form $L_z = A(r) + \cos(2\varphi)B(r)$, then it has two maxima on the horizontal axis at $\varphi = 0$ and $\varphi = \pi$ and two minima on the vertical axis at $\varphi = \pi/2$ and $\varphi = 3\pi/4$. It can be shown that the main contribution is made by terms containing integrals (1.37) with zero first index, that is, expression (1.46) can be written approximately as follows:

$$\begin{aligned} L_z \approx & \, nI_{0,n}^2 + 2\cos(2\varphi)I_{0,n}\left((n+1)I_{2,n+2} + (n-1)I_{2,n-2}\right) \\ & nI_{0,n}\left(I_{0,n} + 4I_{2,n+2}\cos(2\varphi)\right). \end{aligned} \tag{1.47}$$

It can be seen from (1.47) that at $\varphi = 0$ and $\varphi = \pi$ there are regions with $L_z > 0$ in focus, and at $\varphi = \pi/2$ and $\varphi = 3\pi/2$ there are regions with $L_z < 0$. That is, there is a spatial separation of the OAM with different signs at the focus of field (1.35). Moreover, the location in the focal plane of these 4 regions with centers on the horizontal and vertical axes correlates with areas of elliptical polarization with different signs (1.43). It should be noted that in the initial plane (1.35) the OAM axial component (1.45) is equal to: $L_z = nW$, where W is the total beam power. If we integrate (1.46) over the entire focus plane, we will get that the terms containing $\cos(2\varphi)$ disappear since the integration over the angle is performed over an integer number of periods. Integration of other terms will lead to the expression:

$$
\begin{aligned}
\hat{L}_z &= \int\limits_0^\infty r\mathrm{d}r \int\limits_0^{2\pi} \mathrm{d}\varphi L_z(r,\varphi) \\
&= \frac{1}{2}\int\limits_0^\infty r\mathrm{d}r \int\limits_0^{2\pi} \mathrm{d}\varphi \big(2nI_{0,n}^2 + 2(n+1)I_{1,n+1}^2 + 2(n-1)I_{1,n-1}^2 \\
&\quad + (n+2)I_{2,n+2}^2 + (n-2)I_{2,n-2}^2\big) \\
&= nW_0 + (n+1)W_1 + (n-1)W_1 + \frac{1}{2}(n+2)W_2 \\
&\quad + \frac{1}{2}(n-2)W_2 = n(W_0 + 2W_1 + W_2) = nW.
\end{aligned} \tag{1.48}
$$

The last equality in (1.48) follows from the power balance of the entire beam and its components at the focus. The balance can be obtained by integrating the intensity distribution over the entire beam cross section. The intensity distribution at the focus follows from (1.38) and is equal to:

$$
\begin{aligned}
I = &\frac{1}{2}\big[2I_{0,n}^2 + I_{2,n+2}^2 + I_{2,n-2}^2 + 2I_{1,n+1}^2 + 2I_{1,n-1}^2 + \\
&+ 2\cos(2\varphi)\big(I_{0,n}I_{2,n+2} + I_{0,n}I_{2,n-2} - 2I_{1,n+1}I_{1,n-1}\big)\big].
\end{aligned} \tag{1.49}
$$

We integrate expression (1.49) for the intensity over the entire beam cross section at the focus and obtain:

$$
\begin{aligned}
W = &\int\limits_0^\infty \int\limits_0^{2\pi} I(r,\varphi) r\mathrm{d}r\mathrm{d}\varphi = \frac{1}{2}\int\limits_0^\infty \int\limits_0^{2\pi} r\mathrm{d}r\mathrm{d}\varphi\big[2I_{0,n}^2 + I_{2,n+2}^2 + I_{2,n-2}^2 + 2I_{1,n+1}^2 \\
&+ 2I_{1,n-1}^2 + 2\cos(2\varphi)\big(I_{0,n}I_{2,n+2} + I_{0,n}I_{2,n-2} - 2I_{1,n+1}I_{1,n-1}\big)\big] \\
&= W_0 + W_2 + 2W_1.
\end{aligned} \tag{1.50}
$$

In (1.50), Eq. (1.40) and the fact that the integration of the term with $\cos(2\varphi)$ over the period gives zero are used. It can be seen from (1.50) that the total beam power is equal to:

$$
W = W_0 + W_2 + 2W_1 \tag{1.51}
$$

Equation (1.51) was used in the last step of (1.48). Thus, we have shown that the longitudinal OAM component averaged over the beam cross section is preserved for field (1.35). Preservation of the full OAM during propagation of the beam (1.35) is the reason for the formation of an even number of regions in the focus, in which the OAM component has a different sign (the orbital Hall effect).

1.3.4 The Longitudinal Component of the AM Vector at the Focus

Next, we compare the longitudinal components of the AM and of the sum of SAM and OAM. The AM is given by the equation [14]:

$$\boldsymbol{J} = (\boldsymbol{r} \times \boldsymbol{P}) \tag{1.52}$$

The longitudinal AM component is determined only by the angular component of the energy flux at the focus (1.39) and is equal to:

$$J_z = rQ(r) = r\left[I_{1,n+1}\left(I_{0,n} + I_{2,n+2}\right) + I_{1,n-1}\left(I_{0,n} + I_{2,n-2}\right)\right] \tag{1.53}$$

From (1.53) it can be seen that the AM longitudinal component on the optical axis is always equal to zero, since the "leverage" is equal to zero. We compare expression (1.53) with the sum of SAM (1.43) and OAM (1.46) for field (1.35) in the focus:

$$\begin{aligned} S_z + L_z = \frac{1}{2}\Big\{ & 2nI_{0,n}^2 + 2(n+1)I_{1,n+1}^2 + 2(n-1)I_{1,n-1}^2 + \\ & + (n+1)I_{2,n+2}^2 + (n-1)I_{2,n-2}^2 + 2\cos(2\varphi)\times \\ & \times \left[(4n+3)I_{0,n}I_{2,n+2} + (4n-3)I_{0,n}I_{2,n-2} - 2nI_{1,n+1}I_{1,n-1}\right]\Big\}. \end{aligned} \tag{1.54}$$

Comparison of (1.53) and (1.54) shows that the AM is not equal to the sum of the SAM and OAM. For example, the angular momentum (1.53) is radially symmetric and does not depend on the angle φ, while the sum of SAM and OAM (1.54) depends on the azimuth angle as $\cos(2\varphi)$. Therefore, there must be a third term X_z, which must be added to the sum (1.54) in order for the equality to hold:

$$J_z = S_z + L_z + X_z \tag{1.55}$$

Several questions arise from this information. What is transferred to the particle and causes it to rotate along a circular path: AM (1.53) or OAM (1.46)? And what should be called the orbital Hall effect: separation of regions with different OAM signs (1.46) or AM signs (1.53)? Most likely, the orbital Hall effect is determined by the different direction of the transverse energy flow (1.39), since the transverse flow "catches" the microparticle and forces it to rotate along the "orbit" [1]. Therefore, the AM, which is proportional to the transverse energy flux $Q(r)$, is responsible for the rotation of the particle along a circular trajectory.

1.3.5 Physical Meaning of the Third Term in the Equation for the AM

In this section, we will show that the terms SAM and OAM in (1.55) are artificially formed, and that only two characteristics are sufficient for the light field. These characteristics are SAM and AM, which are not related to each other. We start with the definition of the AM (1.52) and write out explicitly the quantities included in it:

$$\boldsymbol{J} = (\boldsymbol{r} \times \boldsymbol{P}) = \mathrm{Im}\big(\boldsymbol{r} \times (\boldsymbol{E}^* \times (\nabla \times \boldsymbol{E})\big) \tag{1.56}$$

In (1.56), all dimensional constants are omitted. Further, for definiteness, we consider obtaining the longitudinal component of the SAM vector in Cartesian coordinates. From (1.56) we get:

$$\begin{aligned} J_z &= \mathrm{Im}\Big\{x\Big[E_x^*\left(\frac{\partial E_x}{\partial y} - \frac{\partial E_y}{\partial x}\right) + E_z^*\left(\frac{\partial E_z}{\partial y} - \frac{\partial E_y}{\partial z}\right)\Big] \\ &\quad - y\left[E_y^*\left(\frac{\partial E_y}{\partial x} - \frac{\partial E_x}{\partial y}\right) + E_z^*\left(\frac{\partial E_z}{\partial x} - \frac{\partial E_x}{\partial z}\right)\right]\Big\} \\ &= \mathrm{Im}\Big[x\left(E_x^*\frac{\partial E_x}{\partial y} + E_z^*\frac{\partial E_z}{\partial y} - E_x^*\frac{\partial E_y}{\partial x} - E_z^*\frac{\partial E_y}{\partial z}\right) \\ &\quad - y\left(E_x^*\frac{\partial E_y}{\partial x} + E_z^*\frac{\partial E_z}{\partial x} - E_y^*\frac{\partial E_x}{\partial y} - E_z^*\frac{\partial E_x}{\partial z}\right)\Big]. \end{aligned} \tag{1.57}$$

Let us write in a general form the expression for the OAM longitudinal component (1.45), but using Cartesian coordinates:

$$\begin{aligned} L_z &= \mathrm{Im}\Big[E_x^*\left(x\frac{\partial}{\partial y} - y\frac{\partial}{\partial x}\right)E_x + E_y^*\left(x\frac{\partial}{\partial y} - y\frac{\partial}{\partial x}\right)E_y \\ &\quad + E_z^*\left(x\frac{\partial}{\partial y} - y\frac{\partial}{\partial x}\right)E_z\Big] = \mathrm{Im}\Big[x\left(E_x^*\frac{\partial E_x}{\partial y} + E_y^*\frac{\partial E_y}{\partial y}\right. \\ &\quad \left. + E_z^*\frac{\partial E_z}{\partial y}\right) - y\left(E_x^*\frac{\partial E_x}{\partial x} + E_y^*\frac{\partial E_y}{\partial x} + E_z^*\frac{\partial E_z}{\partial x}\right)\Big]. \end{aligned} \tag{1.58}$$

Comparing (1.57) and (1.58), there are four terms at x and y in (1.57) and three terms at x and y in (1.58). Therefore, in order to form in (1.57) a separate term as in (1.58), we add and subtract two terms in (1.57). Then instead of (1.57) we get:

$$\begin{aligned} J_z &= \mathrm{Im}\Big[x\left(E_x^*\frac{\partial E_x}{\partial y} + E_z^*\frac{\partial E_z}{\partial y} + \left\langle E_y^*\frac{\partial E_y}{\partial y} - E_y^*\frac{\partial E_y}{\partial y}\right\rangle - E_x^*\frac{\partial E_y}{\partial x} - E_z^*\frac{\partial E_y}{\partial z}\right) \\ &\quad - y\left(E_x^*\frac{\partial E_y}{\partial x} + E_z^*\frac{\partial E_z}{\partial x} + \left\langle E_x^*\frac{\partial E_x}{\partial x} - E_x^*\frac{\partial E_x}{\partial x}\right\rangle - E_y^*\frac{\partial E_x}{\partial y} - E_z^*\frac{\partial E_x}{\partial z}\right)\Big]. \end{aligned} \tag{1.59}$$

The added terms in (1.59) are marked with triangular parentheses. They do not change the value of the expression (1.57). Now in (1.59) we group the terms in order to explicitly separate the term equal to L_z (1.58):

$$J_z = L_z - \mathrm{Im}[x\left(E_x^* \frac{\partial E_y}{\partial x} + E_y^* \frac{\partial E_y}{\partial y} + E_z^* \frac{\partial E_y}{\partial z}\right) - y\left(E_x^* \frac{\partial E_x}{\partial x} + E_y^* \frac{\partial E_x}{\partial y} + E_z^* \frac{\partial E_x}{\partial z}\right)]. \tag{1.60}$$

Next, we add and subtract in (1.60) the SAM longitudinal component (1.42) and get:

$$\begin{aligned} J_z &= L_z + S_z - \mathrm{Im}[x\left(E_x^* \frac{\partial E_y}{\partial x} + E_y^* \frac{\partial E_y}{\partial y} + E_z^* \frac{\partial E_y}{\partial z}\right) \\ &\quad - y\left(E_x^* \frac{\partial E_x}{\partial x} + E_y^* \frac{\partial E_x}{\partial y} + E_z^* \frac{\partial E_x}{\partial z}\right) + \left\langle E_x^* E_y - E_y^* E_x \right\rangle] \\ &= L_z + S_z + X_z. \end{aligned} \tag{1.61}$$

Thus, from (1.57) we have obtained (1.55). In (1.61), the difference between the two terms in small triangular brackets is equal to the SAM with the opposite sign. That is, OAM and SAM in the expression for AM are artificially formed by adding and subtracting additional terms. As a result, the third term X_z appeared, which has no meaning in the general case. Although in some cases it can be attributed a certain meaning. For instance, if $L_z + S_z = 0$, then the term X_z is equal to the angular momentum of the light field ($X_z = J_z$). The conclusion from this subsection is presented below. The orbital Hall effect occurs at the focus when the regions with the AM longitudinal component of different signs are separated, that is, the regions appear with a different direction of the transverse energy flow rotation. The spin Hall effect occurs at the focus when the regions with the SAM longitudinal component of different signs are separated from each other, that is, the regions in which the polarization vector rotates in different directions are separated.

1.3.6 Simulation

Figure 1.15 shows the distributions of the intensity, as well as the densities of the SAM, the OAM, and the AM of the beam (1.35) in a tight focus at $n = 1$ (Fig. 1.15a–d), $n = 3$ (Fig. 1.15e–h), $n = 5$ (Fig. 1.15i–l). Figure 1.15 confirms formula (1.43), according to which the maximum and minimum values of the SAM density are achieved on the Cartesian axes. Figure 1.15 also confirms formulas (1.43) and (1.47), according to which the OAM density is symmetric with respect to the Cartesian axes, and the AM density has radial symmetry.

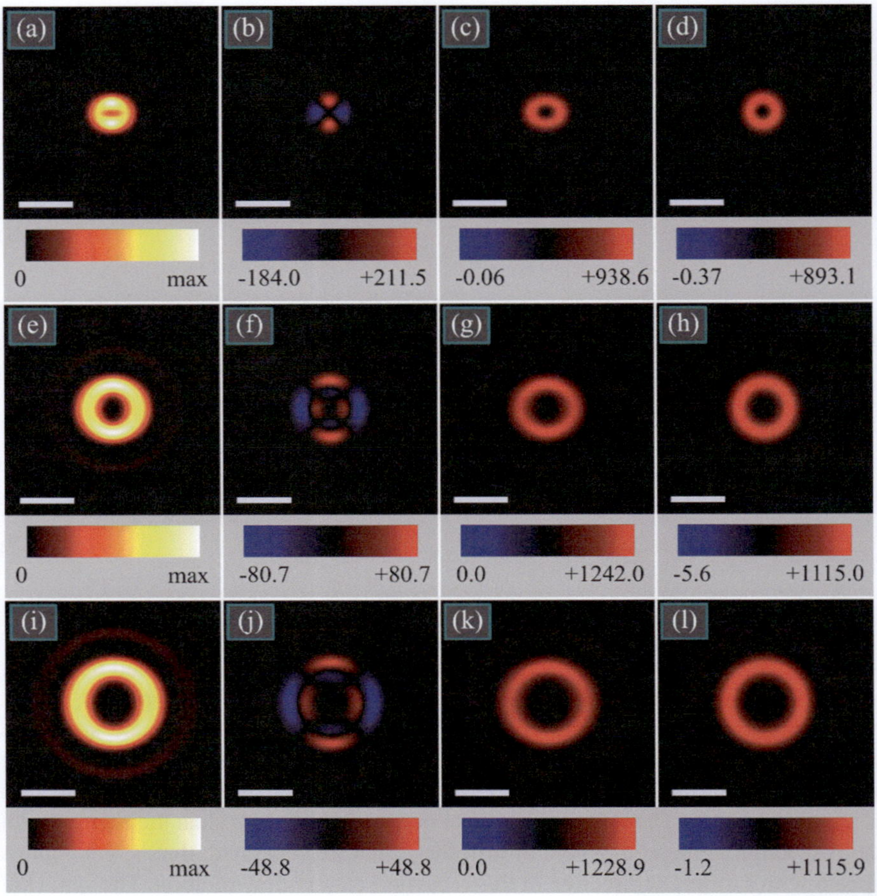

Fig. 1.15 Distributions of the intensity (**a**, **e**, **i**), the SAM density (**b**, **f**, **j**), the OAM density (**c**, **g**, **k**), and the AM density (**d**, **h**, **l**) of the beam (1.35) in a tight focus at $n = 1$ (a–d), $n = 3$ (e–h), $n = 5$ (i–l) and with the following calculation parameters: wavelength $\lambda = 532$ nm, focal length $f = 10$ μm, numerical aperture NA = 0.95, the size of the computational domain is 4×4 μm^2. The scale mark in all figures means 1 μm. The numbers on the color scales below each figure indicate the minimum and maximum values

It follows from Fig. 1.15 that the spin Hall effect occurs at the focus (Fig. 1.15b, f, j), when four local regions with positive and negative (approximately equal in absolute value) SAM are formed at different radii in the focal plane. The orbital Hall effect also takes place at the focus (Fig. 1.15d, h, l). However, firstly, it is radial, and, secondly, it is weakly expressed, since the positive AM distributed over a ring of the one radius is much larger by modulus of a negative AM distributed over a ring of another radius. In Fig. 1.15d, h, l, the blue ring with the negative AM is not visible, but the value of the negative AM is shown on the horizontal color scale.

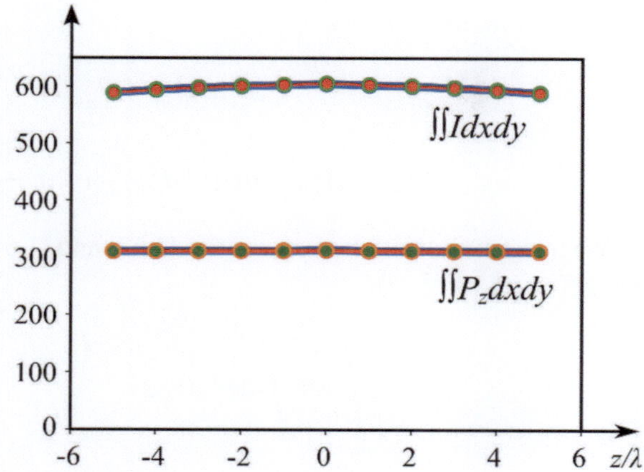

Fig. 1.16 Dependence of the intensity and of the longitudinal component of the Umov-Poynting vector, integrated over the transverse plane, on the distance to the focus z. The upper curve is the total intensity (power), the lower curve is the total longitudinal power flow. Calculation parameters are the same as in Fig. 1.15

Figure 1.16 illustrates the dependences of the total intensity (power) and the total longitudinal power flux on the distance to the focus.

The graphs show the curves for $n = 1$ and $n = 3$, but they are almost coincide with each other. It was assumed that the distribution of the focused light is uniform ($|A(\theta)| \equiv 1$). In this case, the total energy is $2\pi f^2 \approx 628\ \mu\text{m}^2$. Numerically obtained values are approximately equal to $\int\int I\text{d}x\text{d}y \approx 600\ \mu\text{m}^2$. The theoretical value of the total longitudinal power flow is $\pi f^2 \approx 314\ \mu\text{m}^2$. Numerically obtained values are approximately equal to $\int\int P_z\text{d}x\text{d}y \approx 310\ \mu\text{m}^2$.

Figure 1.16 confirms formula (1.51), according to which the total light energy (power) should be equal to $2\pi f^2$ (at $\alpha = \pi/2$ and $|A(\theta)| \equiv 1$) and formula (1.41), according to which the total longitudinal power flow should be equal to πf^2. That is, the representation of the longitudinal power flow through the forward flow, the perpendicular flow, and the reverse flow is also confirmed.

1.3.7 Discussion of Results

In this section, we have shown that the longitudinal SAM and OAM components averaged over the focus plane for an initial optical vortex (with arbitrary radially symmetric real amplitude) with linear polarization are preserved separately. However it is not true in any cases. For example, if we consider the tight focusing of an optical vortex with circular polarization [62], then the averaged axial SAM and OAM components are not conserved. Instead, only their sum is conserved. Indeed, the density of the longitudinal components of the SAM and OAM vectors at the focus of an optical vortex with right-hand circular polarization

$$\boldsymbol{E}(\theta, \varphi) = \frac{1}{\sqrt{2}} \exp(in\varphi) \begin{pmatrix} 1 \\ i \end{pmatrix} \tag{1.62}$$

have the form, respectively:

$$S_z = I_{0,n}^2 - I_{2,n+2}^2, \tag{1.63}$$

$$L_z = nI_{0,n}^2 + (n+2)I_{2,n+2}^2 + 2(n+1)I_{1,n+1}^2. \tag{1.64}$$

We integrate both of these quantities (1.63) and (1.64) over the plane of the focus and get:

$$\begin{aligned} \hat{S}_z &= \int_0^\infty r\mathrm{d}r \int_0^{2\pi} \mathrm{d}\varphi S_z(r,\varphi) \\ &= \int_0^\infty r\mathrm{d}r \int_0^{2\pi} \mathrm{d}\varphi \left(I_{0,n}^2 - I_{2,n+2}^2\right) = (W_0 - W_2), \end{aligned} \tag{1.65}$$

$$\begin{aligned} \hat{L}_z &= \int_0^\infty r\mathrm{d}r \int_0^{2\pi} \mathrm{d}\varphi L_z(r,\varphi) \\ &= \frac{1}{2}\int_0^\infty r\mathrm{d}r \int_0^{2\pi} \mathrm{d}\varphi \left(nI_{0,n}^2 + (n+2)I_{2,n+2}^2 + 2(n+1)I_{1,n+1}^2\right) \\ &= nW_0 + (n+2)W_2 + 2(n+1)W_1 = nW + 2(W_2 + W_1). \end{aligned} \tag{1.66}$$

And the sum of (1.65) and (1.66) is equal to:

$$\hat{S}_z + \hat{L}_z = (n+1)W. \tag{1.67}$$

We obtain the following expressions for the SAM and the OAM in the initial plane:

$$\hat{S}_z = W, \hat{L}_z = nW, \hat{S}_z + \hat{L}_z = (n+1)W. \tag{1.68}$$

It can be seen from (1.67) that in the initial field for an individual photon, the sum of the spin and the OAM for the beam (1.62) was equal to $S_z + L_z = (n+1)\hbar$ while for the entire beam it was $\hat{S}_z + \hat{L}_z = (n+1)W$. During focusing the total spin of the beam (1.62) decreased in the focal plane, while the total OAM increased in it:

$$\begin{aligned} \hat{S}_z &= (W_0 - W_2) = W - 2(W_2 + W_1), \\ \hat{L}_z &= nW + 2(W_2 + W_1). \end{aligned} \tag{1.69}$$

This effect is called SOC [33]. So, if the initial field (1.35) has no spin (no SAM), then there is no SOC in the focus and the total spin is zero (1.44): $\hat{S}_z = 0$. However, the spin Hall effect (1.43) can be formed at the focus. The OAM for the field (1.35) is also preserved (1.48) and is equal to $\hat{L}_z = nW$, and there is an orbital Hall effect (1.46), (1.47) at the focus. If there is SAM (1.63) in the initial field (1.62), then due to SOC it is not kept in focus, but decreases, according to (1.69), partially converting into OAM. The beam (1.62) also has the spin and the orbital Hall effects [62]. However, both of these effects are radial, i.e., the sign of the SAM and the OAM is different at different radii from the optical axis.

In this section, the following results are obtained. It is shown that during tight focusing of an optical vortex with an arbitrary radially symmetric amplitude function and with linear polarization the distribution of the SAM axial component (1.43) in the focal plane depends on the azimuthal angle φ as $\cos(2\varphi)$, and, therefore, for TC $n \neq 0$ the spin Hall effect takes place at the focus [67]. This effect leads to formation of two regions on the vertical and horizontal axes in which the polarization vector rotates in different directions (clockwise and counterclockwise) and SAM has different signs. Similarly, it is derived that the OAM axial component (1.46) depends on the azimuth angle φ as $\cos(2\varphi)$ at the focus. However, we cannot call these four regions with different signs of the OAM longitudinal component a manifestation of the orbital Hall effect, since we do not know how the transverse energy flow behaves in these regions (changes the direction of rotation or not). It is also demonstrated that the transverse energy flux rotates in the plane of focus in opposite directions at different radii from the optical axis (1.39). Such a distribution of the transverse energy flux at the focus can be called the radial-orbital Hall effect, since the energy flux will rotate dielectric microparticles trapped at different radii at the focus clockwise or counterclockwise (the angular tractor [1]).

References

1. B. Richards, E. Wolf, Electromagnetic diffraction in optical systems. II. Structure of the image field in an aplanatic system. Proc. R. Soc. Lond. A **253**, 358–379 (1959)
2. G.H. Yuan, S.B. Wei, X.-C. Yuan, Nondiffracting transversally polarized beam. Opt. Lett. **36**, 3479 (2011)
3. C. Ping, Ch. Liang, F. Wang, Y. Cai,Radially polarized multi-Gaussian Schell-model beam and its tight focusing properties. Opt. Exp. **25**, 32475–32490 (2017)
4. T. Grosjean, I. Gauthier, Longitudinally polarized electric and magnetic optical nano-needles of ultra high lengths. Opt. Commun.Commun. **294**, 333–337 (2013)
5. H. Wang, L. Shi, B. Lukyanchuk, C. Sheppard, C.T. Chong, Creation of a needle of longitudinally polarized light in vacuum using binary optics. Nat. Photonics **2**, 501–505 (2008)
6. J. Lin, R. Chen, P. Jin, M. Cada, Y. Ma, Generation of longitudinally polarized optical chain by 4π focusing system. Opt. Commun. Commun. **340**, 69–73 (2015)
7. J. Zhuang, L. Zhang, D. Deng, Tight-focusing properties of linearly polarized circular Airy Gaussian vortex beam. Opt. Lett. **45**, 296 (2020)
8. Y. Lyu, Z. Man, R. Zhao, P. Meng, W. Zhang, X. Ge, S. Fu, Hybrid polarization induced transverse energy flow. Opt. Commun. Commun. **485**, 126704 (2021)

9. H. Li, C. Wang, M. Tang, X. Li, Controlled negative energy flow in the focus of a radial polarized optical beam. Opt. Express **28**, 18607–18615 (2020)
10. V.V. Kotlyar, S.S. Stafeev, A.G. Nalimov, Energy backflow in the focus of a light beam with phase or polarization singularity. Phys. Rev. A **99**, 033840 (2019)
11. Z. Bomzon, M. Gu, J. Shamir, Angular momentum and geometrical phases in tight-focused circularly polarized plane waves. Appl. Phys. Lett. **89**, 241104 (2006)
12. A. Aiello, P. Banzer, M. Neugebauer, G. Leuchs, From transverse angular momentum to photonic wheels. Nat. Photon. **9**, 789–795 (2015)
13. M. Li, Y. Cai, S. Yan, Y. Liang, P. Zhang, B. Yao, Orbit-induced localized spin angular momentum in strong focusing of optical vectorial vortex beams. Phys. Rev. A **97**, 053842 (2018)
14. Y. Zhao, J.S. Edgar, G.D.M. Jeffries, D. McGloin, D.T. Chiu, Spin-to-orbital angular momentum conversion in a strongly focused optical beam. Phys. Rev. Lett. **99**, 073901 (2007)
15. P.B. Monteiro, P.A. Maia, H.M. Nussenzveig, Angular momentum of focused beams: beyond the paraxial approximation. Phys. Rev. A **79**, 033830 (2009)
16. A.Y. Bekshaev, A simple analytical model of the angular momentum transformation in strongly focusec light beams. Cent. Eur. J. Phys. **8**(6), 947–960 (2010)
17. M.V. Berry, Optical currents. J. Opt. A Pure Appl. Opt. **11**, 094001 (2009)
18. H. Gross, W. Singer, M. Totzeck (eds.), *Handbook of Optical Systems* (Wiley, 2005), vol. 2. ISBN 9783527403783
19. D.L. Golovashkin, N.L. Kazanskiy, Mesh domain decomposition in the finite-difference solution of Maxwell's equations. Math. Model. Comput. Simul. **19**, 48–58 (2007)
20. C. Rockstuhl, H.P. Herzig, Calculation of the torque on dielectric elliptical cylinders. J. Opt. Soc. Am. A **22**, 109–116 (2005)
21. A.G. Nalimov, V.V. Kotlyar, Calculation of the moment of the force acting by a cylindrical Gaussian beam on a cylindrical microparticle. Comput. Opt. **31**, 16–20 (2007)
22. S.S. Stafeev, A.G. Nalimov, A.A. Kovalev, V.D. Zaitsev, V.V. Kotlyar, Circular polarization near the tight focus of linearly polarized light. Photonics **9**, 196 (2022)
23. Z. Man, Z. Xi, X. Yuang, R.E. Burge, P.H. Urbach, Dual coaxial longitudinal polarization vortex structures. Phys. Rev. Lett. **124**, 103901 (2020)
24. C. Schwartz, A. Dogariu, Conversation of angular momentum of light in single scattering. Opt. Express **14**, 8425–8433 (2006)
25. T.A. Nieminen, A.B. Stilgoe, N.R. Heckenberg, N. Rubinsztein-Dunlop, Angular momentum of a strongly focused Gaussian beam. J. Opt. A Pure Appl. Opt. **10**, 115005 (2008)
26. D. Haefner, S. Sukhov, A. Dogariu, Spin Hall effect of light in spherical geometry. Phys. Rev. Lett. **102**, 123903 (2009)
27. O.S. Rodriguez-Herrera, D. Lara, K.Y. Bliokh, E.A. Ostrovskaya, C. Dainty, Optical nanoprobing via spin-orbit interaction of light. Phys. Rev. Lett. **104**, 253601 (2010)
28. A. Bekshaev, K.Y. Bliokh, M. Soskin, Internal flows and energy circulation in light beams. J. Opt. **13**, 053001 (2011)
29. V.V. Koltyar, A.G. Nalimov, S.S. Stafeev, Exploiting the circular polarization of light to obtain a spiral energy flow at the subwavelength focus. J. Opt. Soc. Am. B **36**, 2850–2855 (2019)
30. A.V. Volyar, V.G. Shvedov, T.A. Fadeeva, Structure of a nonparaxial Gaussian beam near the focus. III. Stability, eigenmodes and vortices. Opt. Spectr. **91**, 235–245 (2001)
31. P. Torok, P. Varga, G.R. Booker, Electromagnetic diffraction of light focused through a planar interface between materials of mismatched refractive indices: structure of the electromagnetic field. I. J. Opt. Soc. Am. A **12**, 2136–2144 (1995)
32. Z. Bomzon, M. Gu, Space-variant geometrical phases in focused cylindrical light beams. Opt. Lett. **32**, 3017–3019 (2007)
33. K.Y. Bliokh, E.A. Ostrovskaya, M.A. Alonso, O.G. Rodriguez-Herrera, D. Lara, C. Dainty, Spin-to-orbital angular momentum conversion in focusing, scattering, and imaging systems. Opt. Express **19**, 26132–26149 (2011)
34. B. Roy, N. Ghosh, S.D. Gupta, P.K. Panigrahi, S. Roy, A. Banerjee, Controlled transportation of mesoscopic particles by enhanced spin-orbit interaction of light in an optical trap. Phys. Rev. A **87**, 043823 (2013)

35. B. Roy, N. Ghosh, A. Banerjee, S.D. Gupta, S. Roy, Manifestations of geometric phase and enhanced spin Hall shifts in an optical trap. New J. Phys. **16**, 083037 (2014)
36. V.S. Ignatovsky, Diffraction by a lens having arbitrary opening, in *Transactions of the Optical Institute in Petrograd* (1919), vol. 1, no. 4
37. V.V. Kotlyar, A.A. Kovalev, A.G. Nalimov, Energy density and energy flux in the focus of an optical vortex: reverse flux of light energy. Opt. Lett. **43**, 2921–2924 (2018)
38. M.A. Salem, H. Bagei, Energy flow characteristics of vector X-waves. Opt. Express **19**, 8526–8532 (2011)
39. P. Vaveliuk, O. Martinez-Matos, Negative propagation effect in nonparaxial Airy beams. Opt. Express **20**, 26913–26921 (2012)
40. I. Rondon-Ojeda, F. Soto-Eguibar, Properties of the Poynting vector for invariant beams: negative propagation in Weber beams. Wave Motion **78**, 176–184 (2018)
41. A.V. Novitsky, D.V. Novitsky, Negative propagation of vector Bessel beams. J. Opt. Soc. Am. A **24**(9), 2844–2849 (2007)
42. F.G. Mitri, Reverse propagation and negative angular momentum density flux of an optical nondiffracting nonparaxial fractional Bessel ortex beam of progressive waves. J. Opt. Soc. Am. A **33**, 1661–1667 (2016)
43. S. Chang, S.S. Lee, Optical torque exerted on a homogeneous sphere levitated in the circularly polarized fundamental-mode laser beam. J. Opt. Soc. Am. B **2**, 1853–1860 (1985)
44. V.V. Kotlyar, A.G. Nalimov, S.S. Stafeev, Focusing a vortex laser beam with polarization conversion. Photonics **8**(11), 480 (2021)
45. J.H. Poynting, The wave motion of a revolving shaft, and a suggestion as to the angular momentum in a beam of circularly polarised light. Proc. R. Soc. Lond. ALond. A **82**(557), 560–567 (1909)
46. R.A. Beth, Mechanical detection and measurement of the angular momentum of light. Phys. Rev. **50**(2), 115–125 (1936)
47. L. Allen, M.W. Beijersbergen, R.J.C. Spreeuw, J.P. Woerdman, Orbital angular momentum of light and the transformation of Laguerre-Gaussian laser modes. Phys. Rev. A **45**(11), 8185–8189 (1992)
48. J.-X. Guo, W.-Y. Wang, T.-Y. Cheng, J.-Q. Lü, Interaction of spin-orbit angular momentum in the tight focusing of structured light. Front. Phys. **10**, 1079265 (2022)
49. K.S. Youngworth, T.G. Brown, Focusing of high numerical aperture cylindrical-vector beams. Opt. Express **7**(2), 77 (2000)
50. L. Han, S. Liu, P. Li, Y. Zhang, H. Cheng, J. Zhao, Catalystlike effect of orbital angular momentum on the conversion of transverse to three-dimensional spin states within tightly focused radially polarized beams. Phys. Rev. A **97**(5), 053802 (2018)
51. H. Li, Ch. Ma, J. Wang, M. Tang, X. Li,Spin-orbit Hall effect in the tight focusing of a radially polarized vortex beam, Opt. Express, **29**(24), 39419–39427 (2021)
52. X. Zhang, B. Shen, Z. Zhu, G. Rui, J. He, Y. Cui, B. Gu, Understanding of transverse spin angular momentum in tightly focused linearly polarized vortex beams. Opt. Express **30**(4), 5121–3510 (2022)
53. P. Meng, Z. Man, A.P. Konijnenberg, H.P. Urbach, Angular momentum properties of hybrid cylindrical vector vortex beams in tightly focused optical systems. Opt. Express **27**(24), 35336–35348 (2019)
54. M. Li, S. Yan, B. Yao, Y. Liang, P. Zhang, Spinning and orbiting motion of particles in vortex beams with circular or radial polarizations. Opt. Express **24**(18), 20604–20612 (2016)
55. R. Chen, K. Chew, C. Dai, G. Zhou, Optical spin-to-orbital angular momentum conversion in the near field of a highly nonparaxial optical field with hybrid states of polarization. Phys. Rev. A **96**(5), 053862 (2017)
56. K. Hu, Z. Chen, J. Pu, Tight focusing properties of hybridly polarized vector beams. J. Opt. Soc. Am. A **29**(6), 1099–1104 (2012)
57. S. Huang, G. Zhang, Q. Wang, M. Wang, C. Tu, Y. Li, H.-T. Wang, Spin-to-orbital angular momentum conversion via light intensity gradient. Optica **8**(9), 1231–1236 (2021)

58. P. Yu, Y. Liu, Z. Wang, Y. Li, L. Gong, Interplay between spin and orbital angular momenta in tightly focused higher-order Poincaré sphere beams. Ann. Phys. **532**(8), 2000110 (2020)
59. V.V. Kotlyar, S.S. Stafeev, A.A. Kovalev, V.D. Zaitsev, Spin hall effect before and after the focus of a high-order cylindrical vector beam. Appl. Sci. **12**(23), 12218 (2022)
60. V.V. Kotlyar, S.S. Stafeev, E.S. Kozlova, M.A. Butt, High-order orbital and spin hall effects at the tight focus of laser beams. Photonics **9**(12), 970 (2022)
61. V.V. Kotlyar, S.S. Stafeev, V.D. Zaitsev, A.M. Telegin, Poincaré beams at the tight focus: inseparability, radial spin hall effect, and reverse energy flow. Photonics **9**(12), 969 (2022)
62. V.V. Kotlyar, A.A. Kovalev, A.M. Telegin, Angular and orbital angular momenta in the tight focus of a circularly polarized optical vortex. Photonics **10**(2), 160 (2023)
63. A. Kavokin, G. Malpuech, M. Glazov, Optical spin hall effect. Phys. Rev. Lett. **95**(13), 136601 (2005)
64. C. Leyder, M. Romanelli, J.P. Karr, E. Giacobino, T.C.H. Liew, M.M. Glazov, A.V. Kavokin, G. Malpuech, A. Bramati, Observation of the optical spin Hall effect. Nat. Phys. **3**(9), 628–631 (2007)
65. D. Paul, D.K. Sharma, G.V.P. Kumar, Focused linearly-polarized-light scattering from a silver nanowire: Experimental characterization of the optical spin-Hall effect. Phys. Rev. A **103**(1), 013520 (2021)
66. S.N. Khonina, I. Golub, Vectorial spin Hall effect of light upon tight focusing. Opt. Lett. **47**, 2166–2169 (2022)
67. V.V. Kotlyar, A.A. Kovalev, E.S. Kozlova, A.M. Telegin, Hall effect at the focus of an optical vortex with linear polarization. Micromachines **14**, 788 (2023)

Chapter 2
Spin Hall Effect at the Focus for Light with Circular Polarization

2.1 High-Order Orbital and Spin Hall Effects in a Tight Focus of Laser Radiation

The Hall effect in optics and photonics has been known since 2004 [1]. In [2, 3], the theory of the Hall effect for light was developed. In [4, 5], the Hall effect in optics was experimentally discovered. There are several reviews on the Hall effect in photonics [6, 7]. In optics, the role of electrons with different spins is played by photons with left or right circular polarization. Therefore, instead of the quantum Hall effect, which consists in the spatial separation in the magnetic field of particles with different charges and different spins, the Hall effect in optics is reduced to the spatial separation of light beams with different directions of circular polarization (left and right) or is reduced to the spatial separation of light beams with an orbital angular momentum (OAM) of different signs. The first case is the spin Hall effect [8, 9], and the second one is the orbital Hall effect [10, 11]. Usually, the Hall effect in optics is observed when light is reflected from the interface between media [11], or when it passes through multilayer media [4], crystals [12], or metalens [13]. There are almost no publications on the Hall effect in a tight focus of laser light [14, 15].

In this section, using the Richards-Wolf method [16], it is theoretically and numerically shown that in the tight focus of a vortex laser beam with linear polarization, as well as in the tight focus of a vortex-free beam, which is a superposition of beams with azimuthal polarization of the *m*-th and zero orders, high-order spin and orbital Hall effects are taking place.

V. V. Kotlyar et al., *Optical Hall Effect in the Sharp Focus of Laser Light*,
https://doi.org/10.1007/978-3-031-64683-6_2

2.1.1 The Spin Hall Effect in the Focus of an Optical Vortex with Circular Polarization

We consider a tight focusing of an optical vortex with a circular polarization. In [17], the projections of the electric field of an optical vortex with circular and linear polarization were obtained, but no expressions were obtained for the longitudinal projection of the spin angular momentum (SAM), which indicates the presence of circular (or elliptical) polarization in tight focus. In this section, we will obtain an expression for the longitudinal projection of the SAM for an optical vortex with topological charge m and circular polarization. We consider the initial light electric field with the next Jones vector:

$$\boldsymbol{E} = \frac{A(\theta)\mathrm{e}^{im\varphi}}{\sqrt{2}} \begin{pmatrix} 1 \\ \sigma i \end{pmatrix}, \tag{2.1}$$

where $\sigma = 1$ is right circular, $\sigma = -1$ is left circular and $\sigma = 0$ is linear polarization. For convenience, we take the projections of the electric field near the tight focus of the aplanatic optical system from [17]:

$$\begin{aligned} E_x &= -\frac{i^{m+1}}{\sqrt{2}}\mathrm{e}^{im\varphi}\left(I_{0,m} + \gamma_+\mathrm{e}^{i2\varphi}I_{2,m+2} + \gamma_-\mathrm{e}^{-i2\varphi}I_{2,m-2}\right), \\ E_y &= \frac{i^m}{\sqrt{2}}\mathrm{e}^{im\varphi}\left(\sigma I_{0,m} - \gamma_+\mathrm{e}^{i2\varphi}I_{2,m+2} + \gamma_-\mathrm{e}^{-i2\varphi}I_{2,m-2}\right), \\ E_z &= -2\frac{i^m}{\sqrt{2}}\mathrm{e}^{im\varphi}\left(\gamma_+\mathrm{e}^{i\varphi}I_{1,m+1} - \gamma_-\mathrm{e}^{-i\varphi}I_{1,m-1}\right), \end{aligned} \tag{2.2}$$

where

$$\begin{aligned} I_{\nu,\mu} = \left(\frac{4\pi f}{\lambda}\right)\int\limits_0^{\theta_0} \sin^{\nu+1}\left(\frac{\theta}{2}\right)\cos^{3-\nu}\left(\frac{\theta}{2}\right)\cos^{1/2}(\theta) \\ A(\theta)\mathrm{e}^{ikz\cos\theta}J_\mu(x)\mathrm{d}\theta, \end{aligned} \tag{2.3}$$

In these equations, f is the focal length of the aplanatic system, λ is wavelength, $\mathrm{NA} = \sin\theta_0$ is numerical aperture, $J_\mu(x)$ is the first kind Bessel function of the μ-th order, $x = kr\sin\theta$, (r, φ, z) are cylindrical coordinates, $\gamma_\pm = (1 \pm \sigma)/2$, k is wave number. The functions of Gauss, Bessel-Gauss or a constant value (plane wave) can be used as $A(\theta)$. The spin density vector or the SAM vector is given by:

$$\boldsymbol{S} = \frac{1}{16\pi\omega}\mathrm{Im}(\boldsymbol{E} * \times \boldsymbol{E}), \tag{2.4}$$

where ω is the cyclic frequency of light. Further, the constant $1/(16\pi\omega)$ will be ignored. It can be seen from (2.4) that the longitudinal component of the SAM (without taking into account the constant) coincides with the unnormalized third component of the Stokes vector s_3:

$$s_3 = S_z = 2\mathrm{Im}\big(E_x^* E_y\big). \tag{2.5}$$

Substituting the projections of the electric field (2.2) into (2.5), we obtain:

$$\begin{aligned} S_z &= \big(\sigma I_{0,m}^2 - \gamma_+^2 I_{2,m+2}^2 + \gamma_-^2 I_{2,m-2}^2\big) \\ &\quad + \cos(2\varphi)\big(\gamma_+ I_{2,m+2}(\sigma - 1) + \gamma_- I_{2,m-2}(\sigma + 1)\big). \end{aligned} \tag{2.6}$$

The next expression for the SAM longitudinal projection of the an optical vortex with right circular polarization ($\sigma = 1$, $\sigma = 1$, $\gamma_+^2 = 1$, $\gamma_-^2 = 0$) follows from (2.6):

$$S_{z+} = I_{0,m}^2 - I_{2,m+2}^2. \tag{2.7}$$

It can be seen from (2.7) that near the optical axis $S_{z+} > 0$ (right circular polarization), since $I_{0,m}^2 > I_{2,m+2}^2$, and there will be left circular polarization on those radii where $I_{0,m}^2 < I_{2,m+2}^2$, since $S_{z+} < 0$. Separation of left and right polarizations at different radii from the optical axis is a demonstration of the radial spin Hall effect. It is interesting that the Hall effect will also appear for a vortexless beam ($m = 0$). For the initial left circular polarization, from Eq. (2.6) instead of (2.7) we obtain ($\sigma = -1$, $\gamma_+^2 = 0$, $\gamma_-^2 = 1$):

$$S_{z-} = -I_{0,m}^2 + I_{2,m-2}^2. \tag{2.8}$$

It follows from (2.8) that at m = 0 near the optical axis $S_{z-} < 0$ (left circular polarization), since $I_{0,0}^2 > I_{2,-2}^2 = I_{2,2}^2$, and there will be right circular polarization on those radii where $I_{0,0}^2 < I_{2,2}^2$, since $S_{z-} > 0$. At $m \neq 0$ it is impossible to say unequivocally what polarization will be near the optical axis. For example, at $m = 2$ instead of (2.8) we can write $S_{z-} = -I_{0,2}^2 + I_{2,0}^2 > 0$. That is, although the initial field had left circular polarization, the focused field will have right circular polarization on the optical axis. Such an anomalous behavior of the polarization is due to the fact that at $m = 2$ the energy flux (Poynting vector) has a negative longitudinal projection near the optical axis, i.e., there is a reverse energy flow [18]. The alternation of directions of the polarization vector rotation (the Hall effect) depending on the radial variable also takes place for vortex fields with an initial left circular polarization.

2.1.2 The Spin-Orbital Hall Effect in the Focus of an Optical Vortex with Linear Polarization

In this subsection, we consider a vortex field with linear polarization along the x-axis. In this case, the longitudinal component of the SAM vector does not have circular symmetry ($\sigma = 0$, $\gamma_+^2 = \gamma_-^2 = 1$):

$$S_{zL} = \frac{1}{2}(I_{2,m-2} - I_{2,m+2})(I_{2,m-2} + I_{2,m+2} + \cos(2\varphi)I_{0,m}). \tag{2.9}$$

For definiteness, we write (2.9) for $m = 1$

$$S_{zL1} = -\frac{1}{2}(I_{2,1} + I_{2,3})(-I_{2,1} + I_{2,3} + \cos(2\varphi)I_{0,1}). \tag{2.10}$$

It follows from (2.10) that at $\varphi = 0$ and $\varphi = \pi$ the longitudinal component of the SAM is negative near the optical axis (it is equal to zero on the optical axis) $S_{zL1} < 0$, and at $\varphi = \pi/2$ and $\varphi = 3\pi/2$ it will be positive $S_{zL1} > 0$, since $(I_{2,1} + I_{2,3}) > 0$, $(-I_{2,1} + I_{2,3} - I_{0,1}) < 0$. It should be noted that according to (2.9) there will be linear polarization in the entire focus plane at $m = 0$ (non-vortex light), since $S_{zL0} = 0$. This means that the presence of a minimum optical vortex ($m = 1$) in a beam with linear polarization leads to the formation of 4 subwavelength regions in the focus, in which the polarization changes its sign: the left circular polarization in two regions and the right circular polarization in the other two areas. This distribution of the spin in the focus is another variation of the spin Hall effect. We will show further that simulation confirms theory prediction.

2.1.3 The Spin-Orbital Hall Effect in the Focus of a Superposition of a Cylindrical Vector Beam and a Beam with Linear Polarization

The Hall effect for the superposition of a beam with m-th order radial polarization and a beam with linear polarization along the horizontal axis was shown in [19]. Both beams should be in same phase. It was shown in [19] that, although the initial field of such a superposition has zero longitudinal projection of the SAM (there is no spin), $2m$ subwavelength regions with different rotation directions of the transverse energy flux (clockwise or counterclockwise) are formed in the focus for odd m. In this section, we show that the superposition of the beam with m-th order azimuthal polarization and linearly polarized field in antiphase produces a non-zero longitudinal component of the SAM in the focal plane. In this case, local regions with the opposite rotation direction and polarization, and the transverse energy flow are formed in the focus for an even number m. It is a demonstration of the spin-orbital Hall effect in the focus.

We propose the initial light field with a Jones vector of the form:

$$\boldsymbol{E} = A(\theta)\begin{pmatrix} -\sin(m\varphi) \\ ia + \cos(m\varphi) \end{pmatrix} = A(\theta)\left[\begin{pmatrix} -\sin(m\varphi) \\ \cos(m\varphi) \end{pmatrix} + ia\begin{pmatrix} 0 \\ 1 \end{pmatrix}\right], \operatorname{Im} a = 0. \quad (2.11)$$

It can be seen from (2.11) that the initial field is an axial superposition of the light field with m-th order azimuthal polarization [20] and linear polarization along the y-axis. Further, using the formalism of Richards-Wolf [16], projections of the electric and magnetic fields vectors in the tight focus of the aplanatic system for the initial beam (2.11) can be obtained:

$$\begin{aligned}
E_x &= i^{m+1}\big(I_{0,m}\sin(m\varphi) + I_{2,m-2}\sin((m-2)\varphi)\big) + aI_{2,2}\sin(2\varphi),\\
E_y &= -i^{m+1}\big(I_{0,m}\cos(m\varphi) - I_{2,m-2}\cos((m-2)\varphi)\big) + a\big(I_{0,0} - I_{2,2}\cos(2\varphi)\big),\\
E_z &= -2i^m I_{1,m-1}\sin((m-1)\varphi) - 2iaI_{1,1}\sin(\varphi),\\
H_x &= i^{m+1}\big(I_{0,m}\cos(m\varphi) + I_{2,m-2}\cos((m-2)\varphi)\big) - a\big(I_{0,0} + I_{2,2}\cos(2\varphi)\big),\\
H_y &= i^{m+1}\big(I_{0,m}\sin(m\varphi) - I_{2,m-2}\sin((m-2)\varphi)\big) - aI_{2,2}\sin(2\varphi),\\
H_z s &= -2i^m I_{1,m-1}\cos((m-1)\varphi) + 2iaI_{1,1}\cos(\varphi).
\end{aligned} \quad (2.12)$$

The axial projection of the SAM vector (2.5) for the field (2.12) in the focus can be written as:

$$S_z = \begin{cases} 2a(-1)^p\big[\sin(m\varphi)\big(I_{0,0}I_{0,m} - I_{2,2}I_{2,m-2}\big) \\ +\sin((m-2)\varphi)\big(I_{0,0}I_{2,m-2} - I_{2,2}I_{0,m}\big)\big], m = 2p, \\ 0, m = 2p+1, p = 0, 1, 2, \ldots \end{cases} \quad (2.13)$$

It can be seen from (2.13) that the SAM longitudinal projection of the field (2.11) in the focus is non-zero only for even numbers m, provided that the real parameter a is non-zero. The expressions in parentheses in (2.13) is a constant on a circle with radius r and center on the optical axis, since all functions $I_{\mu,\nu}$ depend only on the radial variable r. Therefore, the SAM projection changes sign $2m$ times around this circle. That is, there will be $2m$ local regions in the focus plane in which the elliptical (or circular) polarization changes the rotation direction. There will be right circular polarization in those areas where $S_z > 0$ and there will be left circular polarization in those areas where $S_z < 0$. Thus, the regions with right and left elliptical or circular polarization are separated in the focus of the field (2.11) with even m. This is a demonstration of the spin Hall effect of the m-th order.

Further we show that the orbital Hall effect of the m-th order also takes place in the focus of the field (2.11). To do this, using the projections of the electric and magnetic fields (2.12), we calculate the Poynting vector transverse projections

$$\boldsymbol{P} = \frac{c}{2\pi}\mathrm{Re}(\boldsymbol{E} * \times \boldsymbol{H}), \tag{2.14}$$

where **E** and **H** are vectors of electric and magnetic fields, * is complex conjugation, $\times$ is vector multiplication, c is the light speed in vacuum. Further, we will ignore the constant $c/(2\pi)$. Substituting (2.12) into (2.14), we obtain transverse projections of the energy flux vector:

$$P_x = \begin{cases} 2a(-1)^p\big[\cos((m-1)\varphi)\big(I_{1,1}I_{2,m-2} - I_{0,0}I_{1,m-1}\big) \\ +\cos((m+1)\varphi)\big(I_{2,2}I_{1,m-1} - I_{1,1}I_{0,m}\big)\big],\ m = 2p, \\ 0,\ m = 2p+1, p = 0, 1, 2, 3, \ldots \end{cases}$$

$$P_y = \begin{cases} 2a(-1)^p\big[-\sin((m-1)\varphi)\big(I_{1,1}I_{2,m-2} - I_{0,0}I_{1,m-1}\big) \\ +\sin((m+1)\varphi)\big(I_{2,2}I_{1,m-1} - I_{1,1}I_{0,m}\big)\big],\ m = 2p, \\ 0,\ m = 2p+1, p = 0, 1, 2, 3, \ldots \end{cases} \tag{2.15}$$

It follows from (2.15) that, both projections of the Poynting vector change sign $2(m + 1)$ around a circle of some radius centered on the optical axis when the expressions in parentheses are constant. This means that the centers of $2m$ local subwavelength regions, in which the transverse energy flux will rotate along a closed trajectory, will lie in focus on a certain circle centered on the optical axis. Moreover, the rotation will be directed in different directions (clockwise and counterclockwise) in neighboring regions. Thus, we have shown that, transverse energy fluxes rotating in different directions are separated in the tight focus of the field (2.11). That is, there is an orbital Hall effect of the m-th order.

2.1.4 Simulation

Hall Effect in the Focus of an Optical Vortex with Linear Polarization

The focusing of an optical vortex with a unit topological charge ($m = 1$) and linear polarization along the x-axis was simulated by the Richards-Wolf integral method [16]. The wavelength is $\lambda = 532$ nm, numerical aperture of aplanatic objective is NA $= 0.95$.

It can be seen from Fig. 2.1 that the spin density has the form of a Maltese cross, in the upper and lower parts of which there is a right circular (elliptical) polarization, since $S_z > 0$, and in the left and right parts of the cross there is a left circular (elliptical) polarization, since $S_z < 0$.

Hall Effect in the Focus of a Beam with Hybrid Polarization

Using the Richards-Wolf formulas, the focusing of a beam (2.11), which is a superposition of a cylindrical vector beam of the m-th order and a plane wave with linear polarization along the y-axis, was simulated. The wavelength is $\lambda = 532$ nm, order

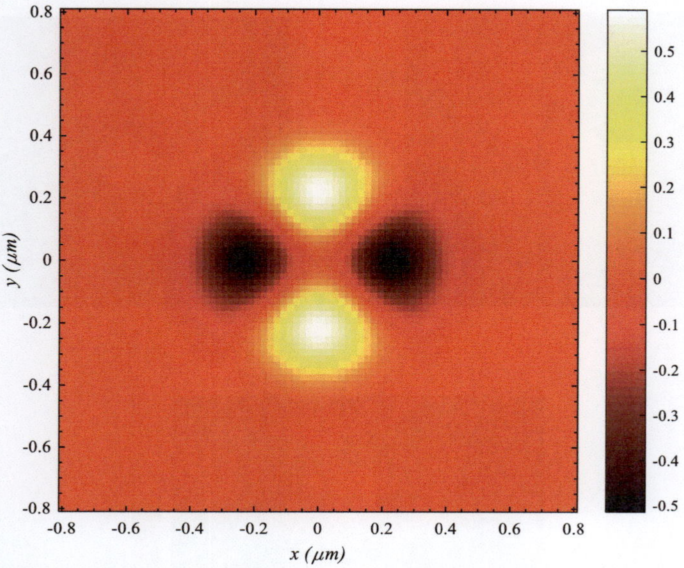

Fig. 2.1 Distribution of the SAM longitudinal component S_z in the focus of an aplanatic objective with NA $= 0.95$ illuminated by an optical vortex ($m = 1$) with linear polarization along the x-axis

of azimuthal polarization is $m = 4$, parameter a is equal to 1. Focusing was proposed by a flat diffractive lens with a numerical aperture NA $= 0.95$.

Figure 2.2a shows the beam intensity (2.11) in the focus. It can be seen that the focal spot is an ellipse extended along the direction of the polarization axis y. Also, the intensity in Fig. 2.2 has 6 side lobes determined by the longitudinal component (Fig. 2.2c). The transverse intensity distribution (Fig. 2.2b) has $2m = 8$ isolated intensity nulls (the energy flux is zero at these points). The transverse energy flow rotates along a closed trajectory (Fig. 2.3)) around these zeros.

Figure 2.3 shows the intensity distribution (same as in Fig. 2.2a) and the direction and magnitude (marked by arrows) of the transverse energy flux $P_x\boldsymbol{e}_x + P_y\boldsymbol{e}_y$, where e_x, e_y are unit vectors along Cartesian coordinates in the focus plane. It can be seen from Fig. 2.3 that $2m = 8$ local subwavelength regions with a diameter of 200–300 nm are formed in the focus. The centers of these areas lie on a certain circle drawn around the optical axis and passing through the intensity zeros surrounding the focal spot. In each of these regions, the transverse energy flow rotates along a closed trajectory. The direction of rotation is different in neighboring regions. That is, the OAM axial projection has a different sign in neighboring regions. Such a separation in space of transverse energy flows rotating in different directions is a demonstration of the orbital Hall effect of the 4-th order.

In this section, using the Richards-Wolf formalism, which adequately describes the behavior of a vector light field in a tight focus, if the focal length is much greater than the wavelength, it is shown that in the focus of an optical vortex with

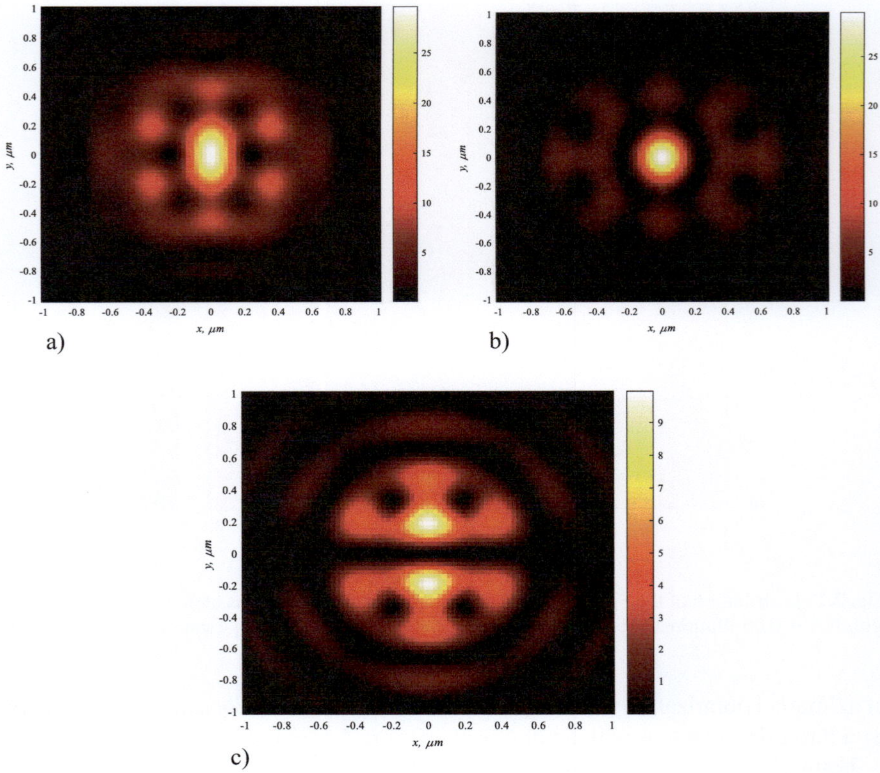

Fig. 2.2 Distribution of the total intensity for the initial field (2.11) $I = |E_x|^2 + |E_y|^2 + |E_z|^2$ (**a**), and its components: transverse $|E_x|^2 + |E_y|^2$ (**b**) and longitudinal $|E_z|^2$ (**c**) in the focus plane ($m = 4, a = 1$)

circular polarization there will be circular polarizations of different signs (left and right circular polarizations) at different radii from the optical axis [21]. That is, photons that fall into focus at different distances from the optical axis will have different spins: either left circular polarization or right circular polarization. Such an effect can be called the radial spin Hall effect. It is also shown by using the Richards-Wolf formalism that four local subwavelength regions about 200 nm in size (the wavelength is 532 nm, numerical aperture is 0.95), in which the spin is directed in different directions in neighboring regions, are formed in the tight focus of an optical vortex with a unit topological charge and linear polarization. That is, in two vertically adjacent areas, the longitudinal projection of the SAM is positive (right elliptical polarization), and in two horizontally located areas, the longitudinal projection of the SAM is negative (left circular or elliptical polarization). Such a spatial separation of regions with different spins in the focus can be called the spin Hall effect of the 2-th order.

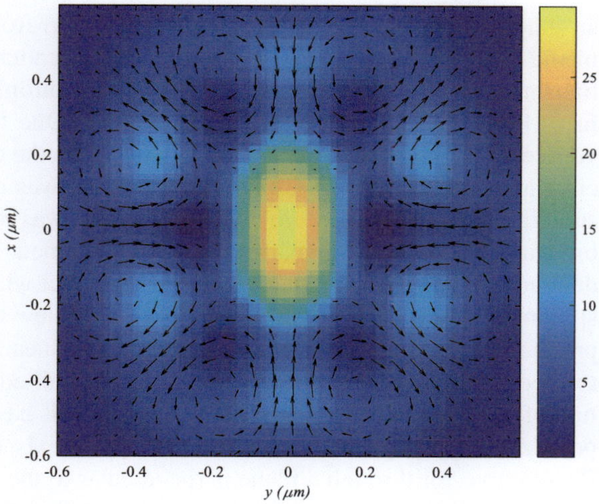

Fig. 2.3 Intensity distribution and direction and magnitude (arrows) of the transverse energy flux in the focus

It is also shown that in the tight focus of the superposition of a cylindrical vector beams with m and zero orders, $2m$ local regions are formed, in neighboring of which the transverse energy flux rotates in different directions. That is, the OAM longitudinal component has different signs in neighboring regions. This is the orbital Hall effect of the m-th order.

These effects arise in the focus due to the conservation of the angular momentum of the beam and due to the spin–orbit conversion. In the first example (an optical vortex with linear polarization), the SAM is zero in the initial plane and the SAM is also zero at the focus. However, in this case an even number of local regions are formed in the focus, in which the SAM is non-zero and has different signs in neighboring regions. In this case, the Hall effect was formed due to the orbital-spin transformation. In the second example (superposition of non-vortex cylindrical beams of the mth and 0th orders), the OAM is equal to zero and the SAM is non-zero in the initial plane. It the focus, $2m$ regions are formed, in which the OAM is non-zero and has different signs in neighboring regions. The total OAM in the focus is still zero. In this case, the Hall effect appeared due to the spin-orbital conversion.

2.2 Spin-Orbital Transformation in a Tight Focus of an Optical Vortex with Circular Polarization

In modern optics, many optical effects are based on the interaction between the polarization of light and the vortex phase of light. A review on spin–orbital conversion (SOC) (interaction, coupling) in photonics can be found in [22]. By SOC, we mean the ability of light with circular polarization to create a rotation of the transverse energy flow. The spin–orbital transformation underlies the spin Hall effect in an inhomogeneous medium [23], upon reflection from the interface between two media [1–3], tight focusing [21, 24–27], light scattering [28–30], and while light transmission through waveguides [31–33], anisotropic crystals [34, 35], metasurfaces [36–38], and in nanostructures [39–41]. Due to SOC, excitation of optical vortices was observed in the nanodisk laser when the disk was excited by light with circular polarization [42]. The spin Hall effect was experimentally observed as a submillimeter Imbert–Fedorov shift when light was reflected from the surface of a birefringent symmetric planar waveguide with a metal shell [43]. Light in free space demonstrates an intrinsic quantum spin Hall effect when surface modes with strong spin–momentum locking are formed [44]. A review of works on SOC in optics is presented in [45]. Due to SOC, when light with linear polarization is reflected from twisted few-layer hyperbolic metasurfaces, two shifted beams with circular polarization of different directions are formed [46]. In this case, when a beam with circular polarization propagates at an angle to the optical axis, then the transverse flow of the Poynting vector through a plane perpendicular to the optical axis will be displaced or separated. In [47], this phenomenon was called the geometric spin Hall effect. If a paraxial vortex beam with cylindrical vector polarization is limited by a sector diaphragm, that is, the circular symmetry of the beam is broken, then regions with left and right circular polarization appear in the transmitted beam [48]. The spin Hall effect also arises if a radially polarized laser beam passes through a sector diaphragm [8]. Then, regions with left and right circular polarization will be formed in the beam after passing the aperture. In the works listed above, the question of what part of the spin angular momentum (SAM) is transferred to the orbital angular momentum (OAM) is not considered. SAM is the spin density in the cross section of the light beam. Light with linear polarization has no spin, i.e., SAM is zero. However, light with circular polarization has a maximum spin denseness, that is, SAM is equal to the power of the beam. The normalized to beam power SAM of circularly polarized light is plus one for right-hand polarization and minus one for left-hand circular polarization. The normalized SAM of light with elliptical polarization is less than one. OAM is a value that indicates the presence of a vortex energy flow in the beam. The normalized to the beam power OAM of vortex laser beams is equal to the topological charge. Therefore, spin–orbit transformation occurs when circularly polarized light that has SAM is converted into light with a vortex energy flow that has OAM.

SOC is a universal physical phenomenon and is present in many processes. For example, it is known that due to SOC, no collapse occurs during Bose–Einstein condensate [49]. That is, due to SOC, the initial Gaussian quantum state does not decrease to zero in a finite time, but transforms into a vortex state and increases in size with time.

In this section, within the Richards-Wolf formalism, exact formulas are obtained for the case of sharp focusing of a circularly polarized optical vortex. Here, we proposed right-hand circular polarization. These formulas show what part of the total longitudinal SAM transforms into the total longitudinal part of the OAM in the focus. It is also shown that the maximum part of the SAM that can convert into the OAM is equal to half of the beam power.

2.2.1 *The Denseness of Lengthwise Projections of the SAM and OAM*

The Jones vector for the electric and magnetic fields of the source beam have the form:

$$\mathbf{E}(\psi,\vartheta)=\frac{\mathbf{A}(\psi)}{\sqrt{2}}\exp(\mathbf{im}\vartheta)\begin{pmatrix}1\\ i\end{pmatrix},\ \mathbf{H}(\psi,\vartheta)=\frac{\mathbf{A}(\psi)}{\sqrt{2}}\exp(\mathbf{im}\vartheta)\begin{pmatrix}-i\\ 1\end{pmatrix}, \quad (2.16)$$

where $A(\psi)$ is an amplitude of the radially symmetric launch field, (ρ,ϑ) are polar coordinates in the cross section of the light beam, $\rho = f\cos\psi, f$ is a focal length of a spherical lens. Figure 2.4 shows the optical scheme, which is studied in this section. The light beam from the laser acquires a linear polarization after the polarizer P_1 and enters the light modulator SLM. The modulator has a transmission in the form of a diffraction grating with a fork. The number of teeth at the fork is equal to the topological charge of the optical vortex, which is formed after SLM. After a quarter wave plate, the optical vortex acquires circular polarization. Aperture D cuts out the working + 1 order of diffraction that is formed by the grating. A microobjective focuses light onto a CCD-camera. Thus, the diagram in Fig. 2.4 makes it possible to register the distribution of light intensity in the focus of the light field (2.16). In Fig. 2.4, for definiteness, a wavelength of 633 nm is used. Simulation results will be also given for this wavelength, but the effect of spin–orbital transformation takes place for any wavelength.

The phase distribution in the form of a grating with a fork (shown in the inset) is generated in the modulator window. The formula after the quarter-wave plate describes the transverse components of the field (2.16).

The amplitudes in the proposed launch field (2.16) are given for right-hand circular polarization. At the source, the lengthwise projection of the SAM $S_z = 2\mathrm{Im}\left(E_x^* E_y\right)$ of the field (2.16) is equal to $S_{z,0} = A^2(\psi)$, and the total spin in the source plane is

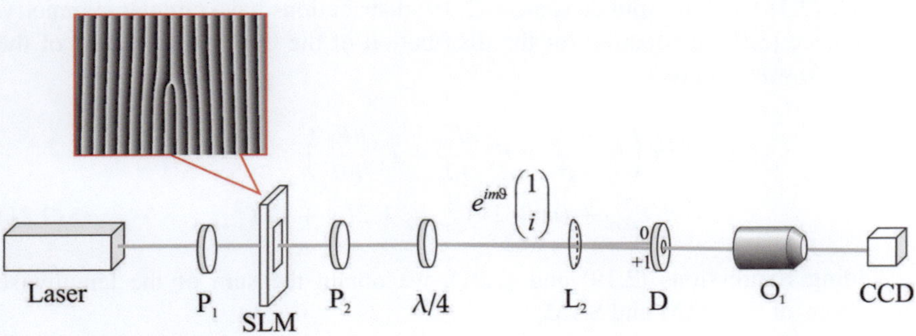

Fig. 2.4 Optical scheme: laser with a wavelength of 633 nm; P_1, P_2 are polarizers; SLM is liquid crystal light modulator; $\lambda/4$ is quarter-wave plate; L_2 is spherical lens; D is aperture: O_1 is microobjective with a numerical aperture NA = 0.95; CCD is camera

equal to the total energy (power) of the beam W:

$$\hat{S}_{z,0} = 2\pi \int_0^\infty \int_0^{2\pi} A^2(\psi)\rho d\rho = W \tag{2.17}$$

In reference [17], using the Richards-Wolf formalism [16], formulas for the intensity of the lengthwise component of the SAM in the focal plane were derived. The distributions of the intensity and the SAM lengthwise component of the focused field (2.16) will have the form:

$$I(\rho, \vartheta) = |E_x|^2 + |E_y|^2 + |E_z|^2 = \Upsilon_{0,m}^2 + \Upsilon_{2,m+2}^2 + 2\Upsilon_{1,m+1}^2, \tag{2.18}$$

$$S_z = 2\mathrm{Im}(E_x^* E_y) = \Upsilon_{0,m}^2 - \Upsilon_{2,m+2}^2. \tag{2.19}$$

Formulas (2.18) and (2.19) include functions $\Upsilon_{\xi,\chi}$ which rely only on the radial variable ρ:

$$\Upsilon_{\xi,\chi} = 2kf \int_0^\beta \sin^{\xi+1}\left(\frac{\psi}{2}\right) \cos^{3-\xi}\left(\frac{\psi}{2}\right) \cos^{1/2}(\psi) A(\psi) e^{ikz\cos\psi} J_\chi(k\rho\sin\psi) d\psi, \tag{2.20}$$

where $k = 2\pi/\lambda$ is the wavenumber of light with the wavelength of λ; f is the focal length; β is the maximum inclination angle of the rays to the optical axis, which determines the numerical aperture of the aplanatic lens $NA = \sin\beta$; $J_\chi(k\rho\sin\psi)$ is the Bessel function of the first kind of χ-th order. In Expression (2.20) and below, the indices χ and ξ can get the next values: $\xi = 0, 1, 2$; $\chi = m-2, m-1, m, m+1, m+2$.

By virtue of the circular polarization of the launch field in the focus, both the intensity (2.18) and the spin denseness (2.19) distributions have circular symmetry. In reference [32], an equation for the distribution of the OAM in the focus of the field (2.16) was obtained:

$$\begin{aligned} L_z &= \mathrm{Im}\left(E_x^* \frac{\partial E_x}{\partial \vartheta} + E_y^* \frac{\partial E_y}{\partial \vartheta} + E_z^* \frac{\partial E_z}{\partial \vartheta}\right) \\ &= m\Upsilon_{0,m}^2 + (m+2)\Upsilon_{2,m+2}^2 + 2(m+1)\Upsilon_{1,m+1}^2. \end{aligned} \tag{2.21}$$

Adding Expressions (2.19) and (2.21), we obtain the sum of the lengthwise projections of the OAM and SAM:

$$J_z = L_z + S_z = (m+1)\left(\Upsilon_{0,m}^2 + \Upsilon_{2,m+2}^2 + 2\Upsilon_{1,m+1}^2\right) = (m+1)I. \tag{2.22}$$

Expression (2.22) shows that the lengthwise component of the sum of orbital and spin angular momenta in the focus of the field (2.1) is equal to the intensity (2.18) of light in the focus multiplied by the sum of the topological charge m and the "spin" of the launch field, which equals to 1.

2.2.2 The Total Lengthwise OAM and SAM Averaged Over the Cross Section of the Beam

In reference [50], an expression was obtained for the partial power of the angular harmonics included in (2.18), (2.19), and (2.22) in the form:

$$W_\xi = 4\pi f^2 \int_0^\beta \sin^{2\xi+1}\left(\frac{\psi}{2}\right)\cos^{5-2\xi}\left(\frac{\psi}{2}\right)|A(\psi)|^2 \mathrm{d}\psi \tag{2.23}$$

Integrals (2.23) can be analytically calculated only in ordinary cases, but, nevertheless, it is possible to estimate the contribution of each angular harmonic. For instance, if a uniform field with a constant amplitude $A(\psi) \equiv 1$ is focused, then [50]:

$$W_0 = 4\pi f^2 \int_0^\beta \sin\left(\frac{\psi}{2}\right)\cos^5\left(\frac{\psi}{2}\right)\mathrm{d}\psi = \frac{4}{3}\pi f^2\left[1 - \cos^6\left(\frac{\beta}{2}\right)\right],$$

$$W_1 = 4\pi f^2 \int_0^\beta \sin^3\left(\frac{\psi}{2}\right)\cos^3\left(\frac{\psi}{2}\right)\mathrm{d}\psi = \frac{2}{3}\pi f^2 \sin^4\left(\frac{\beta}{2}\right)\left[1 + 2\cos^2\left(\frac{\beta}{2}\right)\right],$$

$$W_2 = 4\pi f^2 \int_0^\beta \sin^5\left(\frac{\psi}{2}\right)\cos\left(\frac{\psi}{2}\right)\mathrm{d}\psi = \frac{4}{3}\pi f^2 \sin^6\left(\frac{\beta}{2}\right),$$

$$W = 2\pi f^2[1 - \cos(\beta)]. \tag{2.24}$$

In the limiting case, when the numerical aperture is close to unity, i.e., $\beta \approx \pi/2$, it can be obtained: $W_0 = (7/6)\pi f^2$, $W_1 = (1/3)\pi f^2$, $W_2 = (1/6)\pi f^2$. All energy coincides with the area of the hemisphere: $W = W_0 + W_2 + 2W_1 = 2\pi f^2$. It also should be noted that $W_0 - W_2 = W/2$.

Based on (2.18), (2.19), (2.22), and (2.23), we find the total (averaged over the cross-section of the whole beam) intensity (total beam power), the OAM, and the lengthwise SAM in the focal plane:

$$\hat{I} = W = \int_0^\infty \int_0^{2\pi} I(\rho, \vartheta)\rho \mathrm{d}\rho \mathrm{d}\vartheta = \int_0^\infty \int_0^{2\pi} \rho \mathrm{d}\rho \mathrm{d}\vartheta \left(\Upsilon_{0,m}^2 + \Upsilon_{2,m+2}^2 + 2\Upsilon_{1,m+1}^2\right)$$

$$= W_0 + W_2 + 2W_1. \tag{2.25}$$

$$\hat{S}_z = \int_0^\infty \int_0^{2\pi} S_z \rho \mathrm{d}\rho \mathrm{d}\vartheta = \int_0^\infty \int_0^{2\pi} \rho \mathrm{d}\rho \mathrm{d}\vartheta \left(\Upsilon_{0,m}^2 - \Upsilon_{2,m+2}^2 \right) = W_0 - W_2. \tag{2.26}$$

$$\begin{aligned} \hat{L}_z = \int_0^\infty \int_0^{2\pi} L_z \rho \mathrm{d}\rho \mathrm{d}\vartheta &= \int_0^\infty \int_0^{2\pi} \rho \mathrm{d}\rho \mathrm{d}\vartheta \\ &\times \left(m\Upsilon_{0,m}^2 + (m+2)\Upsilon_{2,m+2}^2 + 2(m+1)\Upsilon_{1,m+1}^2 \right) \\ &= mW_0 + (m+2)W_2 + 2(m+1)W_1. \end{aligned} \tag{2.27}$$

Expressions (2.25)–(2.27) are the main result of this section. It follows from them that the sum of the full longitudinal OAM and SAM is equal to:

$$\begin{aligned} \hat{S}_z + \hat{L}_z &= mW_0 + (m+2)W_2 + 2(m+1)W_1 + W_0 - W_2 \\ &= (m+1)(W_0 + W_2 + 2W_1) = (m+1)W. \end{aligned} \tag{2.28}$$

Equation (2.28) should be supplemented with a similar sum in the initial beam plane (2.16). To do this, we obtain the denseness of the lengthwise OAM component and the total longitudinal OAM in the launch plane from (2.1):

$$\begin{aligned} L_{z,0} &= \mathrm{Im}\left(E_x^* \frac{\partial E_x}{\partial \varphi} + E_y^* \frac{\partial E_y}{\partial \varphi} \right) = m|A(\psi)|^2, \\ \hat{L}_{z,0} &= \int_0^\infty \int_0^{2\pi} L_{z,0} \rho d\rho d\vartheta = mW. \end{aligned} \tag{2.29}$$

Adding (2.17) and (2.29), we obtain the sum of the OAM and SAM in the initial plane of the beam (2.16):

$$\hat{S}_{z,0} + \hat{L}_{z,0} = W + mW = (m+1)W. \tag{2.30}$$

2.2.3 *The Spin-Orbit Conversation upon the Light Focusing*

Comparing (2.28) and (2.30), it can be seen that the sum of the total lengthwise projections of the OAM and SAM is conserved upon focusing. However, separately, the full longitudinal OAM and SAM are not conserved due to the SOC. The total SAM during focusing decreases and partially passes into the full OAM, which, in contrast, increases. It is especially evident when $m = 0$. In the launch plane, in this case, the total SAM is equal to the total beam power (2.16): $\hat{S}_{z,0} = W$, and the total OAM is zero: $\hat{L}_{z,0} = 0$. During the focusing process, the total SAM will decrease and will be equal to (2.26) in the focus: $\hat{S}_z = W_0 - W_2$, and the total OAM will grow and will be equal to (2.27) in the focus: $\hat{L}_z = 2W_2 + 2W_1 = W - (W_0 - W_2)$. Taking into account (2.24) and at $\beta \approx \pi/2$, the full longitudinal SAM will decrease in focus compared to the initial plane by 2 times ($\hat{S}_z = W_0 - W_2 = W/2$), and the total longitudinal OAM will increase in focus compared to the initial plane, also by 2 times ($\hat{L}_z = W/2$). This means, that, in the limiting case ($A(\psi) = 1$, $\beta = \pi/2$) with a plane wave with right-hand circular polarization focusing, half of the total longitudinal SAM will transform into a longitudinal OAM. For a smaller numerical aperture ($\beta < \pi/2$) and for any other real amplitudes $A(\psi) < 1$, with field (2.16) focusing, a part less than half will pass from the SAM to the OAM. From (2.24), it can be found how much the SAM and OAM change due to the SOC for the case with a smaller numerical aperture, for instance, at $\beta = \pi/3$ ($m = 0$, $A(\psi) = 1$). Then, we get that SAM and OAM in the focus will be equal to $\hat{S}_z = 3W/4$ and $\hat{L}_z = W/4$, respectively. That is, at a numerical aperture $\sin\beta = \sqrt{3}/2 \approx 0.87$, the limit value of the OAM that can occur in the focus of the field (2.16) is equal to one-fourth of the total beam power.

It is clear that similar formulas can also be obtained for an optical vortex with left-hand circular polarization. Then, for an optical vortex with topological charge n and left circular polarization, instead of (2.26), (2.27), and (2.28) we get:

$$\begin{aligned} \hat{S}_z &= W_2 - W_0, \\ \hat{L}_z &= mW_0 + (m-2)W_2 + 2(m-1)W_1, \\ \hat{S}_z + \hat{L}_z &= (m-1)(W_0 + W_2 + 2W_1) = (m-1)W. \end{aligned} \tag{2.31}$$

2.2.4 *Transformation of the Longitudinal Energy Flux into the Transverse Energy Flux*

It is interesting to follow how the total longitudinal energy flux (the lengthwise Poynting vector component averaged over the beam cross-section) changes while tight focusing of the beam (2.16). It was shown in [50] that the distribution of the longitudinal Poynting vector component in the focus $P_z = \mathrm{Re}\left(E_x^* H_y - E_y^* H_x\right)$ is equal to the lengthwise projection of the SAM:

$$P_z = S_z = \Upsilon_{0,m}^2 - \Upsilon_{2,m+2}^2. \tag{2.32}$$

Therefore, the total longitudinal energy flux averaged over the beam cross-section in the focus will be equal to (2.26):

$$\hat{P}_z = \int_0^{\infty}\int_0^{2\pi} P_z \rho \mathrm{d}\rho \mathrm{d}\vartheta = W_0 - W_2. \tag{2.33}$$

In the initial plane of the field (2.16), the denseness of the longitudinal energy flow and the total longitudinal energy flux are equal to the expressions:

$$P_{z,0} = |A(\psi)|^2, \hat{P}_{z,0} = W. \tag{2.34}$$

It can be seen from a comparison of (2.33) and (2.34) that, upon focusing, the total longitudinal energy flux decreases in the same way as the total longitudinal SAM. Due to the fact that during focusing, a longitudinal component of the electric field strength E_z appears, a transverse energy flux also appears. This occurs simultaneously with the SOC. That is, a transverse energy flow appears as the SAM decreases and the OAM appears. This means that part of the initial longitudinal energy flux is converted into a transverse energy flux, and the longitudinal flux decreases by exactly the same amount as the longitudinal SAM decreases. It should be noted that that the value of the total longitudinal energy flux does not depend on the value of the topological charge, but depends only on the numerical aperture $\sin\beta$ of the focusing optical system and on the initial radially symmetric amplitude $A(\psi)$ of the light field. These facts follow from (2.23) and (2.33).

2.2.5 Separate Measurement of the OAM and SAM in the Focus

In reference [51], formulas are given that relate the force $\vec{F}$ **F** and the moment of the force **T** $\vec{T}$, which influence on a Rayleigh (smaller than wavelength) microparticle with a complex permittivity ε, with the intensity I, the canonical vector of the energy flow **P** $\vec{P}_0$, and the SAM vector **S** $\vec{S}$:

$$\begin{aligned} \vec{F} &= \frac{\mathrm{Re}(\varepsilon)}{4}\nabla I + \frac{\mathrm{Im}(\varepsilon)}{2}\vec{P}_{\mathbf{0}}, \\ \vec{T} &= \frac{\mathrm{Im}(\varepsilon)}{2}\vec{S}. \end{aligned} \tag{2.35}$$

In (2.35), we use next notation:

$$\begin{aligned} I &= |E_x|^2 + |E_y|^2 + |E_z|^2, \\ \vec{P} &= \vec{P}_0 + \vec{P}_s = \frac{\mathrm{Re}}{8\pi c}\left(\vec{E}^* \times \vec{H}\right) = \frac{\mathrm{Im}}{16\pi c}\left(\vec{E}^* \cdot (\nabla)\vec{E}\right) + \frac{1}{2}\nabla \times \vec{S}, \\ \vec{S} &= \frac{\mathrm{Im}}{16\pi c}\left(\vec{E}^* \times \vec{E}\right), \end{aligned} \tag{2.36}$$

where c is the speed of light in vacuum and $\vec{P}_s$ is the spin flow. Therefore, the method of separate measurement of the SAM and OAM in the focal plane of the field (2.16) is as follows. It follows from (2.18) that at $m = 0$, there is an intensity maximum $I(\rho = 0) = \Upsilon_{0,0}^2$ in the focus center. The microparticle in focus will move to the center, where the intensity is maximal and the gradient is equal to zero. Due to the absorption of the particle (Im $\varepsilon \neq 0$), the moment of the force (2.35) effecting on the particle will rotate it around its center of mass and around the optical axis. The value of this moment of forces will be proportional to the lengthwise projection of the SAM vector: $T_z \sim S_z$. At a large numerical aperture (close to unity), the focal spot is smaller than the wavelength and the particle captured by such a focal spot will rotate under the influence of the total longitudinal component of the SAM equal to $\hat{S}_z = W_0 - W_2 \approx W/2$. Therefore, by measuring the speed of the particle rotation, it is possible to estimate the force moment, which will be proportional to the SAM $\hat{S}_z$. To measure the transverse energy flux P_φ, it is necessary to form a ring of light in the focus. It follows from (2.3) that at $m = 1$ the intensity on the optical axis will be zero in the focus, since $I(\rho = 0) = \Upsilon_{0,1}^2 + \Upsilon_{2,3}^2 + 2\Upsilon_{1,2}^2 = 0$. This means that a light ring will form in the focus, because of the presence of an optical vortex with a topological charge $m = 1$. The particle in focus will shift to the ring due to the minimum intensity gradient on the ring, and the force $F_\vartheta \sim P_\vartheta$ will shift this particle along the light ring. This force will be proportional to the transverse energy flow P_ϑ, which, in turn, is proportional to the longitudinal component of the angular momentum and the lengthwise projection of the OAM: $J_z = L_z = \rho P_\vartheta$. It was shown in [50] that the distribution of the transverse flux in the focal plane of the field (2.16) is described by the expression $P_\vartheta = \Upsilon_{1,m+1}\left(\Upsilon_{0,m} + \Upsilon_{2,m+2}\right)$. The speed of the particle moving along a ring of small radius will be influenced not by the full longitudinal OAM, but by a partial one, since the particle is captured not by the entire circular trajectory, but only by its part. By experimental estimation of the value of the light ring, which falls on the particle ($\gamma < 1$), it is possible to estimate the part of the total longitudinal OAM that is transferred to the particle $\gamma \hat{L}_z = \gamma(2W - W_0 + W_2) \approx 3\gamma W/2$. This value can be found by measuring the average speed of the particle along the light ring in the focus of the field (2.16), since this speed v is proportional to the force (2.35), which is proportional to the transversal energy flow P_ϑ, which is proportional to the longitudinal projection of the OAM $\hat{L}_z$: $v \sim F \sim P_\vartheta \sim \gamma \hat{L}_z$.

2.2.6 Simulation Results

This section presents the results of numerical simulation of light field (2.16) focusing on different topological charges of incident beams. The simulation was performed by using the solution of the Richards-Wolf integral [16]. Focusing of light with a wavelength of 633 nm by aplanatic objectives was considered. This wavelength was chosen because it can be implemented in practice using a conventional helium–neon laser. As an example of the calculated patterns of intensity and SAM, the results of focusing for vortices with topological charges of $m = 0$ (no vortex) and $m = 1$ are presented. Figure 2.5 shows the intensity pattern and its individual components for $m = 0$. Patterns of the SAM vector components for the same topological charge ($m = 0$) are presented in Fig. 2.6. Figures 2.7 and 2.8 show the distribution of intensity and SAM for $m = 1$. Figure 2.9 shows the ratio $\hat{S}_z/\hat{L}_z$ depending on the numerical aperture of the focusing lens. All figures are given for the numerical aperture of the focusing lens equal to NA = 0.95. Microobjectives with 100× magnification have such a numerical aperture. A large numerical aperture was chosen for modeling because the effect of SOC is noticeable starting from a numerical aperture of 0.8 and reaches a maximum at a numerical aperture of 1.

The comparison of Fig. 2.5a and Fig. 2.6c shows that at $m = 0$, the intensity I and the lengthwise SAM projection S_z in the focal plane are almost the same in shape (bright round spot) and in magnitude $I_{\max} \simeq S_{z,\max} \simeq 16$. This follows from Eqs. (2.18) and (2.19), since $I \simeq S_z \simeq \Upsilon_{0,0}^2$, because it follows from (2.24) that $\Upsilon_{0,0}^2 \gg \Upsilon_{2,2}^2$, $\Upsilon_{0,0}^2 \gg 2\Upsilon_{1,1}^2$.

The comparison of Figs. 2.7a and 2.8c shows that at $m = 1$, the intensity I and the lengthwise SAM projection S_z in the focal plane have the form of a ring with approximately the same radius and size, although $S_{z,\max} \sim 4$ is slightly less than $I_{\max}$

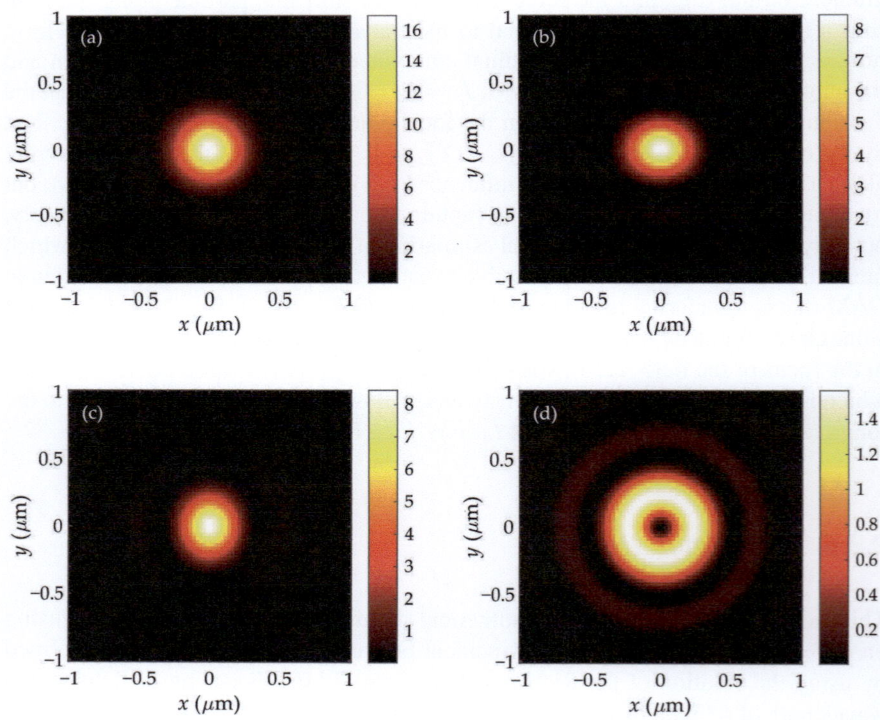

Fig. 2.5 Distribution of intensity I (**a**) and its components I_x (**b**), I_y (**c**), I_z (**d**) in the focus of the beam with circular polarization (without vortex) for NA = 0.95

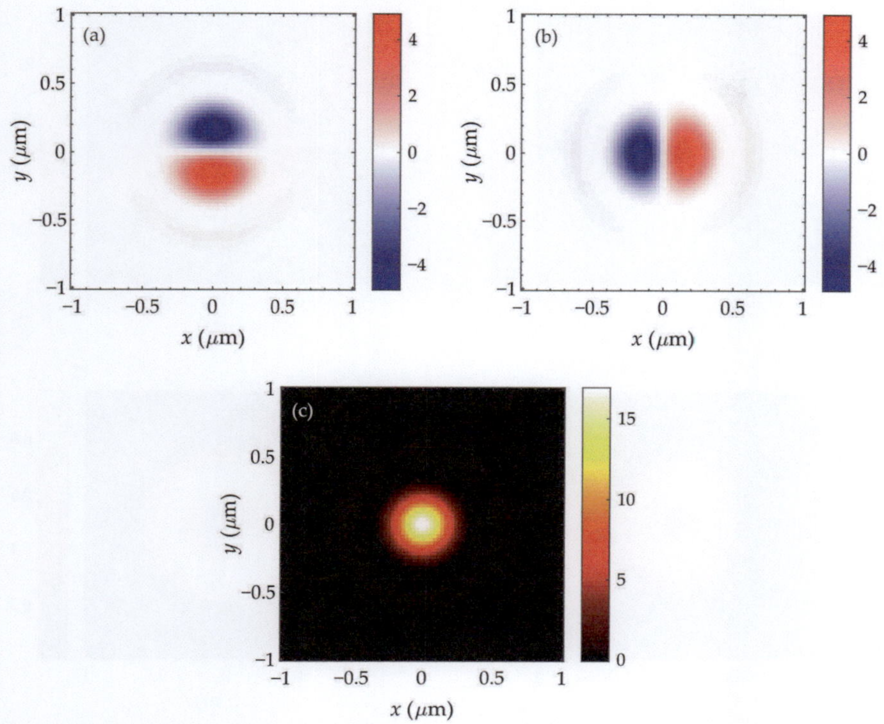

Fig. 2.6 Patterns of the SAM components S_x (**a**), S_y (**b**), S_z (**c**) in the focus of the beam with circular polarization (without vortex) for NA = 0.95

~ 5. This follows from Eqs. (2.18) and (2.19): $I = \Upsilon_{0,1}^2 + \Upsilon_{2,3}^2 + 2\Upsilon_{1,2}^2$, $S_z = \Upsilon_{0,1}^2 - \Upsilon_{2,3}^2$, taking into account (2.24): $\Upsilon_{0,1}^2 \gg \Upsilon_{2,3}^2$, $\Upsilon_{0,1}^2 \gg 2\Upsilon_{1,2}^2$. Figure 2.9 shows the ratio of SAM to OAM $\hat{S}_z/\hat{L}_z$ in the focus of light field (2.16) at different numerical apertures of a spherical lens and for different topological charges of incident light: $m = 0$ (line 1), $m = 1$ (line 2), $m = -1$ (line 3). In Sect. 2.2.3, specific values $\hat{S}_z$ and $\hat{L}_z$ were also obtained for the numerical aperture NA = 0.87. It is obtained that $\hat{S}_z = W_0 - W_2 \approx 3W/4$, $\hat{L}_z \approx W/4$ for $m = 0$, and $\hat{S}_z \approx 3W/4$, $\hat{L}_z = W + W/4 \approx 5W/4$ for $m = 1$. Therefore, it can be seen from Fig. 2.9 that for $m = 0$ $\hat{S}_z/\hat{L}_z = 3$, and for $m = 1$ $\hat{S}_z/\hat{L}_z = 3/5 = 0.6$ at a numerical aperture of 0.87. If $n = -1$, then $\hat{S}_z \approx 3W/4$, $\hat{L}_z = -W + W/4 \approx -3W/4$, that is $\hat{S}_z = -\hat{L}_z$, and their ratio is equal to $\hat{S}_z/\hat{L}_z = -1$ for any numerical aperture. This can be seen from Fig. 2.9 (curve 3). It means that an optical vortex with right circular polarization and a topological charge of −1 has no angular momentum, since the OAM and SAM are equal and have different signs.

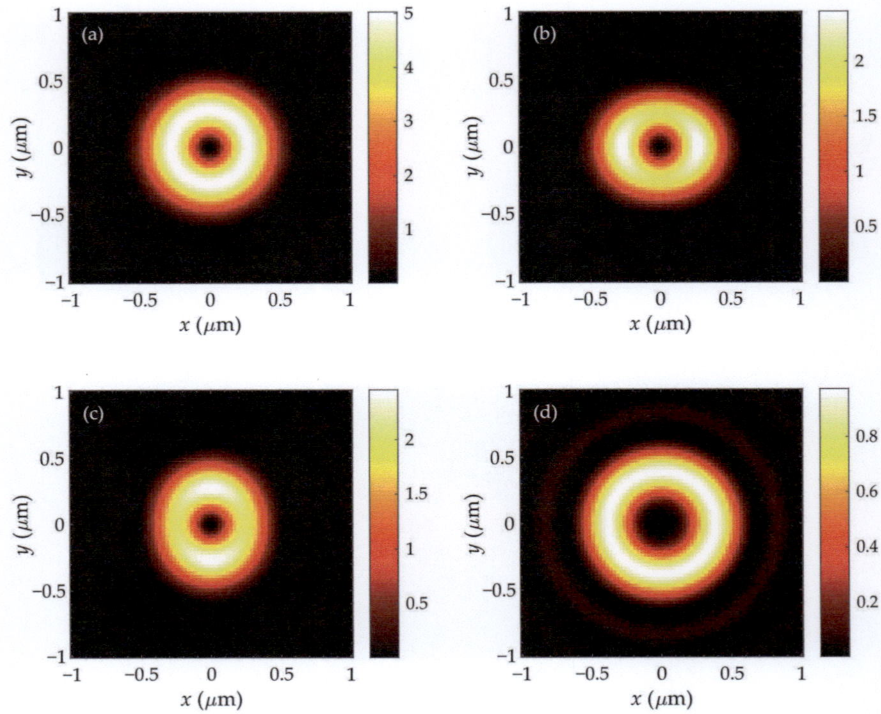

Fig. 2.7 Distribution of intensity I (**a**) and its components I_x (**b**), I_y (**c**), I_z (**d**) in the focus of the optical vortex with unit topological charge and with circular polarization for NA = 0.95

2.2.7 Discussion

In tight focusing, SOC occurs when light passes through a spherical lens. If the optical vortex with the right circular polarization (2.16) before the spherical lens had a full longitudinal SAM $\hat{S}_z$, equal to beam power W (2.17), then the total longitudinal SAM of the beam would decrease immediately after the spherical lens and become equal to (2.26)$\hat{S}_z = W_0 - W_2$. After the spherical lens, beam (2.16) has three vortex harmonics [17]: $\exp(im\vartheta)$, $\exp(i(m+2)\vartheta)$, and $\exp(i(m+1)\vartheta)$, instead of one vortex harmonic $\exp(im\vartheta)$. Therefore, the total beam power W after the spherical lens is distributed between these three harmonics $W = W_0 + W_2 + 2W_1$. The vortex harmonic with the topological charge $m + 1$ has a linear polarization directed along the optical axis z. That is, it propagates perpendicular to the optical axis and does not contribute to the longitudinal projection of the SAM. An optical vortex with topological charge $m+2$ has a left-hand circular polarization, and therefore its contribution to the longitudinal projection of the SAM will be negative. Therefore, the total longitudinal SAM of the beam (2.16) after the spherical lens is reduced by the sum of the powers of the indicated harmonics $\hat{S}_{z,in} - \hat{S}_{z,after} = W - (W_0 - W_2) = 2(W_1 + W_2)$. This difference,

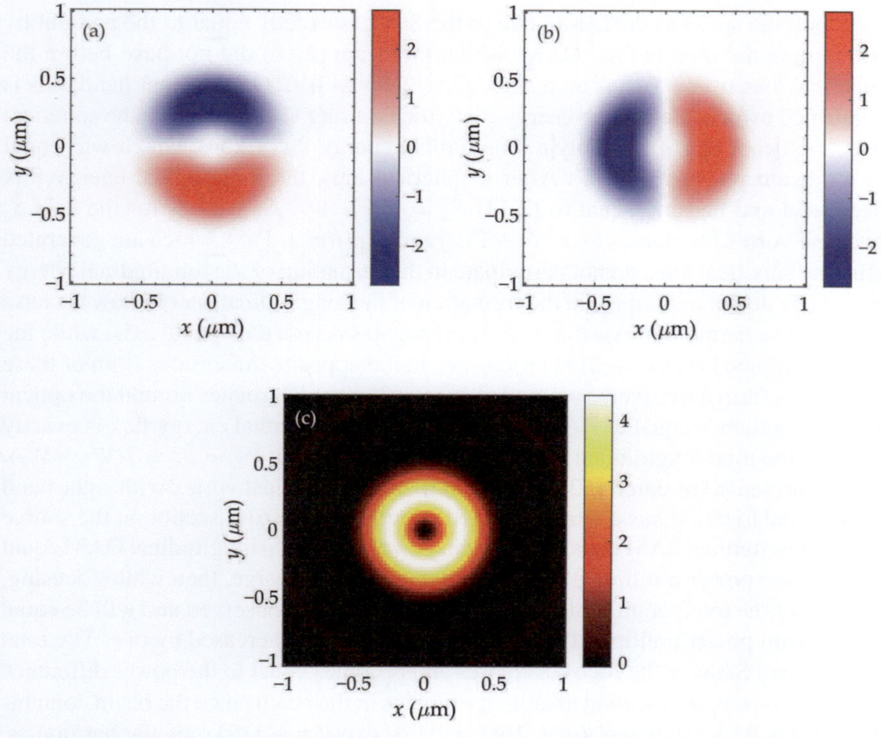

Fig. 2.8 Patterns of the SAM components S_x (**a**), S_y (**b**), S_z (**c**) in the focus of optical vortex with unit topological charge and with circular polarization for NA = 0.95

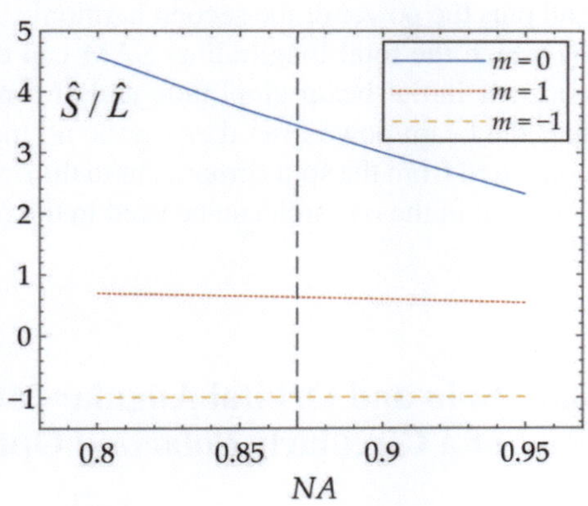

Fig. 2.9 Ratio $\hat{S}_z/\hat{L}_z$ in the focus of light field (2.16) at different numerical apertures of the focusing lens for beams (2.16) with different topological charges: $m = 0$ (without vortex, blue line 1), $m = 1$ (red line 2), $m = -1$ (yellow line 3). The vertical dotted line drawn through the NA = 0.87

by which the spin has decreased due to the SOC, is exactly equal to the amount by which the total longitudinal OAM, which the beam (2.16) did not have before the spherical lens if $m = 0$, has increased: $\hat{L}_z = 2(W_1 + W_2)$. On the other hand, this is confirmed by considering the energy flow (the Pointing vector). Before the spherical lens, the beam (2.16) had only a longitudinal energy flux (2.34), which was equal to the beam power $\hat{P}_z = W$. After a spherical lens, the longitudinal energy flux decreased and became equal to (2.33) $\hat{P}_z = W_0 - W_2$. As well as for the spin $\hat{S}_z$ two new vortex harmonics $\exp(i(m+2)\vartheta)$ and $\exp(i(m+1)\vartheta)$, which are generated after the spherical lens, do not participate in the formation of the longitudinal energy flux. They do not participate in the formation of the longitudinal energy flow because one of these harmonics $\exp(i(m+1)\vartheta)$ propagates across the optical axis, while the other harmonic $\exp(i(m+2)\vartheta)$ propagates in the opposite direction z. Both of these harmonics form a transverse azimuthal energy flux, which rotates around the optical axis, and which is equal to $\hat{P}_\vartheta = 2(W_1 + W_2)$. This azimuthal energy flux is exactly equal to the total longitudinal OAM of the beam at $m = 0$: $\hat{P}_\vartheta = \hat{L}_z = 2(W_1 + W_2)$.

The presented research [52] shows that if an initial optical vortex with right-hand circular polarization has a total (averaged over the beam cross-section in the source plane) longitudinal SAM equal to the beam power and a total longitudinal OAM equal to the beam power multiplied by the beam topological charge, then while focusing, the sum of the total longitudinal OAM and SAM will be conserved and will be equal to the beam power multiplied by the topological charge increased by one. The total longitudinal SAM in the focus decreases and becomes equal to the power difference between the zero and second angular harmonics in the beam since the beam contains zero $\exp(im\vartheta)$, second $\exp(i(m+2)\vartheta)$, and first $\exp(i(m+1)\vartheta))$ angular harmonics. The total longitudinal OAM in the focus will increase and become equal to the beam power multiplied by the topological charge, minus the power of the zero harmonic, and plus the power of the second harmonic. It is also shown that the maximum value by which the total longitudinal SAM can decrease upon focusing is achieved at a constant initial beam amplitude and at a numerical aperture of 1 and is equal to half the beam power. In other words, no more than half of the beam power can be converted from the spin component to the orbital component due to SOC. The results obtained in the research can be used in the problems of microparticles manipulation [53–55].

2.3 Spin and Orbital Angular Momenta in the Tight Focus of a Circularly Polarized Optical Vortex

Angular momentum (AM) of light is an important property of non-paraxial light fields with a different polarization state [56, 68]. Angular momentum of light can be decomposed into the spin angular momentum (SAM) and the orbital angular momentum (OAM) [57, 58]. Of special interest is studying the AM in the tight focus of light [59, 72]. To the angular momentum of light, a recent review is dedicated

[70]. A number of works [51, 60, 69] are dedicated to the spin-orbital conversion, which is inherent to the angular momentum and is especially prominent in the tight focus. In the tight focus, interesting optical phenomena are known, related with the AM and with the spin-orbital conversion: Hall effect [21, 26], "photonic wheels" [61], reverse energy flow [18]. The behavior of light in the tight focus is adequately described by the Richards-Wolf (RW) theory [16]. In [62, 71], the Poynting vector (electromagnetic energy flow) was shown to be the sum of two vectors: orbital flow and spin flow. The spin flow does not carry light energy, although the spin flow generates the SAM vector that can be transferred to an absorbing small particle and to rotate it about its center of mass. The presence of the spin term in the Poynting vector explains the reason of the reverse flow in the focus [18]. If the spin flow has a negative longitudinal component (parallel to the optical axis) and is greater by magnitude than the orbital energy flow (canonical energy flow), then in this area in space the Poynting vector has the negative longitudinal component.

In this section, based on the RW formalism, we derive exact expressions for the longitudinal components of the AM, SAM and OAM vectors in the focal plane of an optical vortex with right circular polarization. It turned out that the sum of the SAM and OAM is not equal to the AM. We found in a literature [63] a general expression for the AM vector via the SAM and OAM vectors. It turned out that only the total (integrated over the whole space) AM is equal to the sum of total momenta SAM and OAM.

2.3.1 Components of the Electric and Magnetic Field Vectors in the Focus

Here we consider an initial light field with the transverse components of the electric field strength vector being described by the following Jones vector:

$$\boldsymbol{E}(\theta,\varphi)=\frac{A(\theta)}{\sqrt{2}}\exp(in\varphi)\begin{pmatrix}1\\ i\end{pmatrix} \tag{2.37}$$

where φ is the angular polar coordinate in the plane of the exit pupil of the aplanatic optical system, θ is the azimuthal angle, describing the tilt of the light rays to the optical axis, $A(\theta)$ is the amplitude of the initial light field (real-valued function), rotationally symmetrical with respect to the optical axis, n is the topological charge of the optical vortex (integer number), $(1, i)$ is the transposed Jones vector for right circular polarization. For the initial field (2.37), tightly focused by an aplanatic system (ideal spherical lens), the RW formalism [16] allows deriving all the components of the electric and magnetic field strength vectors near the tight focus [64]:

$$E_x=\frac{i^{n-1}}{\sqrt{2}}\mathrm{e}^{in\varphi}\left(I_{0,n}+\mathrm{e}^{2i\varphi}I_{2,n+2}\right),$$

$$
\begin{aligned}
E_y &= \frac{i^n}{\sqrt{2}} \mathrm{e}^{in\varphi}\left(I_{0,n} - \mathrm{e}^{2i\varphi} I_{2,n+2}\right), \\
E_z &= -\sqrt{2} i^n \mathrm{e}^{i(n+1)\varphi} I_{1,n+1}, \\
H_x &= \frac{-i^n}{\sqrt{2}} \mathrm{e}^{in\varphi}\left(I_{0,n} + \mathrm{e}^{2i\varphi} I_{2,n+2}\right), \\
H_y &= \frac{i^{n-1}}{\sqrt{2}} \mathrm{e}^{in\varphi}\left(I_{0,n} - \mathrm{e}^{2i\varphi} I_{2,n+2}\right), \\
H_z &= \sqrt{2} i^{n+1} \mathrm{e}^{i(n+1)\varphi} I_{1,n+1}.
\end{aligned}
\tag{2.38}
$$

In Eq. (2.38), $I_{\nu,\mu}$ denote the following integrals, where the first index is the type of the integral ($\nu = 0, 1, 2$) and the second index ($\mu = 0, \pm 1, \pm 2, \ldots$) is equal to the order of the Bessel function $J_\nu(\xi)$ in the integrand:

$$
I_{\nu,\mu} = \left(\frac{4\pi f}{\lambda}\right) \int_0^{\theta_0} \sin^{\nu+1}\left(\frac{\theta}{2}\right) \cos^{3-\nu}\left(\frac{\theta}{2}\right) \cos^{1/2}(\theta) A(\theta) \mathrm{e}^{ikz\cos\theta} J_\mu(\xi) \mathrm{d}\theta \tag{2.39}
$$

where $k = 2\pi/\lambda$ is the wavenumber of light with the wavelength λ, f is the focal distance of an aplanatic system, (r, φ, z) are the cylindrical coordinates, θ_0 is the maximal tilt angle of rays to the optical axis, defining the numerical aperture of the aplanatic lens NA $= \sin(\theta_0)$, $\xi = kr\sin(\theta)$. Expressions (2.38) allow deriving the expressions for all main characteristics of light in the focus (at $z = 0$).

2.3.2 *Intensity of Light, Poynting Vector and the Spin Angular Momentum Vector in the Focus*

From Eq. (2.38), we obtain an expression for the intensity distribution in the focus of the field (2.37):

$$
\begin{aligned}
I(r, \varphi, z = 0) &= |E_x|^2 + \left|E_y\right|^2 + |E_z|^2 \\
&= I_{0,n}^2 + I_{2,n+2}^2 + 2I_{1,n+1}^2.
\end{aligned}
\tag{2.40}
$$

As seen in Eq. (2.40), the intensity in the focus is independent of the polar angle φ, i.e., is circularly symmetric. On the optical axis ($r = 0$), the intensity is non-zero only at $n = 0, -1, -2$. For other values n, the intensity (2.40) on the optical axis is zero, i.e., there is a light ring in the focus.

Further, using Eq. (2.38), we derive expressions for the components of the Poynting vector (energy flow) in the focus of the field (2.37). The Poynting vector **P** is given by [16]:

$$\boldsymbol{P} = \frac{c}{2\pi}\mathrm{Re}\big(\boldsymbol{E}^* \times \boldsymbol{H}\big) \tag{2.41}$$

where **E** and **H** are the strength vectors of the electric and magnetic fields, '*' means complex conjugation, × means the cross product, Re is the real part of a complex number, and c is the speed of light in vacuum. Below we omit the constant $c/(2\pi)$. Then, for the components of the energy flow vector in the focus we get the following expressions:

$$\begin{aligned} P_x &= -Q(r)\sin\varphi, \\ P_y &= Q(r)\cos\varphi, \\ P_z &= I_{0,n}^2 - I_{2,n+2}^2, \\ Q(r) &= I_{1,n+1}\big(I_{0,n} + I_{2,n+2}\big). \end{aligned} \tag{2.42}$$

In the cylindrical coordinates, the Poynting vector in the focus of the field (2.37) has the following components:

$$\begin{aligned} P_r &= 0, \\ P_\varphi &= Q(r), \\ P_z &= I_{0,n}^2 - I_{2,n+2}^2. \end{aligned} \tag{2.43}$$

As seen from Eq. (2.43), the energy flow in the transverse plane is rotated counterclockwise if $Q(r) > 0$, and clockwise if $Q(r) < 0$. The transverse energy flow near the optical axis, for instance, at $n = 0$ ($Q(r) > 0$, $kr \ll 1$), is rotated counterclockwise, similarly to the rotation of the polarization vector of the initial field (2.37). It is interesting that both at $n = -1$ and at $n = -2$, near the optical axis, $Q(r) < 0$ and the transverse energy flow is rotated in the opposite direction (clockwise), whereas at $n = -1$ and at $n = -2$ the energy near the optical axis flows in different directions:

$$\begin{aligned} P_{z,-1} &= I_{0,1}^2 - I_{2,1}^2 = \underbrace{\big(I_{0,1} - I_{2,1}\big)}_{>0}\underbrace{\big(I_{0,1} + I_{2,1}\big)}_{>0} > 0, \\ P_{z,-2} &= I_{0,2}^2 - I_{2,0}^2 < 0, kr << 1. \end{aligned} \tag{2.44}$$

The reverse energy flow in the focus was studied earlier in [18, 64]. Further, we obtain the components of the spin angular momentum (SAM) in the focus of the initial field (2.37). The SAM vector is given by [58, 59, 72]

$$\boldsymbol{S} = \frac{1}{8\pi\omega}\mathrm{Im}\big(\boldsymbol{E}^* \times \boldsymbol{E}\big) \tag{2.45}$$

with Im denoting the imaginary part of a complex number and ω being the angular frequency of light. Below we omit the constant $1/(8\pi\omega)$. Then the expressions for the components of the SAM vector (2.45) in the focus of the initial field (2.37) are quite the similar to the components of the Poynting vector (2.43):

$$
\begin{aligned}
S_r &= 0, \\
S_\varphi &= Q(r), \\
S_z &= I_{0,n}^2 - I_{2,n+2}^2.
\end{aligned}
\tag{2.46}
$$

According to Eq. (2.46), the transverse SAM vector is rotated in the focus counterclockwise or clockwise in the same cases as does the transverse energy flow. It is interesting that at $n = -2$, when there is the reverse energy flow ($P_z < 0$) in the focus on the optical axis, then the longitudinal component of the SAM vector (2.46) is also negative $S_{z,-2} = I_{0,2}^2 - I_{2,0}^2 < 0$. This means that near the optical axis in the focus, the polarization vector rotates clockwise (left circular polarization), although circular polarization in the initial plane (2.37) is right-handed. Such change of the direction of polarization rotation in the focus can be an indicator used for detecting the reverse flow in the focus. Non-zero transverse SAM vector in the focus means that in the longitudinal planes (x, z) and (y, z) near the focal plane, there is right or left circular polarization, i.e., the polarization vectors rotate reminiscing the photonic wheels [61].

2.3.3 *Angular Momentum and Orbital Angular Momentum at the Focus*

Further, we derive the longitudinal component of the angular momentum (AM) and of the orbital angular momentum (OAM) in the focus of the field (2.37). The AM vector is given by the following formula [65]

$$
\boldsymbol{J} = \frac{1}{2c}\mathrm{Re}\big(\boldsymbol{r} \times \big(\boldsymbol{E}^* \times \mathbf{H}\big)\big) = \frac{2\pi}{c^2}(\boldsymbol{r} \times \boldsymbol{P}) \tag{2.47}
$$

where all designations are described above. Below we omit the constant $2\pi/c^2$. Then the longitudinal AM component in the focus of the field (2.37) is equal to

$$
J_z = rP_\varphi = rQ(r) \tag{2.48}
$$

On the other hand, it is known that the AM vector is a sum of the SAM and OAM vectors [65]:

$$
\boldsymbol{J} = \boldsymbol{S} + \boldsymbol{L} = \frac{1}{8\pi\omega}\mathrm{Im}\big(\boldsymbol{E}^* \times \boldsymbol{E}\big) + \frac{1}{8\pi\omega}\sum_{p=x,y,z}\mathrm{Im}\Big(E_p^*(\boldsymbol{r} \times \nabla)E_p\Big) \tag{2.49}
$$

We note that the first work on the angular momentum of light [65] contains the expression (2.49) without the derivation from the expression (2.47).

From now on, we omit the constant $1/(8\pi\omega)$ in Eq. (2.49) for brevity. The first term in Eq. (2.49) in the focus we already derived Eq. (2.46), and now we will obtain the second term in Eq. (2.49). Thus, we get the longitudinal component of the OAM vector in the focus of the field (2.37):

$$\begin{aligned} L_z &= \mathrm{Im}\left(E_x^* \frac{\partial}{\partial\varphi} E_x + E_y^* \frac{\partial}{\partial\varphi} E_y + E_z^* \frac{\partial}{\partial\varphi} E_z \right) \\ &= n I_{0,n}^2 + (n+2) I_{2,n+2}^2 + 2(n+1) I_{1,n+1}^2. \end{aligned} \tag{2.50}$$

As seen from Eq. (2.46), the light field (2.37) in the focus has a non-zero longitudinal OAM component at an arbitrary value n. It is interesting that if $n \geq 0$ then the OAM is positive ($L_z > 0$), and if $n \leq -2$ then the OAM is negative ($L_z < 0$). If $n = -1$, the OAM can be both negative and positive. Even when $n = 0$, when there is no the optical vortex and only circular polarization (2.37) remains, the longitudinal OAM component is non-zero and equals

$$L_{z,0} = 2I_{2,2}^2 + 2I_{1,1}^2 \geq 0. \tag{2.51}$$

According to Eq. (2.51), due to spin-orbital conversion, circular polarization of a vortex-free field (2.37) generates in the focus non-negative longitudinal component of the OAM vector. This component is equal to zero only on the optical axis. Equation (2.51) shows that in the focus, the transverse energy flow is rotated counterclockwise.

Using Eq. (2.46), we can derive an expression for the longitudinal component of the AM vector in the focus of the field (2.37):

$$J_z = S_z + L_z = (n+1)\left(I_{0,n}^2 + I_{2,n+2}^2 + 2I_{1,n+1}^2\right) = (n+1)I. \tag{2.52}$$

Equation (2.52) indicates that the longitudinal AM component in the focus is equal to the light intensity in the focus, multiplied by the sum of the topological charge n and of the «spin» of the initial field equal to 1. Equation (2.52) shows that, in contrast to the OAM (2.50), the longitudinal AM component in the focus is not always non-zero. If $n = -1$, the angular momentum (2.52) in the focus is zero. This means that the optical vortex with the topological charge $n = -1$ compensates the influence of right circular polarization:

$$J_{z,-1} = 0 = S_{z,-1} + L_{z,-1}, \tag{2.53}$$

that is the longitudinal SAM component is equal by magnitude to the longitudinal OAM component and directed oppositely: $S_{z,-1} = -L_{z,-1} = I_{0,1}^2 - I_{2,1}^2$. It can be shown that the longitudinal AM component for an optical vortex with left circular polarization, in difference with Eq. (2.52), is given by

$$J_z = (n-1)\left(I_{0,n}^2 + I_{2,n-2}^2 + 2I_{1,n-1}^2\right) = (n-1)I. \tag{2.54}$$

Comparison of Eqs. (2.52) and (2.54) indicates that the longitudinal AM component in the focus is proportional to the algebraic sum of the normalized OAM of the initial field n and of the spin $\sigma = \pm 1$.

2.3.4 Is the AM a Sum of the SAM and OAM?

In this subsection, we consider some contradiction, which we failed to resolve completely, and for which we have not found an explanation in the literature [51, 59, 60, 69, 70]. The point is that Eq. (2.49) cannot be derived from Eq. (2.47). For the initial light field (2.37), Eq. (2.52) indicates that the AM in the focus is equal to zero at $n = -1$. However, according to Eq. (2.48), the AM in the focus of the same field (2.37) at $n = -1$ is non-zero: $J_{z,-1} = rI_{1,0}\left(-I_{0,1} + I_{2,1}\right) \neq 0$. Indeed, since the Poynting vector $\mathbf{P}$ in Eq. (2.47) can be represented as a sum of two flows [62]: the orbital flow $\mathbf{P_o}$ and the spin flow $\mathbf{P_s}$ ($\mathbf{P} = \mathbf{P_o} + \mathbf{P_s}$), then instead of Eq. (2.47), we get (the constant is omitted):

$$\boldsymbol{J} = (\boldsymbol{r} \times \boldsymbol{P}) = (\boldsymbol{r} \times \boldsymbol{P}_o) + (\boldsymbol{r} \times \boldsymbol{P}_s) = \boldsymbol{L} + \boldsymbol{S} \tag{2.55}$$

In Eq. (2.55), the orbital flow and the spin equal to flow are [71]

$$\begin{aligned} \boldsymbol{P}_0 &= \frac{1}{2\omega}\mathrm{Im}\left(\boldsymbol{E}^* \cdot (\nabla)\boldsymbol{E}\right), \\ \boldsymbol{P}_s &= \frac{1}{4\omega}\nabla \times \mathrm{Im}\left(\boldsymbol{E}^* \times \boldsymbol{E}\right), \end{aligned} \tag{2.56}$$

with ω being the angular frequency of monochromatic light. Omitting the constant $1/(2\omega)$, it can be shown that

$$\boldsymbol{L} = \boldsymbol{r} \times \boldsymbol{P}_0 = \mathrm{Im}\left(\boldsymbol{E}^* \cdot (\boldsymbol{r} \times \nabla)\boldsymbol{E}\right) = \sum_{p=x,y,z} \mathrm{Im}\left(E_p^*(\boldsymbol{r} \times \nabla)E_p\right), \tag{2.57}$$

but we failed to show that

$$\boldsymbol{S} = (\boldsymbol{r} \times \boldsymbol{P}_s) = \frac{1}{2}\left(\boldsymbol{r} \times \nabla \times \mathrm{Im}\left(\boldsymbol{E}^* \times \boldsymbol{E}\right)\right) = \mathrm{Im}\left(\boldsymbol{E}^* \times \boldsymbol{E}\right) \tag{2.58}$$

The physical reason is that despite the spin flow $\mathbf{P_s}$ presents in the expression for the Poynting vector, it does not carry energy. Since the divergence of the curl is zero, the spin flow $\mathbf{P_s}$ does not contribute to the differential conservation law of light energy (without currents and charges): $\partial W/\partial t = \mathrm{div}\,\mathbf{P}$, where W is the light energy density, t is time. However, the spin flow $\mathbf{P_s}$ generates the SAM $\mathbf{S}$, which is included into the expression for the angular momentum (2.52) and can be physically observed in the rotation of a small absorbing particle around its center of mass. We found the only work in the internet [63], where the above stated problem is formally solved. It is shown in [63] that the following integral identity is fulfilled:

$$\int \mathrm{Re}\big(\boldsymbol{r} \times (\boldsymbol{E}^* \times \mathbf{H})\big)dv = \int \mathrm{Im}\big(\boldsymbol{E}^* \times \boldsymbol{E}\big)\mathrm{d}v \\ + \int \sum_{p=x,y,z} \mathrm{Im}\Big(E_p^*(\boldsymbol{r} \times \nabla)E_p\Big)\mathrm{d}v. \tag{2.59}$$

The integrals in Eq. (2.59) are evaluated over the whole three-dimensional space, since the derivation of Eq. (2.59) supposes that the amplitude of the light field **E** tends to zero at the infinity. Indeed, if we replace in Eq. (2.59) the magnetic field by the curl of the electric field, we get

$$\widehat{\boldsymbol{J}} = \mathrm{Im} \iiint \big(\boldsymbol{r} \times (\boldsymbol{E}^* \times \mathrm{rot}\boldsymbol{E})\big)\mathrm{d}^3\boldsymbol{r} \tag{2.60}$$

For decomposing the summary AM into the SAM and the OAM, we use the following vectorial identity

$$\boldsymbol{A} \times \mathrm{rot}\boldsymbol{B} = \nabla_{\boldsymbol{B}}(\boldsymbol{A} \cdot \boldsymbol{B}) - (\boldsymbol{A} \cdot \nabla)\boldsymbol{B} \tag{2.61}$$

where $\nabla_{\boldsymbol{B}}$ is the Feynman subscript notation, which considers only the variation due to the vector field **B** [66]. Then we get

$$\widehat{\boldsymbol{J}} = \mathrm{Im} \iiint \left[\sum_{p=x,y,z} E_p^*(\boldsymbol{r} \times \nabla)E_p - \big(\boldsymbol{E}^* \cdot \nabla\big)(\boldsymbol{r} \times \boldsymbol{E}) + \big(\boldsymbol{E}^* \times \boldsymbol{E}\big)\right]\mathrm{d}^3\boldsymbol{r} \tag{2.62}$$

The first and the third terms in this expression are exactly the summary OAM $\widehat{\boldsymbol{L}}$ and the summary SAM $\widehat{\boldsymbol{S}}$. The second term can be transformed by integrating by parts:

$$\widehat{\boldsymbol{J}} = \widehat{\boldsymbol{L}} + \mathrm{Im} \iiint \big[(\nabla \cdot \boldsymbol{E}^*)(\boldsymbol{r} \times \boldsymbol{E})\big]\mathrm{d}^3\boldsymbol{r} + \widehat{\boldsymbol{S}} \tag{2.63}$$

where

$$\widehat{\boldsymbol{L}} = \iiint \boldsymbol{L}\mathrm{d}^3\boldsymbol{r},\ \widehat{\boldsymbol{S}} = \iiint \boldsymbol{S}\mathrm{d}^3\boldsymbol{r} \tag{2.64}$$

The second term in the integrand in Eq. (2.63) for the summary AM is proportional to the divergence of the electric field and is thus equal to zero. Then, the total AM is equal to the sum of the total SAM and OAM, i.e., equal to the expression (2.59). Thus, the second equation in Eq. (2.55) for the AM density is not valid. Instead, only the integral identity (2.59) is valid. The correct expression for the AM density follows from Eq. (2.62) and reads as

$$\begin{aligned} \boldsymbol{J} &= \mathrm{Im}\left\{ \sum_{p=x,y,z} E_p^*(\boldsymbol{r} \times \nabla)E_p - \left(\boldsymbol{E}^* \cdot \nabla\right)(\boldsymbol{r} \times \boldsymbol{E}) + \left(\boldsymbol{E}^* \times \boldsymbol{E}\right) \right\} \\ &= \boldsymbol{L} - \mathrm{Im}\left\{\left(\boldsymbol{E}^* \cdot \nabla\right)(\boldsymbol{r} \times \boldsymbol{E})\right\} + \boldsymbol{S}. \end{aligned} \tag{2.65}$$

The second term in Eq. (2.65) does not have a physical meaning since it vanishes in the integration (2.62).

2.3.5 *Light Field in the Focus, Obtained by the Richards-Wolf Theory, is a Solution of the Maxwell's Equations*

In this subsection, we obtain expressions for the magnetic components of the field (2.38) from the expressions for the electric field components by using the Maxwell's equations for a monochromatic light. The magnetic field strength vector of a monochromatic light is related with the electric field strength vector by the well-known expression:

$$\boldsymbol{H} = \frac{1}{i\omega\mu_0\mu}\mathrm{rot}\boldsymbol{E} \tag{2.66}$$

where μ_0 and μ are respectively the magnetic permeability of free space and of the material. For the longitudinal component of the magnetic field, we get:

$$\begin{aligned} H_z &= \frac{1}{i\omega\mu_0\mu}(rotE)_z = \frac{1}{i\omega\mu_0\mu}\left[\frac{1}{r}E_\varphi + \frac{\partial E_\varphi}{\partial r} - \frac{1}{r}\frac{\partial E_r}{\partial \varphi}\right] \\ &= \frac{k}{\omega\mu_0\mu}\left[\sqrt{2}i^{n+1}\mathrm{e}^{i(n+1)\varphi}I_{1,n+1}\right]. \end{aligned} \tag{2.67}$$

In derivation, we used the following auxiliary expressions:

$$\begin{aligned} \frac{\partial}{\partial r}\left(I_{0,n} - I_{2,n+2}\right) &= \frac{n}{r}I_{0,n} + \frac{n+2}{r}I_{2,n+2} - 2kI_{1,n+1}, \\ \frac{\partial E_\varphi}{\partial r} &= \frac{i^n}{\sqrt{2}}e^{i(n+1)\varphi}\left(\frac{n}{r}I_{0,n} + \frac{n+2}{r}I_{2,n+2} - 2kI_{1,n+1}\right). \end{aligned} \tag{2.68}$$

Similarly, we derive expressions for the transverse components of the magnetic vector in the polar and then in the Cartesian coordinates:

$$H_r = \frac{1}{i\omega\mu_0\mu}(\mathrm{rot}E)_r = \frac{1}{i\omega\mu_0\mu}\left(\frac{1}{r}\frac{\partial E_z}{\partial \varphi} - \frac{\partial E_\varphi}{\partial z}\right)$$

$$
\begin{aligned}
&= \frac{1}{i\omega\mu_0\mu}\frac{-k}{\sqrt{2}}i^{n+1}e^{i(n+1)\varphi}\left(I_{0,n} + I_{2,n+2}\right),\\
H_\varphi &= \frac{1}{i\omega\mu_0\mu}(\mathrm{rot}E)_\varphi = \frac{1}{i\omega\mu_0\mu}\left(\frac{\partial E_r}{\partial z} - \frac{\partial E_z}{\partial r}\right)\\
&= \frac{1}{i\omega\mu_0\mu}\frac{k}{\sqrt{2}}i^n e^{i(n+1)\varphi}\left(I_{0,n} - I_{2,n+2}\right),\\
H_x &= H_r\cos\varphi - H_\varphi\sin\varphi = \frac{k}{\omega\mu_0\mu}\frac{-i^n}{\sqrt{2}}e^{in\varphi}\left(I_{0,n} + e^{2i\varphi}I_{2,n+2}\right),\\
H_y &= H_\varphi\cos\varphi + H_r\sin\varphi = \frac{k}{\omega\mu_0\mu}\frac{i^{n-1}}{\sqrt{2}}e^{in\varphi}\left(I_{0,n} - e^{2i\varphi}I_{2,n+2}\right).
\end{aligned}
\tag{2.69}
$$

If the constant multiplier $k/(\omega\mu_0\mu)$ is omitted, then the last two expressions in Eq. (2.69) and the expression (2.67) coincide with the Cartesian components of the magnetic field in Eq. (2.38). For deriving the expressions (2.69), we used auxiliary expressions, similar to Eq. (2.38):

$$
\frac{\partial}{\partial z}\left(I_{0,n} - I_{2,n+2}\right) = ikI_{0,n} + ikI_{2,n+2} - 2i\frac{n+1}{r}I_{1,n+1}.
\tag{2.70}
$$

$$
\frac{\partial}{\partial z}\left(I_{0,n} + I_{2,n+2}\right) = ikI_{0,n} - ikI_{2,n+2} - 2i\frac{\partial}{\partial r}I_{1,n+1}.
\tag{2.71}
$$

Since the components of the light field near the focus (2.38), derived by the Richards-Wolf theory, are exact solutions of the Maxwell's equations, then this theory exactly describes the light field in the tight focus in free space.

2.3.6 Explaining Some Experiments on Microparticles Rotation

It is interesting that the above developed theory explains the experiment described in [27]. In this work, an absorbing microparticle is rotated along a light ring in the tight focus of an optical vortex with the topological charge $n = 1$ and with right circular polarization. When right circular polarization was changed to left circular polarization, the particle continued rotation in the same direction (counterclockwise), but with a lower speed. This phenomenon can be explained the following way. For comparison, the transverse energy flow in the focus of an optical vortex with right (2.43) and left circular polarization is given by

$$
\begin{aligned}
P_{\varphi,R} &= I_{1,n+1}\left(I_{0,n} + I_{2,n+2}\right),\\
P_{\varphi,L} &= I_{1,n-1}\left(I_{0,n} + I_{2,n-2}\right).
\end{aligned}
\tag{2.72}
$$

For the topological charge $n = 1$, instead of Eq. (2.72), we get

$$P_{\varphi,R,1} = I_{1,2}(I_{0,1} + I_{2,3}) > 0,$$
$$P_{\varphi,L,1} = I_{1,0}(I_{0,1} - I_{2,1}) > 0, kr << 1. \quad (2.73)$$

According to Eq. (2.73), the transverse energy flow for a vortex with left and with right circular polarization does not change sign, i.e., the light energy in the focal plane is rotated along a ring counterclockwise, but has a different magnitude. For left circular polarization, the energy flow is lower than that for right circular polarization. Besides, the longitudinal SAM component for right and left circularly polarized vortices

$$S_{z,R} = I_{0,n}^2 - I_{2,n+2}^2,$$
$$S_{z,L} = -(I_{0,n}^2 - I_{2,n-2}^2), \quad (2.74)$$

has different signs at $n = 1$: the SAM is positive for right circular polarization and negative for left circular polarization:

$$S_{z,R,1} = I_{0,1}^2 - I_{2,3}^2 > 0,$$
$$S_{z,L,1} = -(I_{0,1}^2 - I_{2,1}^2) < 0. \quad (2.75)$$

Thus, for left circular polarization, the OAM and the SAM rotate a particle in different directions. However, in the experiment, the particle is rotated by the left circularly polarized light still counterclockwise. This means that the OAM of light more affects the particle, than the SAM. It is this that explains why the particle is rotated in the same direction both for left and right circular polarization, but with a speed, lower for left circular polarization. We note that there are no such detailed explanations of the experiment results in [27].

The above obtained expressions also explain another experiment described in [67]. In this work, in the focus of a circularly polarized light beam, an aspherical dielectric microparticle was rotated around the optical axis. When the polarization handedness was changed (from left-handed to right-handed), the particle also changed its rotation direction (from clockwise rotation to counterclockwise rotation). This can be explained by the transverse energy flow in the focus of a Gaussian beam with left and right polarizations, which is directed in different sides and has equal by magnitude value:

$$P_{\varphi,R} = I_{1,1}(I_{0,0} + I_{2,2}),$$
$$P_{\varphi,L} = -I_{1,1}(I_{0,0} + I_{2,2}). \quad (2.76)$$

The longitudinal SAM component is also of different sign for left and right circular polarization:

$$S_{z,R} = I_{0,0}^2 - I_{2,2}^2,$$
$$S_{z,L} = -(I_{0,0}^2 - I_{2,2}^2). \quad (2.77)$$

Therefore, a particle in the focus is rotated with the same velocity, but in different directions for the light with left and right circular polarization.

2.3.7 Simulation

Numerical simulation was conducted by computing the Debye integrals within the framework of the Richards-Wolf formalism. Figure 2.10 illustrates the intensity distributions I (2.40), as well as distributions of the longitudinal components of the SAM S_z (2.46), OAM L_z (2.50), second term in Eq. (2.55), and full AM J_z (2.48) in the tight focus of a vortex field with right circular polarization and with homogeneous unit initial amplitude $A(\theta) = 1$ (2.1) for $n = 0, -1, -2$ at the following parameters: wavelength $\lambda = 532$ nm, focal length of the focusing lens $f = 10$ μm, numerical aperture NA $= \sin\theta_0 = 0.95$, computation domain 2×2 μm.

As seen from Fig. 2.10f–j, at $n = -1$, the SAM (2.46) and the OAM (2.50) indeed compensate each other, their sum is zero, and the total AM (2.48) consists only one component equal to the second term in Eq. (2.65). At $n = -2$ (Fig. 2.10k–o), the SAM (2.46) and the OAM (2.50) are similar by shape, but no longer compensate each other, the OAM is greater by magnitude, and thus contributes to the total AM

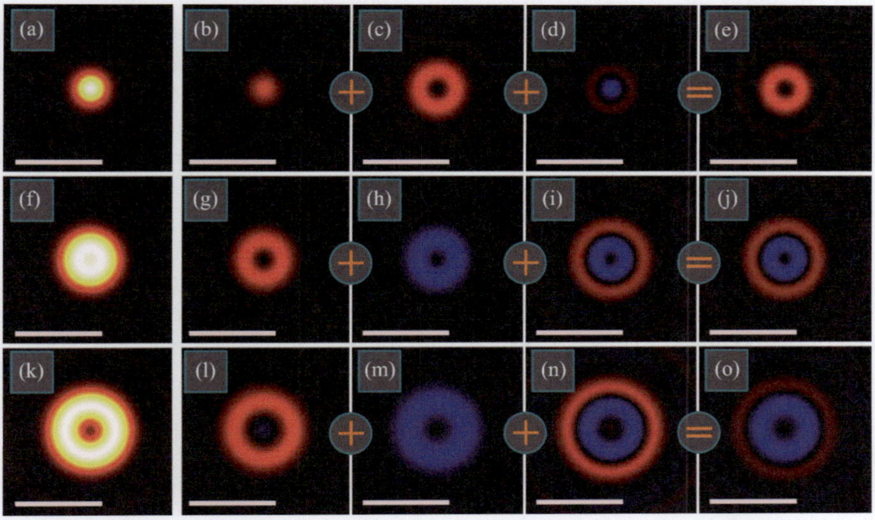

Fig. 2.10 Intensity distributions (**a**, **f**, **k**), as well as distributions of the SAM (2.46) (**b**, **g**, **l**), OAM (2.50) (**c**, **h**, **m**), second term in Eq. (2.65) (**d**, **i**, **n**), and full AM (2.48) (**e**, **j**, **o**) in the tight focus of a plane vortex field ($A(\theta) = 1$) with right circular polarization (2.37) for $n = 0$ (a-e), $n = -1$ (**f–j**), $n = -2$ (**k–o**) at the following parameters: wavelength $\lambda = 532$ nm, focal length of the focusing lens $f = 10$ μm, numerical aperture $\sin\theta_0 = 0.95$, computation domain 2×2 μm. Red and blue colors mean respectively positive and negative values. Scale mark in all figures denotes 1 μm

along with the component defined by the second term in Eq. (2.65). This component includes an inner light ring with a negative AM density and an outer ring with a positive AM density. Due to the summation with the OAM, the outer ring in the total AM is weaker than the inner one. At $n = 0$ (Fig. 2.10a–e), the SAM has a shape of a spot, while the OAM has a shape of a ring, and both they are positive. The component, defined by the second term in Eq. (2.65), contains a central spot with a negative AM and a ring with a positive AM. The central spot compensates the SAM, whereas the ring with the positive AM is summated with the positive OAM, thus generating a ring-shaped distribution of the total positive AM. Therefore, Fig. 2.10 indicates that the summary longitudinal AM component (integrated over the beam cross-section) is positive at $n = 0$ (Fig. 2.10e) and will rotate a nearly 1-μm-size particle with a center on the optical axis counterclockwise. The summary longitudinal AM component at $n = -2$ is negative (Fig. 2.10o) and will rotate a small particle clockwise. We note that initial polarization of the field (2.37) is right-handed and the polarization vector rotates counterclockwise. Nevertheless, the particle in the focus should be rotated clockwise. Summary longitudinal AM component at $n = -1$ is almost zero (Fig. 2.10j), and therefore a small particle with the center on the optical axis, placed into the focus, will not be rotated.

In this section, based on the Richards-Wolf formalism, simple analytical expressions have been derived for the longitudinal component of the AM density vector (2.48) of light in the focus of an optical vortex with a topological charge n and with right circular polarization [50]. In addition, expressions have been obtained for the SAM (2.46) and for the longitudinal OAM component (2.50). We have demonstrated that the sum of the SAM and OAM densities is not equal to the AM density. It has been shown earlier that the sum of only the total (integrated over the whole space) SAM and OAM is equal to the total AM of the light beam (2.59). We have also shown in a general form that the AM density is equal to the sum of three terms (2.65): SAM, OAM, and a certain third term, that does not contribute into the total AM (2.59). It is also shown in this work that the expressions (2.38) for the strength vectors of the electric and magnetic fields near the tight focus, derived by using the RW formalism, are exact solutions of the Maxwell's Eqs. (2.67)–(2.69). Thus, the RW theory exactly describes the behavior of light near the tight focus in free space. The simulation results (Fig. 2.10) agree with the theoretical outcomes.

References

1. M. Onoda, S. Marakami, N. Nagaosa, Hall effect of light. Phys. Rev. Lett. **93**, 083901 (2004)
2. K.Y. Bliokh, Y.P. Bliokh, Topological spin transport of photons: the optical Magnus effect and Berry phase. Phys. Lett. A **333**, 181–186 (2004)
3. K.Y. Bliokh, Y.P. Bliokh, Conservation of angular momentum, transverse shift, and spin Hall effect in reflection and refraction of an electromagnetic wave packet. Phys. Rev. Lett. **96**, 073903 (2006)
4. A. Kavokin, G. Malpuech, M. Glazov, Optical Spin Hall Effect. Phys. Rev. Lett. **95**(13), 136601 (2005)

5. O. Hosten, P. Kwiat, Observation of the spin Hall effect of light via weak measurements. Science **319**, 787–790 (2008)
6. X. Ling, X. Zhou, K. Huang, Y. Liu, C. Qiu, H. Luo, S. Wen, Recent advances in the spin Hall effect of light. Rep. Prog. Phys. **80**, 066401 (2017)
7. S. Liu, S. Chen, S. Wen, H. Luo, Photonics spin Hall effect: fundamentals and emergent applications. Opto-Electr. Sci. **1**, 220007 (2022)
8. X. Ling, X. Yi, X. Zhou, Y. Liu, W. Shu, H. Luo, S. Wen, Realization of tunable spin-dependent splitting in intrinsic photonic spin Hall effect. Appl. Phys. Lett. **105**, 151101 (2014)
9. X. Yin, Z. Ye, J. Rho, Y. Wang, X. Zhang, Photonic spin Hall effect at metasurfaces. Science **339**, 1405 (2013)
10. R.N. Kumar, Yatish, S.D. Gupta, N. Ghosh, A. Banerjee, Probing the rotational spin-Hall effect in a structured Gaussian beam. Phys. Rev. A **105**, 023503 (2022)
11. J. Zhang, X. Zhou, X. Ling, S. Chrn, H. Luo, S. Wen, Orbit-orbit interaction and photonics orbital Hall effect in reflection of a light beam. Chin. Phys. B **23**, 064215 (2014)
12. S. Fu, C. Guo, G. Liu, Y. Li, H. Yin, Z. Li, Z. Chen, Spin-orbit optical Hall effect. Phys. Rev. Lett. **123**, 243904 (2019)
13. F. Zhang, Y. Guo, M. Pu, X. Li, X. Ma, X. Luo, Metasurfaces enabled by asymmetric photonic spin-orbit interactions. Opto-Electr. Eng. **47**, 200366 (2020)
14. W. Shu, C. Lin, J. Wu, S. Chen, X. Ling, X. Zhou, H. Luo, S. Wen, Three-dimensional spin Hall effect of light in tight focusing. Phys. Rev. A **101**, 023819 (2020)
15. K.Y. Bliokh, E.A. Ostrovskaya, M.A. Alonso, O.G. Rodriguez-Herrera, D. Lara, C. Dainty, Spin-to-orbital angular momentum conversion in focusing, scattering, and imaging systems. Opt. Express **19**, 26132–26149 (2011)
16. B. Richards, E. Wolf, Electromagnetic diffraction in optical systems. II. Structure of the image field in an aplanatic system. Proc. R. Soc. Lond. A **253**, 358–379 (1959)
17. V.V. Koltyar, A.G. Nalimov, S.S. Stafeev, Exploiting the circular polarization of light to obtain a spiral energy flow at the subwavelength focus. J. Opt. Soc. Am. B **36**, 2850–2855 (2019)
18. V.V. Kotlyar, A.A. Kovalev, A.G. Nalimov, Energy density and energy flux in the focus of an optical vortex: reverse flux of light energy. Opt. Lett. **43**, 2921–2924 (2018)
19. V. Kotlyar, S. Stafeev, V. Zaitsev, E. Kozlova, Spin-orbital conversion with the tight focus of an axial superposition of a high-order cylindrical vector beam and a beam with linear polarization. Micromachines **13**, 1112 (2022)
20. V.V. Kotlyar, A.A. Kovalev, S.S. Stafeev, A.G. Nalimov, S. Rasouli, Tightly focusing vector beams containing V-point polarization singularities. Opt. Las. Tech. **145**, 107479 (2022)
21. V.V. Kotlyar, S.S. Stafeev, E.S. Kozlova, M.A. Butt, High-order orbital and spin hall effects at the tight focus of laser beams. Photonics **9**(12), 970 (2022)
22. K. Bliokh, F. Rodríguez-Fortuño, F. Nori, A.V. Zayats, Spin–orbit interactions of light. Nat. Photon. **9**, 796–808 (2015)
23. K.Y. Bliokh, Geometrical optics of beams with vortices: berry phase and orbital angular momentum hall effect. Phys. Rev. Lett. **97**, 043901 (2006)
24. N. Bokor, Y. Iketaki, T. Watanabe, M. Fujii, Investigation of polarization effects for high-numerical-aperture first-order Laguerre-Gaussian beams by 2D scanning with a single fluorescent microbead. Opt. Express **13**, 10440–10447 (2005)
25. A.A. Kovalev, V.V. Kotlyar, Spin hall effect of double-index cylindrical vector beams in a tight focus. Micromachines **14**, 494 (2023)
26. V. Kotlyar, S.S. Stafeev, A.A. Kovalev, V.D. Zaitsev, Spin hall effect before and after the focus of a high-order cylindrical vector beam. Appl. Sci. **12**(23), 12218 (2022)
27. Y. Zhao, J.S. Edgar, G.D.M. Jeffries, D. McGloin, D.T. Chiu, Spin-to-orbital angular momentum conversion in a strongly focused optical beam. Phys. Rev. Lett. **99**, 073901 (2007)
28. A. Dogariu, C. Schwartz, Conservation of angular momentum of light in single scattering. Opt. Express **14**, 8425–8433 (2006)
29. D. Haefner, S. Sukhov, A. Dogariu, Spin Hall effect of light in spherical geometry. Phys. Rev. Lett. **102**, 123903 (2009)

30. K.Y. Bliokh, E.A. Ostrovskaya, M.A. Alonso, O.G. Rodríguez-Herrera, D. Lara, C. Dainty, Spin-to-orbit angular momentum conversion in focusing, scattering, and imaging systems. Opt. Express **19**, 26132–26149 (2011)
31. N.B. Baranova, A.Y. Savchenko, B.Y. Zel'dovich, Transverse shift of a focal spot due to switching of the sign of circular-polarization, JETP Lett. **59**, 232–234 (1994)
32. B.Y. Zel'dovich, N.D. Kundikova, L.F. Rogacheva, Observed transverse shift of a focal spot upon a change in the sign of circular polarization. JETP Lett. **59**, 766–769 (1994)
33. M.Y. Darsht, B.Y. Zel'dovich, I.V. Kataevskaya, N.D. Kundikova, Formation of an isolated wavefront dislocation. JETP Lett. **80**, 817–821 (1995)
34. E. Brasselet, Y. Izdebskaya, V. Shvedov, A.S. Desyatnikov, W. Krolikowski, Y.S. Kivshar, Dynamics of optical spin-orbit coupling in uniaxial crystals. Opt. Lett. **34**, 1021–1023 (2009)
35. M.V. Berry, M.R. Jeffrey, M. Mansuripur, Orbital and spin angular momentum in conical diffraction. J. Opt. A Pure Appl. Opt. **7**, 685–690 (2005)
36. Z. Bomzon, G. Biener, V. Kleiner, E. Hasman, Space-variant Pancharatnam-Berry phase optical elements with computer-generated subwavelength gratings. Opt. Lett. **27**, 1141–1143 (2002)
37. G. Biener, A. Niv, V. Kleiner, E. Hasman, Formation of helical beams by use of Pancharatnam-Berry phase optical elements. Opt. Lett. **27**, 1875–1877 (2002)
38. L. Marrucci, C. Manzo, D. Paparo, Optical spin-to-orbital angular momentum conversion in inhomogeneous anisotropic media. Phys. Rev. Lett. **96**, 163905 (2006)
39. S.S. Kruk, M. Decker, I. Staude, S. Schlecht, M. Greppmair, D.N. Neshev, Y.S. Kivshar, Spin-polarized photon emission by resonant multipolar nanoantennas. ACS Photon. **1**, 1218–1223 (2014)
40. L.T. Vuong, A.J.L. Adam, J.M. Brok, P.C.M. Planken, H.P. Urbach, Electromagnetic spin-orbit interactions via scattering of subwavelength apertures. Phys. Rev. Lett. **104**, 083903 (2010)
41. S.-H. Gong, F. Alpeggiani, B. Sciacca, E.C. Garnett, L. Kuipers, Nanoscale chiral valley-photon interface through optical spin-orbit coupling. Science **359**, 443–447 (2018)
42. J. Ni, S. Liu, Y. Chen, G. Hu, Y. Hu, W. Chen, J. Li, J. Chu, C.-W. Qiu, D. Wu, Direct observation of spin-orbit interaction of light via Chiroptical responses. Nano Lett. **22**, 9013–9019 (2022)
43. H. Dai, L. Yuan, C. Yin, Z. Cao, X. Chen, Direct visualizing the spin hall effect of light via ultrahigh-order modes. Phys. Rev. Lett. **124**, 053902 (2020)
44. K.Y. Bliokh, D. Smirnova, F. Nori, Quantum spin hall effect of light. Science **348**, 1448–1451 (2015)
45. X. Wenhao, S. Yichang, L. Hailu, Spin-orbit interaction of light. Chin. J. Quantum Electron. **39**, 159–181 (2022)
46. W. Zhang, Y. Wang, D. Xu, H. Luo, Spin-orbit interaction of light: When twisted light meets twisted metasurfaces. Phys. Rev. A **107**, 043502 (2023)
47. A. Aiello, N. Lindlein, C. Marquardt, G. Leuchs, Transverse angular momentum and geometric spin hall effect of light. Phys. Rev. Lett. **103**, 100401 (2009)
48. Y. Zhang, P. Li, S. Liu, J. Zhao, Unveiling the photonic spin Hall effect of freely propagating fan-shaped cylindrical vector vortex beams. Opt. Lett. **40**, 4444–4447 (2015)
49. S. Mardonov, E.Y. Sherman, J.G. Muga, H.W. Wang, Y. Ban, X. Chen, Collapse of spin-orbit-coupled Bose-Einstein condensates. Phys. Rev. A **91**, 043604 (2015)
50. V.V. Kotlyar, A.A. Kovalev, A.M. Telegin, Angular and orbital angular momenta in the tight focus of a circularly polarized optical vortex. Photonics **10**(2), 160 (2023)
51. K. Bliokh, A. Bekshaev, F. Nori, Extraordinary momentum and spin in evanescent waves. Nat. Commun. **5**, 3300 (2014)
52. V.V. Kotlyar, S.S. Stafeev, V.D. Zaitsev, A.M. Telegin, E.S. Kozlova, Spin–orbital transformation in a tight focus of an optical vortex with circular polarization. Appl. Sci. **13**, 8361 (2023)
53. A.V. Arzola, L. Chvátal, P. Jákl, P. Zemnek, Spin to orbital light momentum conversion visualized by particle trajectory. Sci. Rep. **9**, 4127 (2019)
54. P. Chandravati, Numerical study of spin–orbit interaction of light in nonparaxial focusing of Gaussian beams. Optik **228**, 166199 (2021)

55. Q. Wang, C. Tu, H. He, Z. Xia, X. Hou, Y. Li, H. Wang, Local angular momentum induced dual orbital effect. APL Photon. **7**, 086102 (2022)
56. R.A. Beth, Direct detection of the angular momentum of light. Phys. Rev. **48**, 471 (1935)
57. S.M. Barnett, On the six components of optical angular momentum. J. Opt. **13**, 064010 (2011)
58. M. Mansuripur, Spin and orbital angular momenta of electromagnetic waves in free space. Phys. Rev. A **84**, 033838 (2011)
59. K.Y. Bliokh, A.Y. Bekshaev, F. Nori, Dual electromagnetism: helicity, spin, momentum, and angular momentum. New J. Phys. **15**, 033026 (2013)
60. K.Y. Bliokh, M.A. Alonso, E.A. Ostrovskaya, A. Aiello, Angular momenta and spin-orbit interaction of nonparaxial light in free space. Phys. Rev. A **82**, 063825 (2010)
61. O. Graydon, Photonic wheel. Nat. Photon. **7**, 672 (2013)
62. A.Y. Bekshaev, M.S. Soskin, Transverse energy flows in vectorial fields of paraxial beams with singularities. Opt. Commun. **271**, 332–348 (2007)
63. K.T. McDonald, *Orbital and Spin Angular Momentum of Electromagnetic Fields* (March 2009, update Oct. 2021). http://kirkmcd.princeton.edu/examples/spin.pdf
64. V.V. Kotlyar, S.S. Stafeev, A.A. Kovalev, Reverse and toroidal flux of light fields with both phase and polarization higher-order singularities in the sharp focus area. Opt. Express **27**, 16689–16702 (2019)
65. J. Humblet, Sur le moment d'impulsion d'une onde électromagnétique. Physica **10**, 585–603 (1943)
66. R.P. Feynman, R.B. Leighton, M. Sands, *The Feynman Lectures on Physics* (Addison-Wesley, 1964), vol. II, pp. 27–24. ISBN 0-8053-9049-9
67. V.V. Kotlyar, A.G. Nalimov, A.A. Kovalev, A.P. Porfirev, S.S. Stafeev, Spin-orbit and orbit-spin conversion in the sharp focus of laser light: Theory and experiment. Phys. Rev. A **102**, 033502 (2020)
68. R.A. Beth, Mechanical detection and measurement of the angular momentum of light. Phys. Rev. **50**(2), 115–125 (1936)
69. A. Bekshaev, K.Y. Bliokh, M. Soskin, Internal flows and energy circulation in light beams. J. Opt. **13**, 053001 (2011)
70. J.-X. Guo, W.-Y. Wang, T.-Y. Cheng, J.-Q. Lü, Interaction of spin-orbit angular momentum in the tight focusing of structured light. Front. Phys. **10**, 1079265 (2022)
71. M.V. Berry, Optical currents. J. Opt. A Pure Appl. Opt. **11**, 094001 (2009)
72. X. Zhang, B. Shen, Z. Zhu, G. Rui, J. He, Y. Cui, B. Gu, Understanding of transverse spin angular momentum in tightly focused linearly polarized vortex beams. Opt. Express **30**(4), 5121–3510 (2022)

Chapter 3
Focusing of Cylindrical Vector Beams and Their Modifications

3.1 Tightly Focusing Vector Beams Containing V-Point Polarization Singularities

In recent years, high-order vector light fields, whose linear polarization vector varies across the beam cross section, have been at the focus of research [1–6]. Such beams can be produced with a variety of techniques, including components with optical metasurfaces [7]. The vector beams feature a robust intensity profile on propagation through turbulence [8] and polarization singularity points [9–11] that, in many respects, are similar to phase singularity points of vortex fields [12]. Polarization singularity points (V-points) are intensity nulls in a vector field where the linear polarization vector is indefinite. The V-points are characterized [10] by a Poincare-Hopf index denoted by η, which equals the number of integer phase steps by 2π when making a full circle around the V-point. The phase is understood as the argument of a complex field composed of transverse E-field components, $E_x + iE_y$. This definition is similar to a relationship utilized in Ref. [12] to calculate the topological charge (TC) of a scalar vortex field with complex amplitude $E(x, y)$. V-points can also be characterized using a Stokes index σ, which is defined through the Poincare-Hopf index η as $\sigma = 2\eta$ and also equals the number of integer phase steps by 2π of a complex Stokes field when making a full circle around the V-point. With the unit Stokes vector $\mathbf{S} = (S_1, S_2, S_3)$ [13] having three components, the complex Stokes field is composed of the first two components: $S_c = S_1 + iS_2$. The phase of the complex Stokes field is the argument of a complex number S_c.

In this section, we derive the Poincare-Hopf and Stokes indices η and σ for nth-order cylindrical vector beams. We show that in the source plane of the beams (where the on-axis field component is zero), fields of linear polarization vectors are formed centered at the V-points, which look like a 'flower' or a 'web', with the number of petals depending on the vector field order n. Using Richards-Wolf formulae, we derive expressions for E-vector components at the tight focus for three types of vector fields, namely, for nth-order radial polarization (n is positive), –nth-order radial polarization

V. V. Kotlyar et al., *Optical Hall Effect in the Sharp Focus of Laser Light*,
https://doi.org/10.1007/978-3-031-64683-6_3

(–n is negative), and nth-order azimuthal polarization. Relying on the expressions derived for the complex E-field amplitudes, we deduce expressions for transverse intensity profiles of the fields of interest. Based on the expressions derived, we obtain a major finding of this work, showing that the number of petals of the 'polarization flower' of the initial vector field equals the number of local intensity maxima at the focal plane. We also show that a V-point of an nth-order vector field is 'disintegrated' at the tight focus into several first-order points with no petals around them.

3.1.1 Vector Field Polarization Index in the Source Plane

Let us analyze an nth-order azimuthally polarized source field whose Jones vector takes the form [14, 15]:

$$\mathbf{E_n}(\varphi) = \begin{pmatrix} -\sin n\varphi \\ \cos n\varphi \end{pmatrix} \tag{3.1}$$

where (r, φ) are the polar coordinates at the source plane. At the field center (at $r = 0$), there is a singular V-point, where the linear polarization vector is indefinite. According to Ref. [10], field (3.1) can be characterized by a singularity index similar to the TC of scalar optical vortices. V-points are described using a Poincare-Hopf index η, which can be calculated for field (3.1) similar to the TC of a complex field

$$E_{c,n}(\varphi) = E_x + iE_y = -\sin n\varphi + i\cos n\varphi = i\exp(in\varphi) \tag{3.2s}$$

The index of field (3.1) and a V-point equal TC of field (3.2): $\eta = n$. On the other hand, vector field (3.1) can be characterized using Stokes parameters $\mathbf{S} = (S_1, S_2, S_3)$ [13], where

$$S_1 = \frac{|E_x|^2 - |E_y|^2}{|E_x|^2 + |E_y|^2},\ S_2 = \frac{2\mathrm{Re}(E_x^* E_y)}{|E_x|^2 + |E_y|^2},\ S_3 = \frac{2\mathrm{Im}(E_x^* E_y)}{|E_x|^2 + |E_y|^2}, \tag{3.3}$$

with Re and Im stand for the real and imaginary parts of a number. From (3.3), the Stokes vector is seen to be of unit length: $S_1^2 + S_2^2 + S_3^2 = 1$. For the field (1), the Stokes parameters from (3.3) are given by

$$S_1 = -\cos(2n\varphi),\ S_2 = -\sin(2n\varphi),\ S_3 = 0. \tag{3.4}$$

Since $S_3 = 0$ in Eq. (3.4), we can infer that the field (3.1) is linearly polarized at any point, excepting the V-point, where polarization is indefinite. The complex Stokes field for the vector (3.4) takes the form:

$$S_c = S_1 + iS_2 = -\cos(2n\varphi) - i\sin(2n\varphi) = -\exp(i2n\varphi) \tag{3.5}$$

The Stokes index for the field (3.1) equals TC of the field (3.5): $\sigma = 2\eta = 2n$. Thus, the Stokes index is twice as large as the Poincare-Hopf index. For a radially polarized nth-order field with the Jones vector

$$E_{1,n}(\varphi) = \begin{pmatrix} \cos n\varphi \\ \sin n\varphi \end{pmatrix} \tag{3.6}$$

the Poincare-Hopf index of the central V-point ($r = 0$) also equals $\eta = n$. The V-point singularity index has the opposite sign ($\eta = -n$) for a vector field

$$E_{2,n}(\varphi) = \begin{pmatrix} \cos n\varphi \\ -\sin n\varphi \end{pmatrix} \tag{3.7}$$

3.1.2 Number of Local Intensity Maxima at the Focus of a Vector Field

Interestingly, vector field (3.6) produces a 'flower'-shaped pattern of linear polarization vectors composed of $2(n-1)$ petals. Actually, a petal is inscribed between the vector found at an angle $\varphi = 0$ and the vector rotated by an angle $\varphi = \pi + \varphi_0$. From the first to the second angle, the phase of the field (3.6) changes by $n\varphi_0$ rad. Equating $\pi + \varphi_0 = n\varphi_0$, we find the angle for a single petal to be $\varphi_0 = \pi/(n-1)$. In total, there are N petals: $2\pi = N\varphi_0$. Hence, we find that $N = 2(n-1)$. A similar reasoning suggests that a polarization 'web' composed of linear polarization vectors around the V-point of field (3.7) has $N = 2(n+1)$ cells.

Next, we demonstrate that a 'flower' of linear polarization vectors composed of $2(n-1)$ petals formed by the field (3.6) in the source plane is transformed at the tight focus into a 'flower'-shaped intensity pattern with $2(n-1)$ local maxima. Actually, using Richards-Wolf formulae [17], which describe the electromagnetic field components in the tight focus neighborhood, the E-field components can be derived in the form:

$$\begin{aligned} E_x &= -i^{n+1}\left(I_{0,n}\cos n\varphi + I_{2,n-2}\cos(n-2)\varphi\right), \\ E_y &= -i^{n+1}\left(I_{0,n}\sin n\varphi - I_{2,n-2}\sin(n-2)\varphi\right), \\ E_z &= 2i^n I_{1,n-1}\sin(n-1)\varphi, \end{aligned} \tag{3.8}$$

where

$$I_{\nu,\mu} = \left(\frac{4\pi f}{\lambda}\right) \int_0^{\theta_0} \sin^{\nu+1}\left(\frac{\theta}{2}\right) \cos^{3-\nu}\left(\frac{\theta}{2}\right) \cos^{1/2}(\theta) A(\theta) e^{ikz\cos\theta} J_\mu(x) d\theta, \quad (3.9)$$

where λ is the wavelength of light, f is the focal length of an aplanatic optical system, $x = kr\sin\theta$, $J_\mu(x)$ is the first-kind Bessel function, and $NA = \sin\theta_0$ is the numerical aperture. The initial amplitude function $A(\theta)$ (herein assumed to be real) may be either constant (a plane wave) or in the form of a Gaussian beam

$$A(\theta) = \exp\left(\frac{-\gamma^2 \sin^2\theta}{\sin^2\theta_0}\right) \quad (3.10)$$

where γ is constant. The transverse intensity (without regard for the longitudinal component of the field (3.8)) is given by

$$I_t = |E_x|^2 + |E_y|^2 = I_{0,n}^2 + I_{2,n-2}^2 + 2I_{0,n}I_{2,n-2}\cos(2(n-1)\varphi). \quad (3.11)$$

From (3.11), the transverse intensity profile is seen to have $2(n-1)$ local intensity maxima centered on the optical axis, each being located on a ray $\varphi = 2\pi p/(2n-2)$, $p = 1, 2, 3, \ldots, 2(n-1)$. Now we will determine an index of the V-point at the focus of the vector field (3.8). For this purpose, an equivalent complex field and its amplitude can be expressed as

$$\begin{aligned} E_{c,n} &= \left(I_{0,n}\cos n\varphi + I_{2,n-2}\cos(n-2)\varphi\right) \\ &\quad + i\left(I_{0,n}\sin n\varphi - I_{2,n-2}\sin(n-2)\varphi\right) \\ &= I_{0,n}\exp(in\varphi) + I_{2,n-2}\exp(-i(n-2)\varphi). \end{aligned} \quad (3.12)$$

In the general case, the index of the field (3.8) is undefined, because while at certain radii r coefficients in one exponential function can be larger than those in another one, the situation may be opposite at other radii. In the complex field of Eq. (3.12), TC depends on asymptotic properties of integrals (3.9). For instance, putting $A(\theta) = \delta(\theta - \theta_0)$, the integrals in (3.9) are replaced by Bessel functions, so that (3.12) is rearranged to

$$E_{c,n} = AJ_n(\alpha r)\exp(in\varphi) + BJ_{n-2}(\alpha r)\exp(-i(n-2)\varphi), \quad (3.13)$$

with $\alpha = kr\sin\theta_0$ and

$$\begin{aligned} A &= \left(\frac{4\pi f}{\lambda}\right) \sin\left(\frac{\theta_0}{2}\right) \cos^3\left(\frac{\theta_0}{2}\right) \cos^{1/2}\theta_0, \\ B &= \left(\frac{4\pi f}{\lambda}\right) \sin^3\left(\frac{\theta_0}{2}\right) \cos\left(\frac{\theta_0}{2}\right) \cos^{1/2}\theta_0. \end{aligned}$$

While from (3.13), the index is still seen to be undefined, near the optical axis the amplitude of a lower-order Bessel function is larger than that of a higher-order Bessel function, which means that, similar to the TC of a superposition of two optical vortices [16], the near-axis index equals $\eta = -(n-2)$. In a particular case of $n = 1$ (conventional radial polarization) Eq. (3.12) suggests that

$$E_{c,1} = (I_{0,1} - I_{2,1})\exp(i\varphi). \tag{3.14}$$

In this case, the V-point index is unit ($\eta = 1$) and, considering that $n = 1$, the source field index remains the same at the focus. This clearly follows from the fact that a singular point with unit index is unable to disintegrate into a number of V-points with smaller indices. In a similar way, a scalar optical vortex with TC = 1 remains robust following stochastic amplitude and phase distortions.

For an nth-order azimuthally polarized vector source field of Eq. (3.1), $2(n-1)$ local intensity maxima will also occur at the focus, though being located on other rays. Hence, a focal 'flower' composed of local intensity maxima will be rotated by an angle of $\pi/(2n-2)$. Using the angle magnitude, it becomes possible to distinguish nth-order radial polarization from nth-order azimuthal one. Meanwhile, the number of 'flower's petals' enables a cylindrical polarization order to be determined. Actually, for a source field (3.1), E-vector components in the focal plane take a form similar to Eq. (3.8):

$$\begin{aligned} E_x &= i^{n+1}\left(I_{0,n}\sin n\varphi + I_{2,n-2}\sin(n-2)\varphi\right), \\ E_y &= i^{n+1}\left(-I_{0,n}\cos n\varphi + I_{2,n-2}\cos(n-2)\varphi\right), \\ E_z &= -2i^n I_{1,n-1}\sin(n-1)\varphi. \end{aligned} \tag{3.15}$$

For the source field (3.1), the transverse intensity distribution in the focus is

$$I_t = |E_x|^2 + \left|E_y\right|^2 = I_{0,n}^2 + I_{2,n-2}^2 - 2I_{0,n}I_{2,n-2}\cos(2(n-1)\varphi). \tag{3.16}$$

From (3.16), $2(n-1)$ local maxima are seen to reside on a circle centered at the optical axis and on the rays outgoing from the center at angles $\varphi = (\pi + 2\pi p)/(2n-2)$, $p = 0, 1, 2, \ldots, 2(n-2)$. To find indices of V-points at the focal spot of the vector field (3.1), we can express an equivalent complex field with the amplitude:

$$\begin{aligned} E_{c,n} &= \left(I_{0,n}\sin n\varphi + I_{2,n-2}\sin(n-2)\varphi\right) \\ &\quad + i\left(-I_{0,n}\cos n\varphi + I_{2,n-2}\cos(n-2)\varphi\right) \\ &= -iI_{0,n}\exp(in\varphi) + iI_{2,n-2}\exp(-i(n-2)\varphi). \end{aligned} \tag{3.17}$$

In the general case, the index of the field (3.17) is undefined, because while at certain radii r coefficients in one exponential function can be larger than those in another one, the situation may be opposite at other radii. However, at $n = 1$ (ordinary azimuthal polarization), from Eq. (3.17) it follows that

$$E_{c,1} = -i(I_{2,1} + I_{0,1})\exp(i\varphi). \tag{3.18}$$

In this case, the V-point index is unit ($\eta = 1$), meaning that the index of initial field (3.1) remains unchanged at the focus.

A vector 'web' of source field (3.7) with $2(n + 1)$ cells, centered on the V-point polarization singularity is transformed at the focus into an intensity pattern with $2(n + 1)$ local maxima. Actually, for the source field in (3.7), projections of the E-vector are given by ($n > 0$)

$$\begin{aligned}
E_x &= i^{n-1}\big(I_{0,n}\sin n\varphi + I_{2,n+2}\sin(n+2)\varphi\big),\\
E_y &= i^{n-1}\big(I_{0,n}\cos n\varphi - I_{2,n+2}\cos(n+2)\varphi\big),\\
E_z &= -2i^n I_{1,n+1}\sin(n+1)\varphi.
\end{aligned} \tag{3.19}$$

For the field (3.19), the transverse intensity distribution at the focus is given by

$$I_t = |E_x|^2 + \left|E_y\right|^2 = I_{0,n}^2 + I_{2,n+2}^2 - 2I_{0,n}I_{2,n+2}\cos(2(n+1)\varphi). \tag{3.20}$$

From (3.20), the intensity distribution is seen to have $2(n + 1)$ local intensity maxima at the focus on an axis-centered circle of a certain radius. Hence, the vector 'web' in the source field of Eq. (3.7) can be identified based on the number of petals of an nth-order vector 'flower'.

Putting $n = -1$ in Eq. (3.7) for the source field, we may infer from (3.14) that the V-point index changes sign at the focus, because based on Eq. (3.19) for the E-vectors at the focus, we find that

$$E_{c,1} = -i(I_{2,1} + I_{0,1})\exp(i\varphi). \tag{3.21}$$

Aiming to determine the V-point index at the focus of the vector field (3.7) and using Eq. (3.19), we form an equivalent complex field with the amplitude:

$$\begin{aligned}
E_{c,n} &= \big(I_{0,n}\sin n\varphi + I_{2,n+2}\sin(n+2)\varphi\big)\\
&\quad + i\big(I_{0,n}\cos n\varphi - I_{2,n+2}\cos(n+2)\varphi\big)\\
&= iI_{0,n}\exp(-in\varphi) - iI_{2,n+2}\exp(i(n+2)\varphi).
\end{aligned} \tag{3.22}$$

Just like in Eq. (3.17), the index of field (3.19) is undefined, but like in Eq. (3.13), it can be asserted that at the focus the near-axis V-point index is equal to a lesser number of the Bessel function, i.e., $\eta = -n$. That is, given the source field of Eq. (3.7), the near-axis V-point index at the focus is the same as in the source plane.

3.1.3 Polarization Singularity Index for a Generalized Vector Field

Obviously, the above reasoning cannot be automatically applied to a generalized vector field as it has different orders on the different axes. For such a field, the Jones vector is [10]

$$\mathbf{E}_{\mathbf{2,n}}(\varphi) = \begin{pmatrix} \cos n\varphi \\ \sin m\varphi \end{pmatrix} \tag{3.23}$$

Although the field (3.23) may also be said to have a central V-point, its index can be defined analytically only in some cases. Actually, the complex field equivalent to the field (3.23) is given by

$$E_{c,n}(\varphi) = E_x + iE_y = \cos n\varphi + i \sin m\varphi \tag{3.24}$$

In the topic-related work [10], it was not specified in which way the index of such a field could be determined if $n \neq m$. In this work, we propose that the V-point index of the vector field (3.23) should be calculated in a similar way to calculating the TC of scalar optical vortices using the Berry's formula [12]:

$$TC = \frac{1}{2\pi} \lim_{r\to\infty} \operatorname{Im} \int_0^{2\pi} d\varphi \frac{\partial E(r,\varphi)/\partial\varphi}{E(r,\varphi)}. \tag{3.25}$$

Then, according to (3.25), the Poincare-Hopf index for vector field (3.24) is given by

$$\eta = \frac{1}{2\pi} \lim_{r\to\infty} \operatorname{Im} \int_0^{2\pi} d\varphi \frac{-n \sin n\varphi + im \cos m\varphi}{\cos n\varphi + i \sin m\varphi} =$$

$$\frac{1}{2\pi} \int_0^{2\pi} d\varphi \frac{n \sin n\varphi \sin m\varphi + m \cos m\varphi \cos n\varphi}{\cos^2 n\varphi + \sin^2 m\varphi}. \tag{3.26}$$

From (3.26), it follows that at $m = n$, $\eta = n$, whereas at $m = -n$, $\eta = -n$. However, at $n \neq \pm m$, the integral in Eq. (3.26) is not reduced to reference integrals. In separate cases, Eq. (3.26) can be calculated analytically, but in other cases it needs to be calculated numerically. Below, we deduce some properties of polarization singularity index (3.26), including properties of parity, symmetry, reciprocity, and multiplicity. The parity property is expressed in the fact that for different-parity m and n (i.e., $m + n$ is odd), polarization singularity index (3.26) equals zero. Actually, the first integral in (3.26) can be broken down in two (with the range of integration in the second integral shifted from [π, 2 π] to [0, π]):

$$\eta_{n,m} = \frac{1}{2\pi}\text{Im}\left\{\int_0^{\pi} \frac{-n\sin n\varphi + im\cos m\varphi}{\cos n\varphi + i\sin m\varphi}d\varphi + \int_0^{\pi} \frac{-n(-1)^n\sin n\varphi + im(-1)^m\cos m\varphi}{(-1)^n\cos n\varphi + i(-1)^m\sin m\varphi}d\varphi\right\}. \tag{3.27}$$

Multiplying the numerator and denominator of the first integral by $(-1)^n$ and taking into account that $(-1)^{m+n} = -1$, we get a sum of two complex conjugated numbers whose imaginary part equals zero. Thus, we obtain a symmetry property of the Poincare-Hopf index for vector field (3.23). In this way, it stands to reason that when n changes sign, the first integral in (3.26) does not change:

$$\eta_{-n,m} = \frac{1}{2\pi}\text{Im}\int_0^{2\pi} d\varphi \frac{-(-n)\sin(-n\varphi) + im\cos m\varphi}{\cos(-n\varphi) + i\sin m\varphi} = \eta_{n,m} \tag{3.28}$$

On the contrary, when m changes sign, the integrand becomes complex conjugated and, hence, the imaginary part changes sign:

$$\eta_{n,-m} = -\eta_{n,m} \tag{3.29}$$

Shifting the range of integration by π /2, we get the following relationships between the indices:

$$\eta_{n,m} = \begin{cases} \dfrac{1}{2\pi}\text{Im}\displaystyle\int_0^{2\pi} d\varphi \frac{-n(-1)^{\frac{n}{2}}\sin n\varphi + im(-1)^{\frac{m}{2}}\cos m\varphi}{(-1)^{\frac{n}{2}}\cos n\varphi + i(-1)^{\frac{m}{2}}\sin m\varphi}, \\ \text{if } n, m \text{ are even,} \\ \dfrac{1}{2\pi}\text{Im}\displaystyle\int_0^{2\pi} d\varphi \frac{-n(-1)^{\frac{n-1}{2}}\cos n\varphi + im(-1)^{\frac{m+1}{2}}\sin m\varphi}{(-1)^{\frac{n+1}{2}}\sin n\varphi + i(-1)^{\frac{m-1}{2}}\cos m\varphi}, \\ \text{if } n, m \text{ are odd,} \end{cases}$$
$$= \begin{cases} (-1)^{(m-n)/2}\eta_{n,m}, \text{ if } n, m \text{ are even,} \\ (-1)^{(m-n)/2}\eta_{m,n}, \text{ if } n, m \text{ are odd.} \end{cases} \tag{3.30}$$

This can be termed as a reciprocity property because it enables the indices to be swapped if they are odd. From (3.30), it also follows that if n and m are even, but $(m - n)/2$ is odd, then $\eta = 0$.

If the orders m and n have a common divisor, i.e., $m = p\mu$ and $n = p\nu$, then, performing a change of variables $\varphi = \theta/p$ in (3.26), we obtain a multiplicity property:

$$\eta_{p\nu,p\mu} = p\frac{1}{2\pi}\mathrm{Im}\int\limits_0^{2\pi p}\frac{-\nu\sin\nu\theta + i\mu\cos\mu\theta}{\cos\nu\theta + i\sin\mu\theta}\frac{d\theta}{p} = p\eta_{\nu,\mu} \tag{3.31}$$

For instance, at $m = 2n$, polarization singularity index equals zero thanks to the multiplicity and parity properties: $\eta_{n,2n} = n\eta_{1,2} = 0$. In a simple case, we determine the index $\eta_{1,2p+1}$ analytically through the use of residues. If we denote $\zeta = \cos\varphi + i\,(2p+1)\sin\varphi$, the integral (3.26) can be written as (at $n = 1$)

$$\eta = \frac{1}{2\pi}\mathrm{Im}\oint\limits_{\Gamma}\frac{d\zeta}{\zeta} \tag{3.32}$$

where Γ is the oriented closed contour in the complex plane drawn by the variable ζ, when $0 \le \varphi \le 2\pi$. Figure 3.1 illustrates this contour for $p = 0, 1, 2$.

If $p = 0$, this contour is a simple unit-radius circle. Otherwise, Γ has self-intersections and the integral over Γ can be replaced by a sum of the integrals over several simple contours without the self-intersections [red dashed contours in Fig. 3.1b, c]. The only pole of the integrand in (3.32) is $\zeta = 0$. If $p = 0$, this pole is within the unit-radius circle and, according to the residues theorem, applied to the integral (3.32), $\eta_{11} = 1$. For $p > 0$, only one simple contour contains the pole (Fig. 3.1b, c). Thus, integration over the other simple contours yields 0. If $p = 1$ (and at other odd p), the pole is bypassed clockwise and, therefore, the integration yields $\eta_{13} = -1$. Similarly, if $p = 2$ (and at other even p), the pole is bypassed counter-clockwise and, therefore, the integration yields $\eta_{15} = +1$. Thus, we can write a general rule for the Poincare-Hopf index η_{nm} at $n = 1$ and odd m:

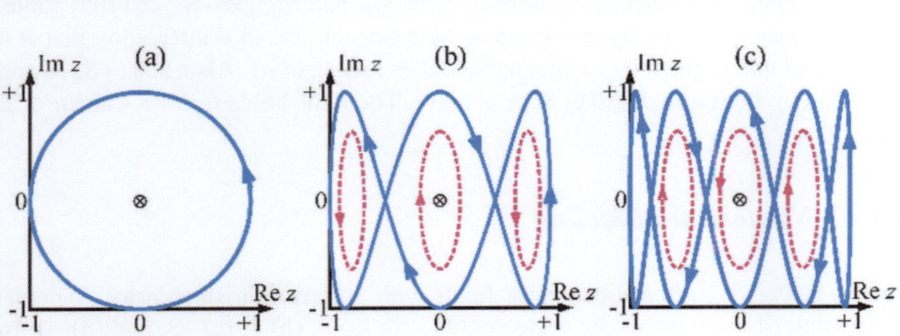

Fig. 3.1 Calculation of the Poincare-Hopf index. Contours Γ (blue solid curves) in the complex plane defined as $\zeta = \cos\varphi + i\,(2p + 1)\sin\varphi$ $(0 \le \varphi \le 2\pi)$ for $p = 0$ (**a**), $p = 1$ (**b**), $p = 2$ (**c**) Red dashed ellipses show the simple contours without self-intersections, into which the contour Γ can be split. The cross in the center denotes $\zeta = 0$, the only pole of the integrand in (3.32)

Table 3.1 Poincare-Hopf index η of vector field (3.23): *n* shown on the horizontal lines and *m*—on the vertical

m	*N*										
	0	1	2	3	4	5	6	7	8	9	10
0	0	0	0	0	0	0	0	0	0	0	0
1	0	1	0	1	0	1	0	1	0	1	0
2	0	0	2	0	0	0	2	0	0	0	2
3	0	–1	0	3	0	–1	0	–1	0	3	0
4	0	0	0	0	4	0	0	0	0	0	0
5	0	1	0	1	0	5	0	1	0	1	0
6	0	0	–2	0	0	0	6	0	0	0	–2
7	0	–1	0	–1	0	–1	0	7	0	–1	0
8	0	0	0	0	0	0	0	0	8	0	0
9	0	1	0	–3	0	1	0	1	0	9	0
10	0	0	2	0	0	0	2	0	0	0	10

$$\eta_{1m} = (-1)^{(m-1)/2} \tag{3.33}$$

or, using the reciprocity property:

$$\eta_{n,1} = 1 \tag{3.34}$$

All the properties of index (3.26) for field (3.23) derived herein can be verified using the Table 3.1.

Table 3.1 below gives values of η, which were calculated using Eq. (3.26) for vector field (3.23), with the orders *m* and *n* being varied from 0 to + 10 (for negative *m* and *n*, symmetry rules can be used: $\eta_{-n,m} = \eta_{n,m}$ and $\eta_{n,-m} = -\eta_{n,m}$). From Table. 3.1, polarization singularity index can be only integer. It is also interesting that at $n = 1, -1$ and any *m*, the η index is equal to either 1, or 0, or –1. Also, at $n = 8, -8$ and any *m*, the η index equals either 8, or 0, or –8. The same holds for $n = 4$ and $n = 2$.

3.1.4 Numerical Modeling

Shown in Fig. 3.2 are source vector fields with polarization singularity (V-point) at the center for the *n*th-order vector field (3.6): (a) 3, (b) 4, (c) –3, and (d) –4. In compliance with the theoretical predictions, the vector fields in Fig. 3.2a, b are shaped as 'flowers' with the number of petals equal to (a) $2(n - 1) = 4$ and (b) $2(n - 1) = 6$. Whereas two other vector fields in Fig. 3.2c, d produce 'lattice' patterns with the number of cells equal to (c) $2(n + 1) = 8$ and (d) $2(n + 1) = 10$.

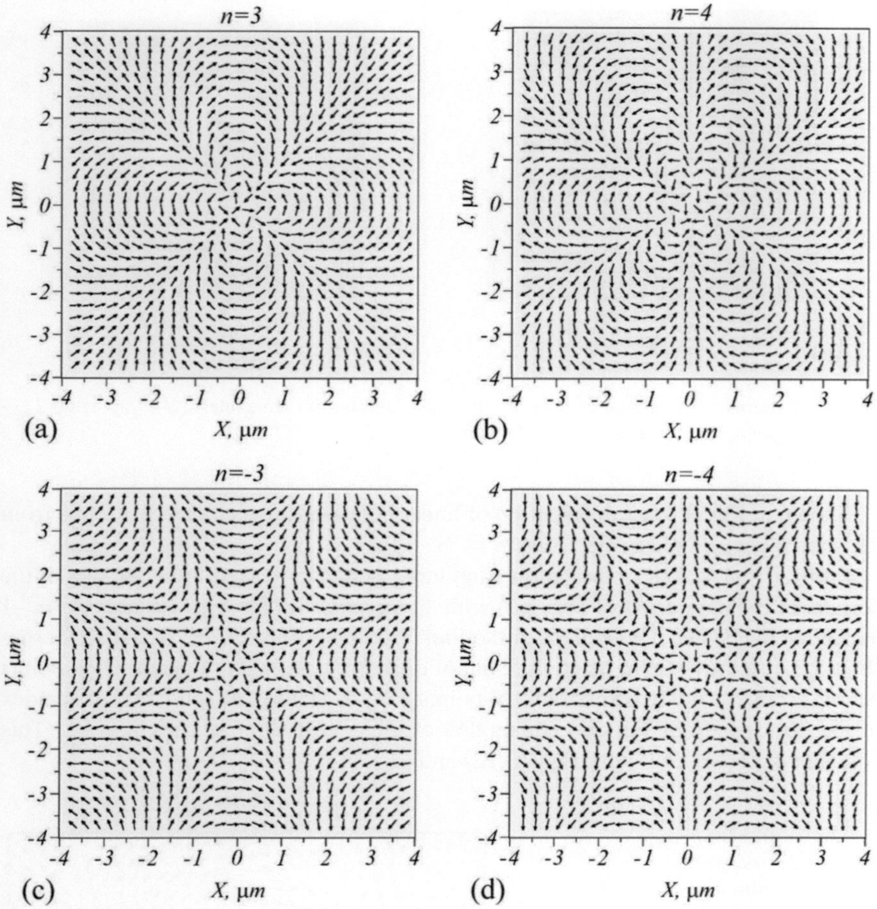

Fig. 3.2 Vector field (3.6) (arrows mark linear polarization vectors at particular points), whose order n coincides with the index of the V-point polarization singularity (Poincare-Hopf index η) at the field center and equals: **a** 3, **b** 4, **c** –3, and **d** –4

Source vector fields of type (3.6) in Fig. 3.2 are transformed at the focal plane into vector fields (3.8), (3.15), and (3.19), which have several points of polarization singularities. Shown in Fig. 3.3 are the total intensity (Fig. 3.3a) and the transverse intensity (Fig. 3.3b) for a source vector field with the index $n = 3$ of Fig. 3.2a. The numerical modeling of focusing vector fields was conducted using Richards-Wolf formulae [17] for wavelength 532 nm and numerical aperture NA = 0.95.

In accordance with theoretical predictions [Eq. (3.11)], there occur $2(n - 1) = 4$ local maxima of the total and transverse intensities at the focus. Due to the longitudinal intensity components, the coordinates of four local maxima in Fig. 3.3a are different from those of the transverse intensity in Fig. 3.3b.

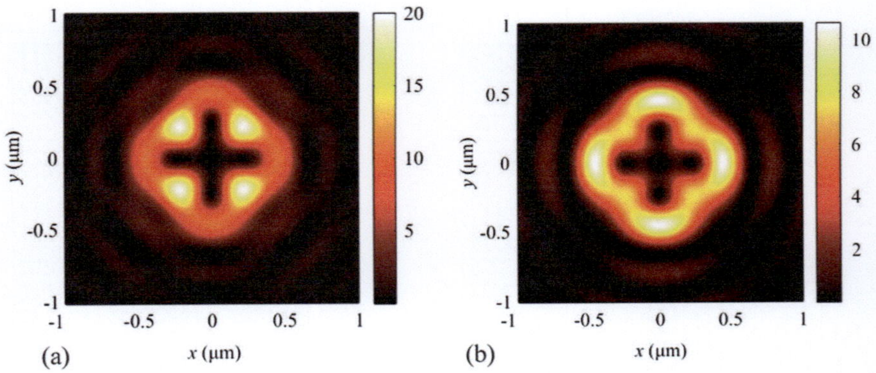

Fig. 3.3 Patterns of the **a** total intensity $I_x + I_y + I_z$ and **b** transverse intensity components $I_x + I_y$ from the source vector field of Fig. 3.2a at $n = 3$

Shown in Fig. 3.4 is a distribution of linear polarization vectors at the focus from the source vector field in Fig. 3.2a ($n = 3$).

From Fig. 3.4, four polarization singularity centers are seen to be located at the corners of the dark cross of Fig. 3.3, with an on-axis V-point with the index $\eta = -1$ located at the center. The indices of the four V-points at the corners of the dark cross (Fig. 3.3) are the same in magnitude but of different sign, with two vertical V-points having $\eta = +1$, and two horizontal V-points $\eta = -1$. Hence, the total near-axis index of the vector field of Fig. 3.4 equals that of the central V-point, i.e., $\eta = -1$. This conclusion agrees well with Eqs. (3.12) and (3.13): $\eta = -(n - 2) = -1$.

Fig. 3.4 Pattern of linear polarization vectors at the focal plane from the source vector field in Fig. 3.2a ($n = 3$)

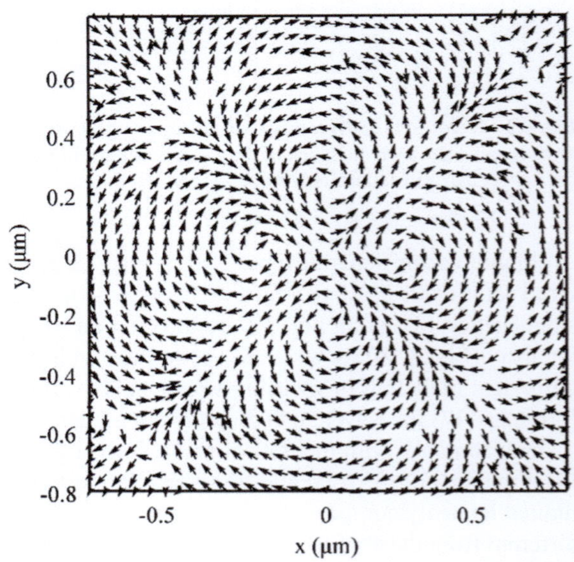

Figure 3.5 depicts numerically simulated patterns for the (a) total intensity and (b) transverse intensity from the source vector field with $n = 4$ (Fig. 3.2b). From Fig. 3.5, the theoretical relation (3.11) is again seen to be corroborated, with $2(n - 1) = 6$ local maxima in the intensity pattern found symmetrically to the optical axis being observed.

Figure 3.6 shows a pattern of linear polarization vectors at the focus from a source vector field with $n = 4$ (Fig. 3.2b). From Fig. 3.6, a set of V-points with indices $\eta = +1, -1$ are seen to form at the 'vertices of a dark six-point star' of Fig. 3.5. Equation (3.12) suggests that an on-axis V-point with $\eta = -2$ is found at the center.

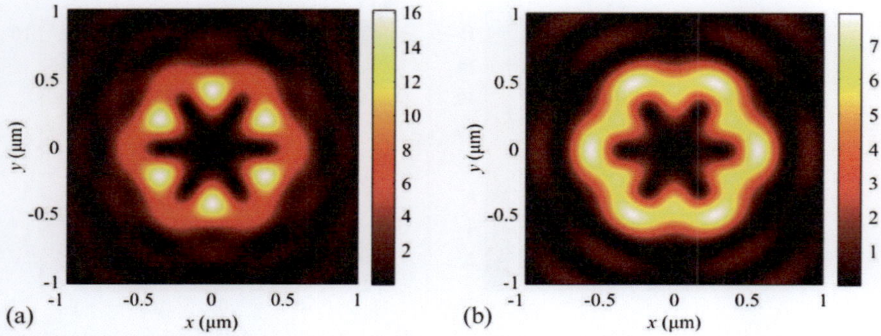

Fig. 3.5 Patterns of the **a** total intensity $I_x + I_y + I_z$ and **b** transverse intensity $I_x + I_y$ component at the focal plane (NA = 0.95) from the source vector field with the index $n = 4$ (Fig. 3.2b)

Fig. 3.6 Pattern of linear polarization vectors for the source vector field of Fig. 3.2b with the index $n = 4$

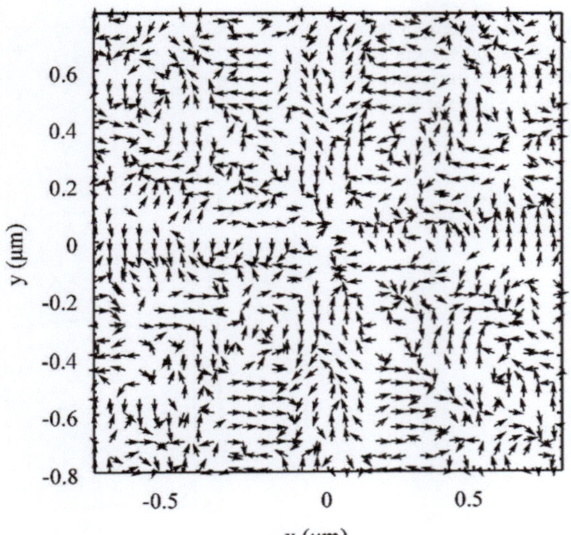

Figure 3.7 depicts patterns for the total (Fig. 3.7a) and transverse (Fig. 3.7b) component of the intensity at the focal plane (NA = 0.95) from the source vector field with $n = -3$ of Fig. 3.2c. Figure 3.7 shows that in compliance with theoretical predictions, there are $2(n+1) = 8$ local intensity maxima in the intensity distribution.

Shown in Fig. 3.8 is a pattern of linear polarization vectors at the focus from the source vector field of Fig. 3.2c at $n = -3$. From Fig. 3.8, eight V-points are seen to be located on a circle (at the vertices of a 'dark eight-point star'), with four of them having the index $\eta = +1$ and four having the index $\eta = -1$. Equation (3.22) suggests that at the center of the focal spot there is a V-point with $\eta = 3$.

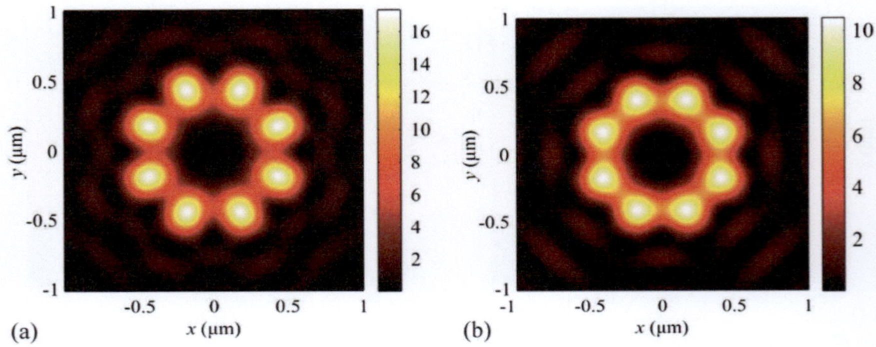

Fig. 3.7 Patterns of the **a** total intensity $I_x + I_y + I_z$ and **b** transverse intensity $I_x + I_y$ component at the focal plane (NA = 0.95) for the source vector field with the index $n = -3$ of Fig. 3.2c

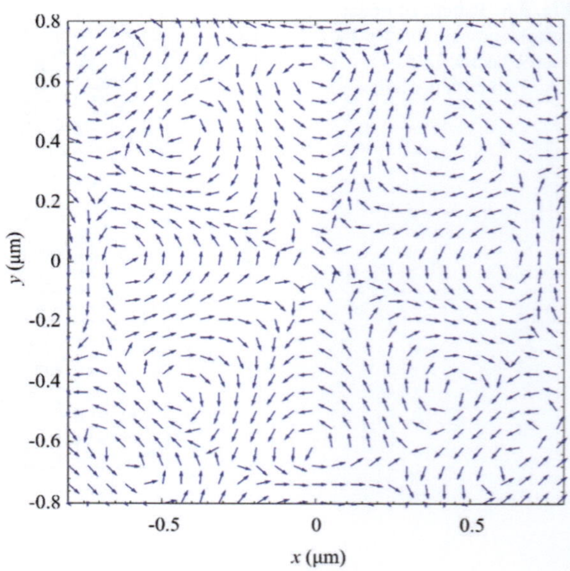

Fig. 3.8 Pattern of linear polarization vectors at the focus from the source field with the index $n = -3$ of Fig. 3.2c

Figure 3.9 presents patterns of linear polarization vectors for the source field of Eq. (3.23) at different values of (n, m): (a) (2,1), (b) (3, –7), (c) (9, –3), and (d) (6,2). Using the Table 3.1 above, the Poincare-Hopf indices η for the said vector fields can be found to be (a) 0, (b) 1, (c) –3, and (d) 2. By looking at Fig. 3.9, indices of the V-points of such complex vector fields would be difficult to determine. The pattern for linear polarization vectors at the focus would be even more complicated (not presented here). Shown in Fig. 3.10 is an intensity pattern at the focus of an aplanatic objective with NA = 0.95 when focusing vector beams with $n = 2$, $m = 1$ (Fig. 3.10a) and $n = 3$, $m = -7$ (Fig. 3.10b).

Figure 3.10 suggests that a source field with $\eta = 0$ (Fig. 3.9a) produces neither an intensity null nor a V-point at the center of the focal spot (Fig. 3.10a), whereas a source vector field with $\eta = 1$ (Fig. 3.10b) produces at the center an intensity null and a V-point.

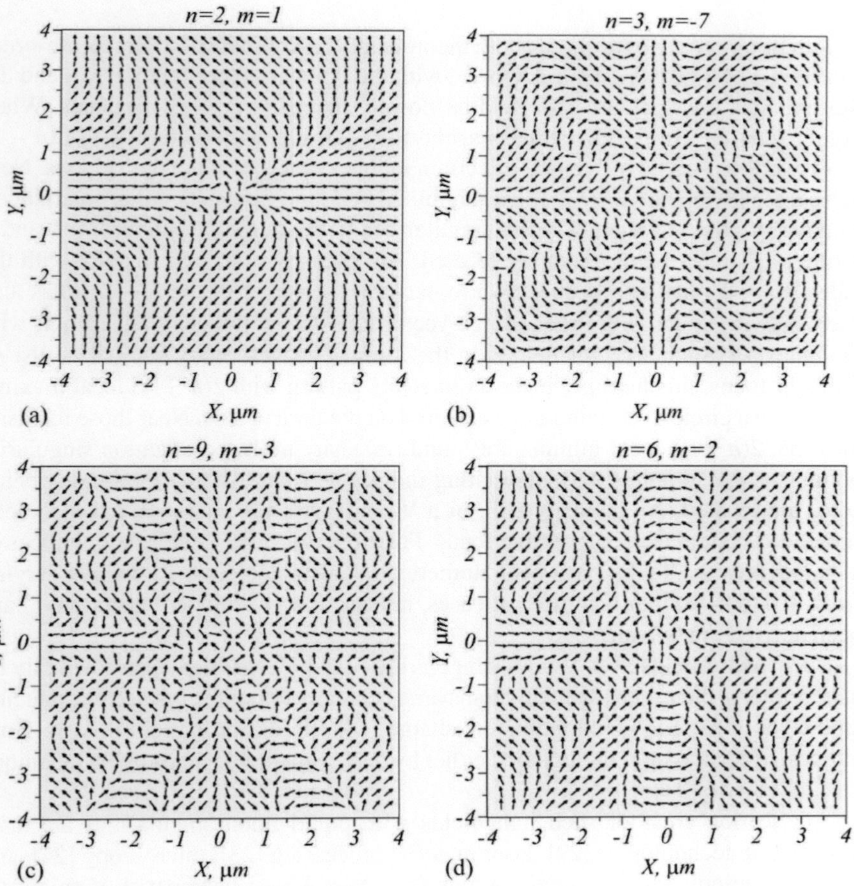

Fig. 3.9 Source vector fields (3.23) at different n and m

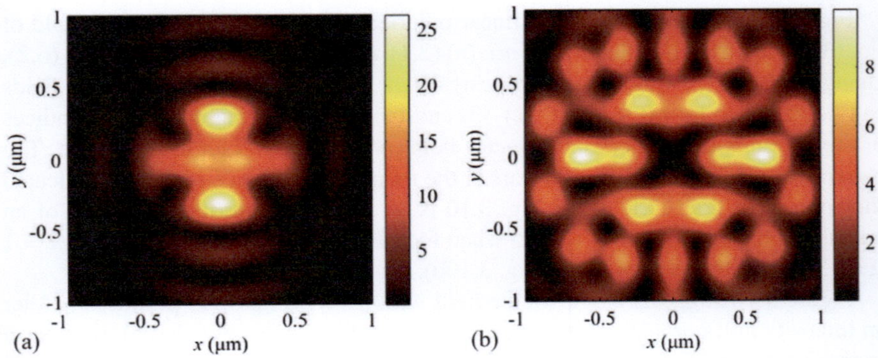

Fig. 3.10 Intensity patterns at the focus of vector beams with **a** $n = 2, m = 1$ (a 'butterfly') and **b** $n = 3$, $m = -7$ (a 'dragon mouth')

Summing up, we have shown both theoretically and numerically that an nth-order source vector field has a central V-point with the Poincare-Hopf index $\eta = n$ and the Stokes index $2n$. Such a vector field is 'flower'-shaped with $2(n - 1)$ petals. When tightly focused, this field produces at the focus an intensity pattern with $2(n - 1)$ local maxima located on a circle of certain radius, centered on the optical axis. Near those intensity maxima, $2(n - 1)$ local minima are found (intensity nulls, polarization singularity points), where V-point singularities with alternating indices $+ 1$ and $- 1$ (the total index being zero) are located. An intensity null, or a V-point with the index $-(n - 2)$, has also been shown to occur at the center of the focus. It has also been shown that an $-n$th-order source vector field has at the center a V-point with the index $-n$. Such a vector field is in the form of a 'web' with $2(n + 1)$ cells. At the tight focus, this field produces an intensity pattern with $2(n + 1)$ local maxima located on a circle of certain radius centered on the optical axis. Near those intensity maxima, $2(n + 1)$ local minima are found (intensity nulls, polarization singularity points), where V-points with alternating indices $+ 1$ and -1 (the total index being zero) are located. An intensity null, or a V-point with the index $-n$, has also been shown to occur at the center of the focus. For an (n, m)-order vector field, indices of V-points (see the Table) have been numerically calculated for the numbers varying from -10 to $+ 10$. For a number of cases, indices of a generalized vector field have been derived analytically.

Such vector fields with V-point singularities can be generated experimentally by using either q-plates, i.e., specially transversely patterned liquid crystal cells inducing an integer or semi-integer topological charge [18, 19], or, for higher Poincare-Hopf indices, by spatial light modulators: either by one, with double modulation technique [20], or by two [21].

Application areas of such light fields with polarization singularities are laser information technologies [22], laser material processing [23], microscopy [24], and particle manipulation or optical trapping [25]. Vector laser beams with V-points are considered in [26].

3.2 Spin Hall Effect Before and After the Focus of a High-Order Cylindrical Vector Beam

In optics, cylindrical vector beams (CVB) are well known [14, 27], including the high-order beams. When the order is $n = 1$, these beams have specific names—radially polarized beam [28] and azimuthally polarized beam [29]. In addition, a lot of works studied sharp focusing of the CVBs of the first order [30, 33], of the high order [4, 31], of the fractional order [32], as well as focusing of the vortex beams with high-order azimuthal polarization [34].

It is known that both in the initial plane and in the focus of a CVB of an arbitrary order, polarization is inhomogeneous and linear in every point of the beams cross section. The beam order n is equal to the number of full rotations of the polarization vector when passing along a closed contour around the optical axis. On the optical axis in the initial plane, such beams have a point of a polarization singularity (V-point), where the direction of linear polarization vector [26] is undefined. The polarization index of this V-point is equal to the beam order n. It is also known, that in the sharp focus of the nth-order CVB, the intensity distribution has $2(n-1)$ local maxima [26]. The third component of the Stokes vector is zero: $S_3 = 2\text{Im}(E_x^* E_y) = 0$, i.e., the spin angular momentum (SAM) of a CVB in the initial plane is zero. Equal to zero is also the orbital angular momentum (OAM) of the CVB in the initial plane and in the focus.

In this section, using the Debye integrals and a numerical simulation, we show that near the sharp focus (before the focus and beyond the focus), local subwavelength areas are generated with elliptic and circular polarization of different sign.

3.2.1 *Spin Angular Momentum Before and Beyond the Focus*

If the initial light field is an nth-order cylindrical vector beam, the Jones vectors of the electric and magnetic fields are given by

$$E_n(\varphi) = \begin{pmatrix} \cos n\varphi \\ \sin n\varphi \end{pmatrix}, H_n(\varphi) = \begin{pmatrix} -\sin n\varphi \\ \cos n\varphi \end{pmatrix}, \tag{3.35}$$

with (r, φ) being the polar coordinates in the transverse initial plane of the beam. Adopting the Debye integrals [17], we derive all the Cartesian components of the strength vectors of the electric and magnetic fields in the sharp focus of the light field with polarization (3.35):

$$\begin{aligned}
E_x(r, \varphi) &= i^{n-1}\left[\cos(n\varphi)I_{0,n} + \cos((n-2)\varphi)I_{2,n-2}\right], \\
E_y(r, \varphi) &= i^{n-1}\left[\sin(n\varphi)I_{0,n} - \sin((n-2)\varphi)I_{2,n-2}\right], \\
E_z(r, \varphi) &= 2i^n \cos((n-1)\varphi)I_{1,n-1},
\end{aligned}$$

$$\begin{aligned} H_x(r,\varphi) &= -i^{n-1}\left[\sin(n\varphi)I_{0,n} + \sin((n-2)\varphi)I_{2,n-2}\right], \\ H_y(r,\varphi) &= -i^{n-1}\left[-\cos(n\varphi)I_{0,n} + \cos((n-2)\varphi)I_{2,n-2}\right], \\ H_z(r,\varphi) &= -2i^n \sin((n-1)\varphi)I_{1,n-1}. \end{aligned} \tag{3.36}$$

In Eq. (3.36), the functions $I_{\nu,\mu}$ depend only on the radial and longitudinal coordinates r and z and equal to:

$$I_{\nu,\mu} = \left(\frac{4\pi f}{\lambda}\right)\int\limits_0^{\theta_0} \sin^{\nu+1}\left(\frac{\theta}{2}\right)\cos^{3-\nu}\left(\frac{\theta}{2}\right)\cos^{1/2}(\theta)A(\theta)e^{ikz\cos\theta}J_\mu(x)d\theta \tag{3.37}$$

where $k = 2\pi/\lambda$ stands for the wavenumber of monochromatic light with the wavelength λ, f is the focal distance of an aberration-free spherical lens that focuses the beam, z is the propagation axis ($z = 0$ is the focal plane), $x = kr\sin\theta$ is a dimensionless argument of the μth-order Bessel function of the first kind $J_\mu(x)$, $NA = \sin\theta_0$ is the numerical aperture of an aplanatic focusing system, whereas an arbitrary real-valued function $A(\theta)$ describes the amplitude of an input field with a cylindrical symmetry (plane wave, Gaussian beam, Bessel-Gaussian beam). In the functions $I_{\nu,\mu}$ in Eq. (3.37), the first index $\nu = 0,1,2$ describes the first part of the integrand, whereas the second index $\mu = 0,1,2,\ldots,n$ defines the order of the Bessel function.

To derive the focal distribution of the spin density or the spin angular momentum (SAM) for the beam with polarization (3.35), we use the general expression for the spin angular momentum vector given in [35]:

$$\mathbf{S} = \frac{1}{16\pi\omega}\mathrm{Im}\left(\mathbf{E}^* \times \mathbf{E}\right), \tag{3.38}$$

where ω is the angular frequency of light. From now on, the constant factor $1/(16\pi\omega)$ will be omitted. It is seen in Eq. (3.38) that the longitudinal SAM component (without the constant factor) is coinciding with the non-normalized third Stokes parameter S_3:

$$S_3 = S_z = 2\mathrm{Im}\left(E_x^* E_y\right). \tag{3.39}$$

It is known that the third Stokes component indicates circular or elliptic polarization of a light field [36]. If $S_3 = 0$, then the field has only linear polarization. If we substitute the electric field components from Eq. (3.36) into Eq. (3.39) and take into account that the integrals (3.37) are complex-valued near the focus (but not in the focus itself), we get the longitudinal component of the SAM vector:

$$S_z = \mathrm{Im}\left[I_{0,n}I_{2,n-2}^* \sin(n\varphi)\cos((n-2)\varphi) - I_{0,n}^* I_{2,n-2}\sin((n-2)\varphi)\cos((n-2)\varphi)\right], \tag{3.40}$$

where the asterisks "*" mean the complex conjugation. Separating the real and imaginary parts in the integrals (3.37), and using the linear approximation $\exp(ikz\cos\theta) \approx 1 + ikz\cos\theta$ near the focus ($kz << 1$), instead of Eq. (3.40) we get:

$$S_z \simeq 2kz\sin(2(n-1)\varphi)(I_0R_2 - I_2R_0), \tag{3.41}$$

where we use the following designations:

$$\begin{aligned}
&R_0 = I_{0,n}(z=0), I_0 = \overline{I}_{0,n},\\
&R_2 = I_{2,n-2}(z=0), I_2 = \overline{I}_{2,n-2},\\
&\overline{I}_{\nu,\mu} = \left(\frac{4\pi f}{\lambda}\right)\int\limits_0^{\theta_0}\sin^{\nu+1}\left(\frac{\theta}{2}\right)\cos^{3-\nu}\left(\frac{\theta}{2}\right)\cos^{3/2}(\theta)A(\theta)J_\mu(x)\mathrm{d}\theta.
\end{aligned} \tag{3.42}$$

As seen from Eq. (3.41), in the focus itself ($z = 0$), $S_3 = 0$, and, therefore, in each point of the focal plane, polarization is linear. However, at small defocusing ($kz << 1$), $S_3 \neq 0$, and areas with elliptic and circular polarization appear, if $n \neq 1$. In the areas where before the focus ($z < 0$) the SAM was negative ($S_3 < 0$), after the focus ($z > 0$) it becomes positive ($S_3 > 0$), and vice versa. According to Eq. (3.41), near the focal plane, on a certain radius circumference with the center on the propagation axis, centers of $4(n – 1)$ local subwavelength areas reside, with elliptic and circular polarization. In such neighboring areas, polarization vector is rotating in opposite directions (clockwise or counter-clockwise). A similar result was obtained in [37], but for $n = 0$. Since, at $n \neq 1$, near the focus of the light field (1), areas with left and right circular polarization (areas with a different "spin") are spatially separated, we can conclude that near the focus (before and after it), the spin Hall effect occurs, although in the focal plane itself, this effect vanishes.

3.2.2 Transverse Energy Flow Before and Beyond the Focus

Further we derive the transverse components of the energy flow vector. The Poynting vector reads as [17]

$$\mathbf{P} = \frac{c}{4\pi}\mathrm{Re}\big(\mathbf{E}^* \times \mathbf{H}\big). \tag{3.43}$$

In Eq. (3.43), vectors **E** and **H** stand respectively for the electric and magnetic fields, the asterisk * means the complex conjugation, $\times$ is the cross product, c is the free-space speed of light. From now on, the constant factor $c/(4\,\pi)$ will be omitted. Substituting the expressions (3.36) for the electromagnetic field components in the focus into Eq. (3.43), we obtain the transverse components of the Poynting vector near the focus:

$$
\begin{aligned}
P_x &= 2\mathrm{Im}\Big\{-I_{0,n}^{*}I_{1,n-1}\sin(n\varphi)\sin((n-1)\varphi)+I_{2,n-2}^{*}I_{1,n-1}\sin((n-2)\varphi)\sin((n-1)\varphi)\\
&\quad +I_{1,n-1}^{*}I_{0,n}\cos(n\varphi)\cos((n-1)\varphi)-I_{1,n-1}^{*}I_{2,n-2}\cos((n-2)\varphi)\cos((n-1)\varphi)\Big\},\\
P_y &= 2\mathrm{Im}\Big\{I_{1,n-1}^{*}I_{0,n}\sin(n\varphi)\cos((n-1)\varphi)+I_{1,n-1}^{*}I_{2,n-2}\sin((n-2)\varphi)\cos((n-1)\varphi)\\
&+ I_{0,n}^{*}I_{1,n-1}\cos(n\varphi)\sin((n-1)\varphi)+I_{2,n-2}^{*}I_{1,n-1}\cos((n-2)\varphi)\sin((n-1)\varphi)\Big\}.
\end{aligned}
\tag{3.44}
$$

Separating the real part and the imaginary part in the integrals (3.37), and using the approximation $\exp(ikz\cos\theta)\approx 1+ikz\cos\theta$ near the focus ($kz< <1$), instead of Eq. (3.44) we get:

$$
\begin{aligned}
P_x &= 2kz\cos\varphi[R_1(I_0-I_2)+I_1(R_2-R_0)],\\
P_y &= 2kz\sin\varphi[R_1(I_0-I_2)+I_1(R_2-R_0)].
\end{aligned}
\tag{3.45}
$$

In Eq. (3.45), the following designations are introduced for the real and imaginary parts of the integrals (3.37):

$$
\begin{aligned}
R_0 &= I_{0,n}(z=0),\, I_0=\overline{I}_{0,n},\, R_1=I_{1,n-1}(z=0),\\
I_1 &= \overline{I}_{1,n-1},\, R_2=I_{2,n-2}(z=0),\, I_2=\overline{I}_{2,n-2},\\
\overline{I}_{\nu,\mu} &= \left(\frac{4\pi f}{\lambda}\right)\int\limits_0^{\theta_0}\sin^{\nu+1}\left(\frac{\theta}{2}\right)\cos^{3-\nu}\left(\frac{\theta}{2}\right)\cos^{3/2}(\theta)A(\theta)J_\mu(x)\mathrm{d}\theta.
\end{aligned}
\tag{3.46}
$$

For demonstrative purpose, we now move to the polar coordinates. Then, instead of Eq. (3.45), we derive the radial and azimuthal components of the transverse Poynting vector near the focus:

$$
\begin{aligned}
P_r &= 2kzQ(r),\\
P_\varphi &= 0,\\
Q(r) &= R_1\left(\overline{I}_0-\overline{I}_2\right)+\overline{I}_1(R_2-R_0).
\end{aligned}
\tag{3.47}
$$

The radial and azimuthal components of the transverse Poynting vector describe respectively the convergence/divergence of the light field and its transverse rotation. Expression (3.47) illustrates that, regardless of the beam order n, the energy flow near the focus of the beam (1) is diverging or converging from the optical axis along the radial paths. At $z=0$ (in the focus), the energy flow is parallel to the optical axis. Before the focus ($z<0$), transverse energy flow is converging along the radii to the optical axis, whereas after the focus ($z>0$) it is diverging. Since the sign of the function $Q(r)$ in Eq. (3.47) can change at some distances from the optical axis, then at certain circles centered on the propagation axis, transverse energy flow is diverging before the focus and converging after the focus.

Now we show for the field with polarization (3.35) that its longitudinal component of the angular momentum vector is equal to zero near the focus, as in the initial plane. Indeed, the angular momentum vector of light is given by [38]:

$$\mathbf{J} = \frac{1}{2c}\mathrm{Re}\big(\mathbf{r} \times (\mathbf{E}^* \times \mathbf{H})\big) = \frac{2\pi}{c^2}(\mathbf{r} \times \mathbf{P}), \tag{3.48}$$

whereas its longitudinal component (without the insignificant constant factor) is

$$J_z = rP_\varphi = 0. \tag{3.49}$$

The expression (3.49) follows from Eq. (3.47) since $P\ \varphi = 0$. The angular momentum (3.48) can be represented [39] as a sum of the spin angular momentum **S** and of the orbital angular momentum **L**:

$$\mathbf{J} = \mathbf{S} + \mathbf{L} = \frac{1}{8\pi\omega}\mathrm{Im}(\mathbf{E}^* \times \mathbf{E}) + \frac{1}{8\pi\omega}\sum_{p=x,y,z}\mathrm{Im}\Big(E_p^*(\mathbf{r} \times \nabla)E_p\Big) \tag{3.50}$$

Without the constant factor 1/(8 π ω), the longitudinal SAM component, as follows from Eq. (3.41), is equal to $S_z \simeq 2kz\sin(2(n-1)\varphi)(I_0R_2 - I_2R_0)$. The longitudinal OAM component from Eq. (3.50) is given by

$$L_z = \mathrm{Im}\left[E_x^*\frac{\partial}{\partial\varphi}E_x + E_y^*\frac{\partial}{\partial\varphi}E_y + E_z^*\frac{\partial}{\partial\varphi}E_z\right] \tag{3.51}$$

Substituting the expressions (3.36) for the electric field components into Eq. (3.51), we get that near the focus

$$L_z \simeq -2kz\sin(2(n-1)\varphi)(I_0R_2 - I_2R_0). \tag{3.52}$$

Summation of Eqs. (3.52) and (3.41) proves the validity of the expression (3.49). Thus, near the focus, longitudinal SAM and OAM components are equal by magnitude and directed oppositely ($S_z = -L_z$), and, therefore, the longitudinal component of the angular momentum is equal to zero. It is interesting that, when passing through the focus, the longitudinal OAM component (3.52) changes sign, as does the longitudinal SAM component (3.41).

Thus, since the transverse energy flow near the focus is not rotating ($P\ \varphi = 0$), the longitudinal component of the angular momentum equals zero, but it does not mean that, near the focus of the field (3.35), the longitudinal OAM component is also equal to zero. The field with polarization (3.35) in the focus has a longitudinal OAM component, equal by the absolute value to the longitudinal SAM component, but with the opposite sign. We also note that the full (averaged along the whole beam cross section) SAM and OAM are the quantities that conserve separately on free-space propagation of light [40], so that near the focus of the beam (3.35), the following two integrals should be equal to zero:

$$\int_0^\infty r\mathrm{d}r \int_0^{2\pi} \mathrm{d}\varphi S_z(r,\varphi,z) = 0, \int_0^\infty r\mathrm{d}r \int_0^{2\pi} \mathrm{d}\varphi L_z(r,\varphi,z) = 0. \tag{3.53}$$

This means that local rotation of the polarization vector near the focus (3.41) should change sign in different areas of the beam transverse section so that to compensate this rotation and so that the full SAM and OAM of the beam (3.35) would be equal to zero (3.53).

3.2.3 Numerical Simulation

To confirm the theoretic findings, we performed a numerical simulation. For this purpose, we computed the electric and magnetic field near the sharp focus by using the expressions (3.36) and verified them by computing the field directly with using the Debye double integrals.

Figure 3.11 illustrates the distributions of intensity (column 1), radial component of the Poynting vector (column 2), longitudinal components of the SAM (column 3) and OAM (column 4) vectors of a sharply focused Gaussian beam with 3rd-order ($n = 3$) cylindrical polarization (3.35) in two transverse planes, one before the focus (row 1) and one beyond the focus (row 2). For computation, we used the following parameters: wavelength $\lambda = 532$ nm, focal length $f = 10$ μm, numerical aperture $\theta_0 = 0.49\,\pi$ (NA = 0.999), calculation area is 4×4 μm, the longitudinal coordinate (relative to the focal plane) is $z = -\lambda$ (row 1) and $z = +\lambda$ (row 2). To describe the Gaussian shape of the input field, radial distribution of the complex amplitude was equal to $A(\theta) = \exp[-(\sin\theta/\sin\alpha_0)^2]$ with $\alpha_0 = \pi/3$. Figure 3.12 depicts the same distributions and for the same parameters as in Fig. 3.11, but for another polarization order $n = 4$.

Despite the visual similarity of SAM and OAM distributions (columns 3 and 4 in Figs. 3.11 and 3.12), they were computed quite differently: SAM distributions were computed by Eq. (3.39), whereas the OAM distributions were computed by Eq. (3.51) with numerical differentiation by the angular coordinate φ. Nevertheless, the obtained patterns confirm that the OAM L_z and the SAM S_z compensate each other, and that after passing the focal plane, the rotation of the polarization vector changes its direction (i.e., longitudinal SAM component changes its sign).

Figures 3.11 and 3.12 also confirm that near the focus there are $4(n-1)$ areas with different rotation direction of the polarization vector. Figure 3.11 contains $4(n-1) = 8$ local subwavelength areas with elliptic polarization (column 3), including 4 dark areas with left elliptic polarization and 4 light areas with right elliptic polarization. In Fig. 3.12, there are $4(n-1) = 12$ such subwavelength areas. Thus, the simulation confirms that the areas with a different "spin" are spatially separated near the focus, that is the Hall effect occurs.

It is known that in the cross section of a high-order cylindrical vector beam (CVB), polarization is locally linear. The number of full rotations of the locally

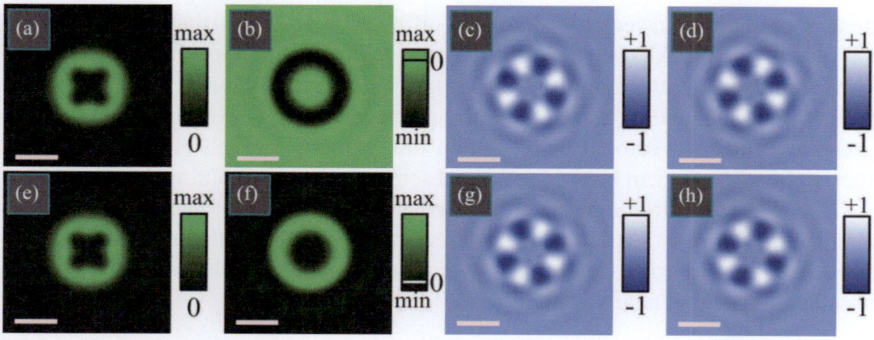

Fig. 3.11 Distributions of intensity (column 1), radial component of the Poynting vector (column 2), normalized-to-maximum longitudinal component of the SAM vector (column 3), and normalized-to-maximum longitudinal component of the OAM vector (column 4) of a sharply focused Gaussian beam with 3rd-order cylindrical polarization before the focus (row 1) and beyond the focus (row 2). In all the figures light and black colors mean respectively maximum and minimum. Scale marks (in the left bottom corner) denote 1 μm

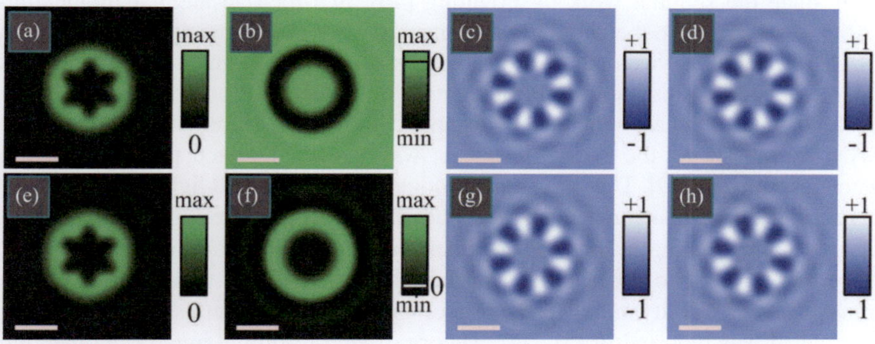

Fig. 3.12 Distributions of intensity (column 1), radial component of the Poynting vector (column 2), normalized-to-maximum longitudinal component of the SAM vector (column 3), and normalized-to-maximum longitudinal component of the OAM vector (column 4) of a sharply focused Gaussian beam with 4th-order cylindrical polarization before the focus (row 1) and beyond the focus (row 2). In all the figures light and black colors mean respectively maximum and minimum. Scale marks (in the left bottom corner) denote 1 μm

linear polarization vector, when passing along a closed contour around the optical axis, is equal to the CVB order, for instance, n. It is also known that both in the initial plane and in the focus, the CVB has neither the spin angular momentum (SAM) nor the orbital angular momentum (OAM). In this section, we demonstrated that near the focal plane of the CVB, for instance, at a distance of wavelength before and beyond the focus, $4(n-1)$ local subwavelength areas are generated, where the polarization vector is rotating in each point. The centers of these local areas reside on a certain radius circle with the center on the optical axis. More of it, in the neighboring areas, the polarization vector is rotating in different directions so that the longitudinal component of the SAM vector in the neighboring areas has opposite signs. The full SAM, obtained by averaging the SAM density over the whole transverse section of the beam, is equal to zero. Such space separation of the left and right rotation of the polarization vectors manifests that the optical spin Hall effect takes place. Such a phenomenon can be used in optical sensorics for determining the CVB order. For example, metasurface-based optical sensors are known [41–44], that detect left-handed and right-handed circular polarizations by separating them in space. In our case, the CVB order n is determined by counting the local areas with left and right circular polarization near the sharp focus [45].

3.3 Spin Angular Momentum at the Tight Focus of a Cylindrical Vector Beam with an Imbedded Optical Vortex

Ever since the seminal work by Poynting [46], in which an angular momentum (AM) was ascribed to the circularly polarized light, optical spin angular momentum (SAM) has actively been studied [47], in particular, at the tight focus. Techniques for measuring Stokes parameters of light fields at different polarization states on the high-order Poincare sphere were discussed in [48]. Measurements of the photon SAM by measuring the optical torque in waveguide integrated photonic devices were described in [49]. The spin angular momentum of a guided light propagating in planar and cylindrical waveguides was studied in [50]. Considering that amplitudes of modes propagating in these waveguides are described by familiar relationships, these can be used to describe all projections of the SAM vector. Measuring the transverse spin density and investigation of a polarization effect using a fluorescent microbead was reported in [51, 52]. Theory of spin-to-orbital angular momentum conversion was proposed in [35, 53], including focusing systems. The study of the transverse spin density in [54] found 'photonic wheels' to occur in the beam. In [55], the spin vector was experimentally shown to form a Möbius strip in the 3D space. The transverse spin angular momentum of tightly focused Poincare beams was discussed in [56]. The optical helicity of electromagnetic waves associated with their polarization states was studied in [57], with the maximum helicity shown to occur for circularly polarized light. Spin angular momentum at the focus of light fields with hybrid polarization states was studied in [60, 61]. Characterization of the SAM of a tightly focused cylindrical vortex beam of the first order was reported in [58]. The study of optical spin properties at the tight focus is of high interest because the SAM vector is partially transferred to the angular momentum of an absorbing microparticle trapped at the beam focus, setting it into rotation around its center of mass [59, 62]. Tight focusing characteristics including SAM were also investigated for specific types of beams, such as Airy beams [63–66] and Pearcey beams [67, 68].

In this section, we analyze the SAM of cylindrical vector beams of an arbitrary integer order. We study the behavior of the SAM, since the non-zero longitudinal SAM is an indicator of elliptical or circular polarization. And we want to know does the focus contain regions with circular and elliptical polarization, if the initial light field has only linear polarization? Using Richards-Wolf theory, we derive an exact expression for the density of the longitudinal SAM vector component at the focus of such beams, revealing that there occur 4(n-1) local subwavelength regions at the focus, with the longitudinal SAM vector component changing sign in the adjacent regions. The total longitudinal component of the SAM vector, i.e., the average value of SAM over the beam cross section, is shown to conserve and equal zero upon focusing. The presence of spatially separated regions with alternating spin sign shows that an optical spin Hall effect occurs at the focus.

3.3.1 Electric and Magnetic Components at the Focus of Light Fields with Phase and Polarization Singularities

Sharp focusing of cylindrical vector beams with an embedded optical vortex was analyzed in [34]. Such beams notably possess both phase and polarization singularities. By simultaneously changing the value of phase singularity (topological charge) and polarization singularity index (Poincare-Hopf index), all characteristics of a laser beam at the sharp focus can be controlled, including the intensity, the energy flow, spin and orbital angular momentums. In this paper, briefly recalling significant relationships from [34] for the readers' convenience, we focus major attention on an analysis of the spin density and the total spin at the focus. We note that the spin angular momentum (SAM) at the focus of the said beams was beyond the scope of Ref. [34]. Below, we demonstrate that the total (averaged over the cross section) longitudinal component of the SAM vector for a cylindrical vector beam of arbitrary integer order with an embedded arbitrary-integer-order optical vortex conserves upon free-space propagation and equals zero. Because of this, an even number of local regions with longitudinal SAM components of alternating sign forms at the focus. In these local regions of alternating spin sign, the elliptic polarization vector rotates clock- or anticlockwise, which is a manifestation of an optical spin Hall effect.

In the original plane, a cylindrical vector beam with an embedded optical vortex is described by the Jones vector:

$$E_{m,n}(\theta,\varphi) = A(\theta)\exp(im\varphi)\begin{pmatrix}\cos n\varphi \\ \sin n\varphi\end{pmatrix}, \tag{3.54}$$

where θ is the polar angle or the angle the optical axis makes with a line drawn from the focal spot center to a point on the spherical wavefront in the original plane, φ is the azimuthal angle at the beam cross section, m is an integer-valued topological charge of the optical vortex, and n is an integer number equal to the Poincare-Hopf polarization singularity index (V-point index), or the order of the cylindrical vector beam. In [34], using Richards-Wolf formalism, we derived relationships for the E- and H-field components at the focus from the original field (3.54). For the convenience of readers, we briefly repeat here the derivation of equations from [34]. The Richards-Wolf equation for calculating the components of the electric and magnetic fields at the focus of a high numerical aperture system is [17]:

$$\begin{aligned}\mathbf{U}(\rho,\psi,z) &= -\frac{if}{\lambda}\int_0^{\alpha}\int_0^{2\pi} B(\theta,\varphi)T(\theta)\mathbf{P}(\theta,\varphi)\\ &\quad\times\exp\{ik[\rho\sin\theta\cos(\varphi-\psi)+z\cos\theta]\}\sin\theta \mathrm{d}\theta\mathrm{d}\varphi\end{aligned} \tag{3.55}$$

where $\mathbf{U}(\rho, \psi, z)$ is the electric or magnetic field, $B(\theta, \varphi)$ is the electric or magnetic field at the input of the high numerical aperture system (θ is the polar angle, φ is the azimuthal angle), $T(\theta)$ is the lens apodization function (in our case, an aplanatic objective was used $T(\theta) = \cos^{1/2}\theta$), f is the focal length, $k = 2\pi/\lambda$ is the wavenumber, λ is the wavelength, α is the maximum polar angle determined by the numerical aperture of the lens (NA = $\sin\alpha$), $\mathbf{P}(\theta, \varphi)$ is the polarization vector, which has the form [69]:

$$\mathbf{P}(\theta,\varphi)=\begin{bmatrix}1+\cos^2\varphi(\cos\theta-1)\\ \sin\varphi\cos\varphi(\cos\theta-1)\\ -\sin\theta\cos\varphi\end{bmatrix}a(\theta,\varphi)+\begin{bmatrix}\sin\varphi\cos\varphi(\cos\theta-1)\\ 1+\sin^2\varphi(\cos\theta-1)\\ -\sin\theta\sin\varphi\end{bmatrix}b(\theta,\varphi) \tag{3.56}$$

where $a(\theta, \phi)$, $b(\theta, \phi)$ are the polarization functions for the x- and y-components of the incident field, i.e., x- and y- components of the Jones vector of the original field (3.54). Field (3.54) can be represented as the sum of two optical vortices with circular polarization:

$$E_{m,n}(\theta,\varphi)=A(\theta)e^{im\varphi}\begin{pmatrix}\cos n\varphi\\ \sin n\varphi\end{pmatrix}=\frac{A(\theta)}{2}e^{i(m+n)\varphi}|L\rangle+\frac{A(\theta)}{2}e^{i(m-n)\varphi}|R\rangle, \tag{3.57}$$

where $|L\rangle = \begin{pmatrix}1\\ -i\end{pmatrix}$; $|R\rangle = \begin{pmatrix}1\\ i\end{pmatrix}$. By substituting Eq. (3.57) into (3.55), we can integrate over the azimuthal angle φ and obtain equations for all components of the electric and magnetic fields in the form [34]:

$$\begin{aligned}E_x &= \frac{i^{m+n-1}}{2}e^{i(m+n)\varphi}\left(I_{0,m+n}+e^{-2i\varphi}I_{2,m+n-2}\right)+\frac{i^{m-n-1}}{2}e^{i(m-n)\varphi}\left(I_{0,m-n}+e^{2i\varphi}I_{2,m-n+2}\right),\\ E_y &= \frac{i^{m+n}}{2}e^{i(m+n)\varphi}\left(-I_{0,m+n}+e^{-2i\varphi}I_{2,m+n-2}\right)+\frac{i^{m-n}}{2}e^{i(m-n)\varphi}\left(I_{0,m-n}-e^{2i\varphi}I_{2,m-n+2}\right),\end{aligned}$$

$$
\begin{aligned}
E_z &= i^{m+n} e^{i(m+n-1)\varphi} I_{1,m+n-1} - i^{m-n} e^{i(m-n+1)\varphi} I_{1,m-n+1},\\
H_x &= \frac{i^{m+n}}{2} e^{i(m+n)\varphi}\left(I_{0,m+n} + e^{-2i\varphi} I_{2,m+n-2}\right) - \frac{i^{m-n}}{2} e^{i(m-n)\varphi}\left(I_{0,m-n} + e^{2i\varphi} I_{2,m-n+2}\right),\\
H_y &= \frac{i^{m+n-1}}{2} e^{i(m+n)\varphi}\left(I_{0,m+n} - e^{-2i\varphi} I_{2,m+n-2}\right) + \frac{i^{m-n-1}}{2} e^{i(m-n)\varphi}\left(I_{0,m-n} - e^{2i\varphi} I_{2,m-n+2}\right),\\
H_z &= i^{m+n+1} e^{i(m+n-1)\varphi} I_{1,m+n-1} + i^{m-n+1} e^{i(m-n+1)\varphi} I_{1,m-n+1},
\end{aligned}
\tag{3.58}
$$

where the functions $I_{\nu,\mu} = I_{\nu,\mu}(r)$ take the form:

$$
I_{\nu,\mu} = 2kf \int\limits_0^{\alpha} \sin^{\nu+1}\left(\frac{\theta}{2}\right) \cos^{3-\nu}\left(\frac{\theta}{2}\right) \cos^{1/2}(\theta) A(\theta) e^{ikz\cos\theta} J_\mu(kr\sin\theta)d\theta, \tag{3.59}
$$

where $J_\mu(\xi)$ is the Bessel function of the first kind and μ-th order, $A(\theta)$ is a real function that defines radially symmetric initial field amplitude, and $\xi = kr\sin\theta$. The first index in (3.59) indicates the function type, $\nu = 0,1,2$, with the second index indicating the number of the Bessel function, $\mu = 0, \pm 1, \pm 2, \ldots$

3.3.2 Distributions of the Intensity, Poynting Vector, and Longitudinal Projection of the SAM

Using Eqs. (3.58), the intensity distribution at the focus of beam (3.54) is found in the form:

$$
\begin{aligned}
I = \frac{1}{2}\left(I_{0,m+n}^2 + I_{0,m-n}^2 + I_{2,m+n-2}^2 + I_{0,m-n+2}^2\right) + I_{1,m+n-1}^2 + I_{0,m-n+1}^2 +\\
(-1)^{n+1}\cos(2(n-1)\varphi)\left(I_{0,m+n} I_{2,m-n+2} + I_{0,m-n} I_{2,m+n-2} - 2I_{1,m+n-1} I_{1,m-n+1}\right).
\end{aligned}
\tag{3.60}
$$

The intensity in (3.60) is defined as the magnitude $I = |E_x|^2 + |E_y|^2 + |E_z|^2$. From (3.60), the intensity pattern is seen to be devoid of circular symmetry, while being symmetric with respect to the Cartesian origin at the focus because the replacement of the angle φ by $\varphi + \pi$ does not lead to a change in the intensity.

Using projections of the **E** and **H** vectors in (3.58), it is also possible to derive projections of the Poynting vector, $\mathbf{P} = \frac{c}{2\pi}\mathrm{Re}(\mathbf{E}^* \times \mathbf{H})$, at the focus of field (3.54) (with the constant $c/(2\pi)$ being dropped below, c is the speed of light):

$$
\begin{aligned}
P_x &= -Q(r)\sin\varphi,\\
P_y &= Q(r)\cos\varphi,\\
P_z &= \frac{1}{2}\left(I_{0,m+n}^2 + I_{0,m-n}^2 - I_{2,m+n-2}^2 - I_{2,m-n+2}^2\right),\\
Q(r) &= I_{1,m+n-1}\left(I_{0,m+n} + I_{2,m+n-2}\right) + I_{1,m-n+1}\left(I_{0,m-n} + I_{2,m-n+2}\right).
\end{aligned}
\tag{3.61}
$$

From (3.61), the longitudinal energy flow is seen to be circularly symmetric, taking both positive ($P_z > 0$) and negative ($P_z < 0$) values. It is also seen from (3.61) and (3.59) that at $m = n$, the longitudinal projection of Poynting vector has a term $I_{0,0}^2$, which is non-zero on the optical axis at $r = 0$, since the term $I_{0,0}^2$ includes the Bessel function, which is equal to one at zero. Accordingly, for $m = n$, the distribution of longitudinal component of Poynting vector in the focal region has the shape of a single spot, while for $m \neq n$, the focus has the shape of a ring.

The transverse energy flow at the focal plane rotates anticlockwise at $Q(r) > 0$ or clockwise at $Q(r) < 0$. Considering that near the optical axis, all the magnitudes entering $Q(r)$ are positive, the transverse energy flow rotates anticlockwise ($m >$ 0). The expression for $Q(r)$ in (3.61) can be shown to change sign following the replacement of $m > 0$ by $m < 0$.

We note that if the topological charge equals zero ($m = 0$, the beam is non-vertical), the transverse flow does not rotate:

$$Q_{m=0}(r) = I_{1,n-1}\big(I_{0,n} + I_{2,n-2}\big) - I_{1,n-1}\big(I_{0,n} + I_{2,n-2}\big) = 0 \tag{3.62}$$

Next, using (8), we can derive the longitudinal component of the SAM vector [70]:

$$\mathbf{S} = \frac{1}{8\pi\omega}\mathrm{Im}\big(\mathbf{E}^* \times \mathbf{E}\big) \tag{3.63}$$

where ω- the angular frequency of light. Below, the constant 1/(8 π ω) is dropped. The density of the longitudinal SAM component at the focus takes the form:

$$\begin{gathered} S_z = \frac{1}{2}\big[I_{0,m-n}^2 - I_{0,m+n}^2 + I_{2,m+n-2}^2 - I_{2,m-n+2}^2 + \\ +2(-1)^n \cos(2(n-1)\varphi)\big(I_{0,m-n}I_{2,m+n-2} - I_{0,m+n}I_{2,m-n+2}\big)\big]. \end{gathered} \tag{3.64}$$

As is the case with intensity (3.60), the pattern of distribution of the longitudinal SAM component from (3.64) is seen to be axially symmetric, i.e., symmetric with respect to the focal spot center, because after replacing φ by $\varphi + \pi$ in (3.64), the value of the SAM remains unchanged. At $m = 0$ (a non-vertical beam), the SAM in (3.64) equals zero at each focal point, as one would expect because at $m = 0$, the cylindrical vector field in (3.54) is linearly polarized everywhere in the focal plane [71]. From (3.64) it is also seen that at $n = 1$ (radial initial polarization) the cosine function equals 1 and the SAM pattern becomes radially symmetric:

$$S_z(n = 1) = \frac{1}{2}\big[(I_{0,m-1} - I_{2,m-1})^2 - (I_{0,m+1} - I_{2,m+1})^2\big]. \tag{3.65}$$

With the terms in the round brackets in (3.65) being subtracted, at definite values of radius r, the SAM can change sign at the focus. Hence, we infer that at the focus a radial optical spin Hall effect occurs when elliptic polarization vectors rotate oppositely at different radii from the optical axis, as they have the opposite sign spins. Note that the optical Hall effect for radially polarized light was analyzed in [72].

For other values of n, the cosine function in (3.64) is dependent on the value of the azimuthal angle. For example, at $n = 0$ (linear polarization), instead of (3.64), we obtain:

$$S_z(n=0) = \frac{1}{2}\left[I_{2,m-2}^2 - I_{2,m+2}^2 + \right.$$
$$\left. +2\cos(2\varphi)\left(I_{0,m}I_{2,m-2} - I_{0,m}I_{2,m+2}\right)\right]. \tag{3.66}$$

Expression (3.66) is identical to a similar expression derived in [71] for a linearly polarized optical vortex. In our case, four subwavelength regions form in the focal spot at a definite distance from the optical axis, with their centers lying on the Cartesian axes and the elliptic polarization vectors alternatively rotating clockwise and anticlockwise. Thus, an optical spin Hall effect occurs at the focus. At $n > 1$, we find that 4(n -1) subwavelength regions occur at the focus, with half of them having negative spin and the other half-positive spin. It is worth noting that at the center of the focal spot, the value of SAM (3.64) takes a positive value of $S_{z,m=n}(r=0) = I_{0,0}^2/2 > 0$ at $m = n$ or a negative value of $S_{z,m=-n}(r=0) = -I_{0,0}^2/2 < 0$ at $m =$ -n. In a similar way, at $m = 2$-n, the SAM at the focus center is positive, but smaller in value, being equal to $S_{z,m=2-n}(r=0) = I_{2,0}^2/2 > 0$. If $m = n$-2, the SAM at the focus center is negative, $S_{z,m=n-2}(r=0) = -I_{2,0}^2/2 < 0$. Hence, by varying the topological charge $m \neq 0$ it is possible to control the speed and direction of rotation about the optical axis of an absorbing microparticle trapped at the focus center. At $m = 0$, SAM (3.64) equals zero, so the microparticle will not rotate. We note that in the original plane, the longitudinal SAM component equals zero.

3.3.3 Longitudinal Projections of the Poynting Vector and Spin Angular Momentum Averaged Over the Beam Cross Section

In this section, we show that the total longitudinal component of SAM equals zero at the focus. In [73], the energy of each vortex harmonic of field (3.64) was shown to be expressed as

$$W_{\nu,\mu} = 2\pi \int_0^\infty \left|I_{\nu\mu}(r)\right|^2 r dr =$$
$$= 4\pi f^2 \int_0^\alpha \sin^{2\nu+1}\left(\frac{\theta}{2}\right)\cos^{5-2\nu}\left(\frac{\theta}{2}\right)|A(\theta)|^2 d\theta = W_\nu. \tag{3.67}$$

Making use of (3.67), we find that SAM (3.64) averaged over the entire focal spot is given by

$$\hat{S}_z = \int_0^\infty \int_0^{2\pi} S_z r dr d\varphi = \frac{1}{2}\int_0^\infty \int_0^{2\pi} r dr d\varphi \left(I_{0,m-n}^2 - I_{0,m+n}^2 + I_{2,m+n-2}^2 - I_{2,m-n+2}^2 +\right)+$$

$$2(-1)^n \int_0^\infty \int_0^{2\pi} r dr d\varphi \cos(2(n-1)\varphi)\left(I_{0,m-n}I_{2,m+n-2} - I_{0,m+n}I_{2,m-n+2}\right) =$$

$$\frac{1}{2}(W_0 - W_0 + W_2 - W_2) = 0. \tag{3.68}$$

The resulting zero in (3.68) is based on Eq. (3.67), leading to the integrals taken of the first two terms being equal to each other. The same is true of the integrals taken of the third and fourth terms, with the integral taken of the cos φ function over its entire period in the last term being also equal to zero. Thus, the total longitudinal projection of SAM turns out to be zero both in the original plane and at the focus, conserving upon focusing, which means that an even number of local regions, 4(n-1) of alternating spin sign is generated at the focus.

Similar to (3.68), we can derive the total energy of the beam at the focus. Averaging the intensity distribution (3.60) over the beam cross section yields:

$$W = \int_0^\infty \int_0^{2\pi} I r dr d\varphi = \frac{1}{2}\int_0^\infty \int_0^{2\pi} \left(I_{0,m+n}^2 + I_{0,m-n}^2 + I_{2,m+n-2}^2 + I_{0,m-n+2}^2\right) r dr d\varphi$$

$$+ \int_0^\infty \int_0^{2\pi} \left(I_{1,m+n-1}^2 + I_{1,m-n+1}^2\right) r dr d\varphi$$

$$+ (-1)^{n+1} \int_0^\infty \int_0^{2\pi} \cos(2(n-1)\varphi)$$

$$\left(I_{0,m+n}I_{2,m-n+2} + I_{0,m-n}I_{2,m+n-2} - 2I_{1,m+n-1}I_{1,m-n+1}\right) r dr d\varphi$$

$$= \frac{1}{2}(W_0 + W_0 + W_2 + W_2) + (W_1 + W_1) = W_0 + W_2 + 2W_1. \tag{3.69}$$

In (3.69), the first two integrals are calculated based on Eq. (3.67), with the last integral of the cosine function being taken over the entire period and equal to zero. It is worth noting that by averaging the longitudinal projection of the energy flow vector in (3.61) we find that not the entire energy of the beam passes the focal plane in the positive z-axis direction. Actually, in view of Eqs. (3.61) and (3.67) and similar to (3.69), we obtain:

$$\hat{P}_z = \int_0^{\infty}\int_0^{2\pi} P_z r dr d\varphi = \frac{1}{2}\int_0^{\infty}\int_0^{2\pi} r dr d\varphi \left(I_{0,m+n}^2 + I_{0,m-n}^2 - I_{2,m+n-2}^2 - I_{2,m-n+2}^2\right)$$
$$= \frac{1}{2}(W_0 + W_0 - W_2 - W_2) = W_0 - W_2 = W - 2W_2 - 2W_1. \tag{3.70}$$

From (3.70), the energy flow traversing the focal plane is smaller than the full energy of the incident beam. The fact is that some energy partially propagates along the focal plane, without crossing it. This portion of energy is transferred by the longitudinal field component at the focus: $I_z = |E_z|^2$. A certain portion of energy is also 'bounced' backwards, in the reverse direction, and referred to as a reverse energy flow [74].

Based on the expression for the transverse energy flow at the focus in (8), we can derive an angular momentum density at the focus from field (3.54). The transverse projection of the energy flow (3.61) can be written in the polar coordinates:

$$\begin{aligned} P_r &= 0, \\ P_\varphi &= Q(r). \end{aligned} \tag{3.71}$$

Then, the longitudinal component of the angular momentum (AM) vector [38]

$$\mathbf{J} = \mathbf{r} \times \mathbf{P} \tag{3.72}$$

is given by

$$J_z = rQ(r) = r\left[I_{1,m+n-1}\left(I_{0,m+n} + I_{2,m+n-2}\right) + I_{1,m-n+1}\left(I_{0,m-n} + I_{2,m-n+2}\right)\right]. \tag{3.73}$$

From (3.73), the longitudinal component of the AM is seen to equal zero on the optical axis, thanks to the 'moment arm' being equal to zero. With the AM being positive near the optical axis, a dielectric microparticle trapped at the beam focus will be drawn into an anticlockwise rotation on a small-radius circle.

3.3.4 Numerical Modeling

Using Richards-Wolf formulae, we numerically simulated sharp focusing of optical vortices with high-order radial polarization, Eq. (3.54), in an aplanatic optical system with NA = 0.95. The wavelength of the focused light was 633 nm and the $A(\theta)$ function was taken to be unit. Figure 3.13a depicts an intensity pattern at the focus from beam (3.54) at $n = m = 1$. Figure 3.13b depicts a distribution pattern of the longitudinal component of the SAM vector. At $n = 1$, Eqs. (3.60) and (3.64) suggest

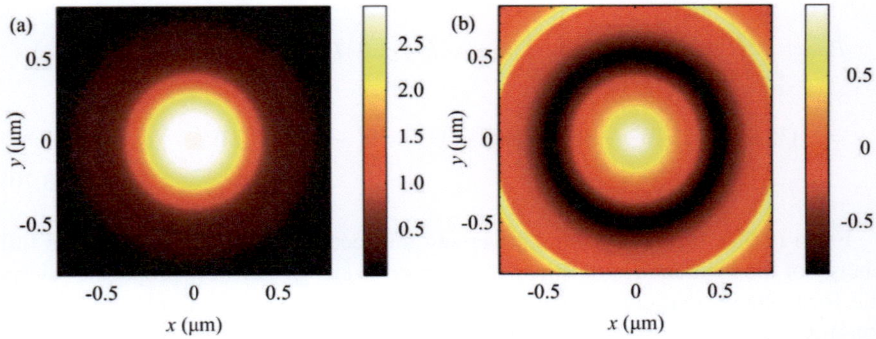

Fig. 3.13 Patterns of **a** intensity $I = I_x + I_y + I_z$ and **b** longitudinal SAM component S_z. Order of phase and polarization singularity: $m = 1, n = 1$

that at the focus the intensity distribution and longitudinal component of SAM vector are radially symmetric and independent of the azimuthal angle φ. Besides, both the intensity and SAM component take maximum positive values on the optical axis at the focus center. Field (3.54) has local linear polarization (longitudinal SAM is equal to zero), however under tight focusing conditions it acquires right-handed elliptical polarization at the focus center, with the polarization vector rotating anticlockwise. On a circle of definite radius (located within a dark annulus in Fig. 3.13b) the spin changes sign and the polarization vector is rotating oppositely (clockwise). That is, Fig. 3.13b demonstrates an optical radial spin Hall effect at the focus of field (3.54), when left and right elliptical polarizations alternate on circles of different radii centered on the optical axis.

Figure 3.14 depicts three Cartesian projections of the Poynting vector. The on-axis projection of the energy flow is seen to have a radially symmetric shape and maximum on the optical axis (Fig. 3.14c). The transverse projections indicate that the energy flow experiences anticlockwise rotation around the axis at the focus (Fig. 3.14a, b), in accordance with the theoretical prediction in (3.61).

Figure 3.15 depicts (a) the intensity and (b) the longitudinal SAM component at the focus of field (3.54) at $n = 2$, $m = 1$. The intensity pattern is seen no more to be circular-shaped, as was the case in Fig. 3.13a, taking the shape of a vertical 'dumb-bell' (Fig. 3.15a) with the longitudinal SAM projection losing its circular symmetry, although four local regions of alternating SAM sign can still be distinguished in Fig. 3.15b.

Figure 3.16 depicts all projections of the Poynting vector at the focus of field (3.54) at $m = 1$ and $n = 2$. The transverse energy flow is seen to rotate around the optical axis within two concentric rings (Fig. 3.16a, b), with the rotation being clockwise in the smaller ring and anticlockwise in the larger ring. Presence of oppositely rotating energy flows at the focus is the manifestation of the radial orbital Hall effect, with the on-axis projection being ring-shaped (Fig. 3.16c). Figure 3.17 depicts (a) the intensity and (b) the longitudinal SAM projection at the focus of field (1) at $n = m = 2$. From Fig. 3.17a, it is seen that although the initial field has a topological

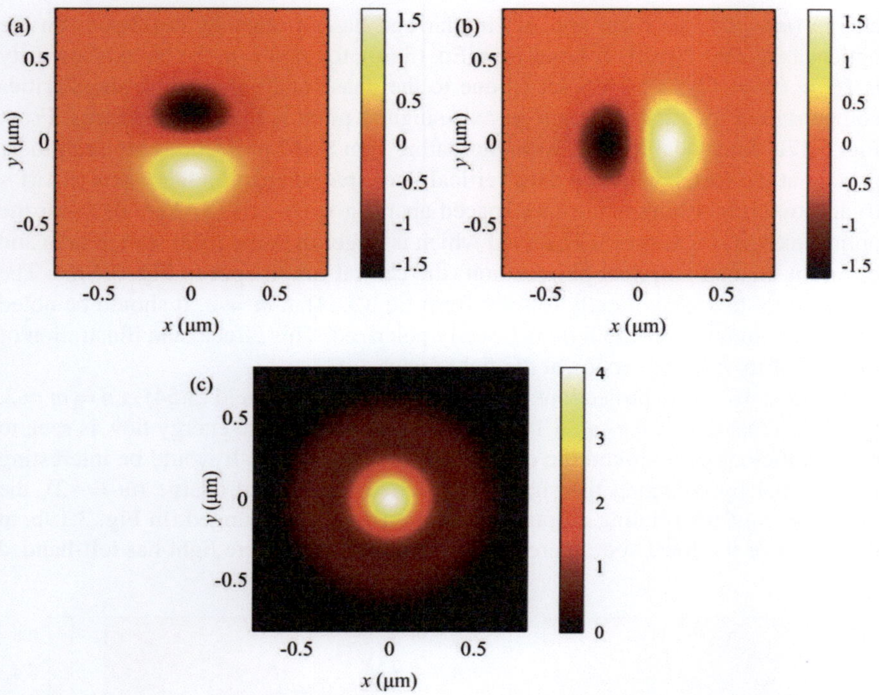

Fig. 3.14 Distribution patterns of the transverse Poynting vector components: **a** P_x and **b** P_y, and **c** the longitudinal Poynting vector component P_z. Order of phase and polarization singularities: $m = 1, n = 1$

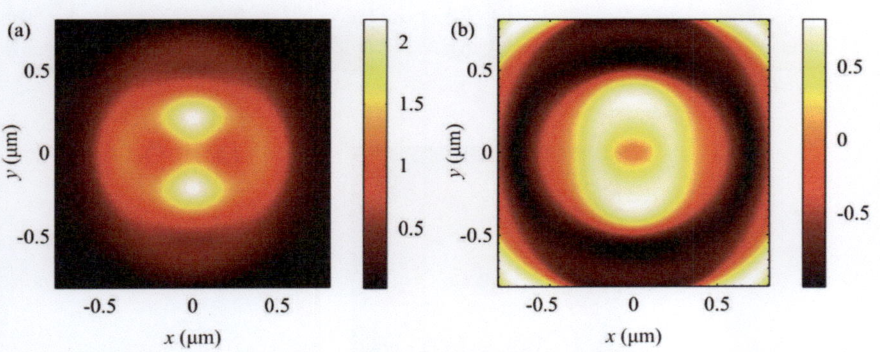

Fig. 3.15 Patterns of **a** intensity $I = I_x + I_y + I_z$ and **b** longitudinal SAM component. Order of phase and polarization singularity: $m = 1, n = 2$

charge ($m = 2$), the focal spot is circular and has an intensity maximum on the optical axis. This directly follows from Eq. (3.60), for at $m = n$, the on-axis intensity is $I(r = 0) = I_{0,0}^2/2 > 0$. That is, due to the phase and polarization singularities compensating for each other, no on-axis singular point is found at the focus. From Fig. 3.17b, four local regions with alternating sign SAM projections are seen to be present at the focus, with two dark vertical lines spaced apart on the x-axis (SAM < 0) and two bright horizontal lines spaced apart on the y-axis (SAM > 0). Near the optical axis, a bright spot is observed which is larger than the focal spot in area and has right-handed elliptical polarization (the central bright spot in Fig. 3.17b). The existence of this spot directly follows from Eq. (3.64) at $m = n$. It should be noted that in the initial plane the light is linearly polarized. This effect is an illustration of the orbital-to-spin conversion at the tight focus.

Figure 3.18 shows projections of the Poynting vector of field (3.54) at $n = m = 2$. Similar to the case of $n = m = 1$ (Fig. 3.14a, b), a transverse energy flow is seen to rotate anticlockwise around the optical axis (Fig. 3.18a, b). It would be interesting to note that by changing the sign of the vortex topological charge (m = -2), the handedness of the rotating elliptic polarization vector is changed. In Fig. 3.19b, at the center of the focal spot, there is a dark elliptic spot where light has left-handed

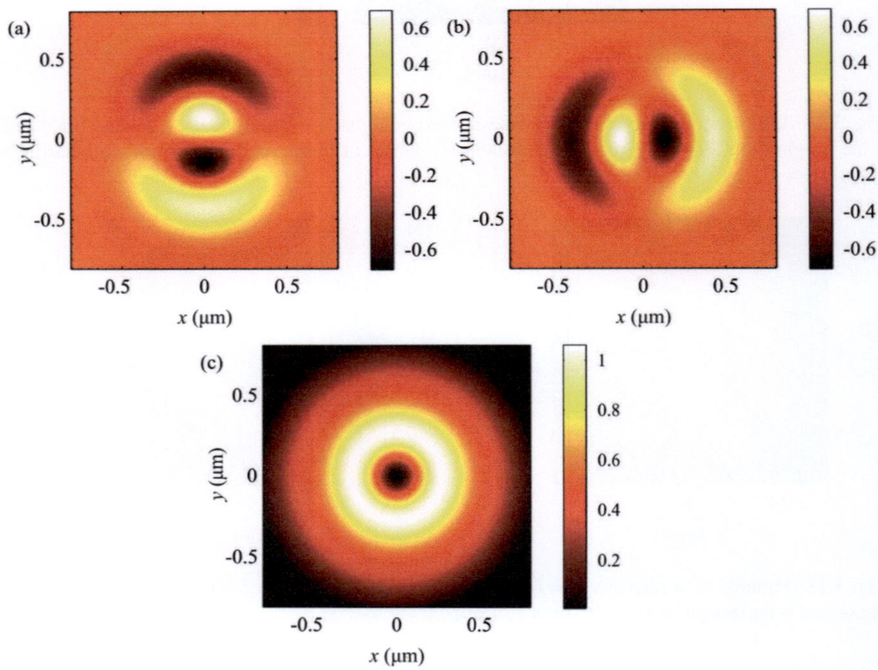

Fig. 3.16 Distribution patterns of the transverse components of the Poynting vector: **a** P_x and P_y and **b** the longitudinal component P_z of the Poynting vector. Order of phase and polarization singularity: $m = 1, n = 2$

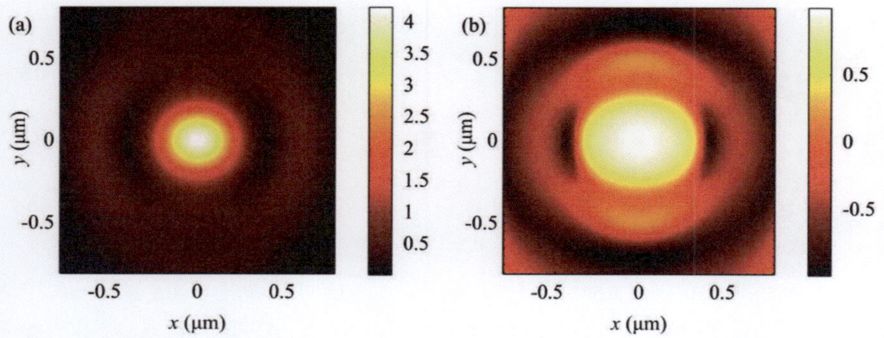

Fig. 3.17 Patterns of **a** intensity $I = I_x + I_y + I_z$ and **b** distribution of the longitudinal SAM component. Order of phase and polarization singularity: $n = 2, m = 2$

elliptic polarization. This effect is the direct consequence of Eq. (3.64) at $n = -m$. Thus, by changing sign of polarization singularity of field (3.54) it is possible to change the handedness of the rotating elliptic polarization vector at the focus near the optical axis.

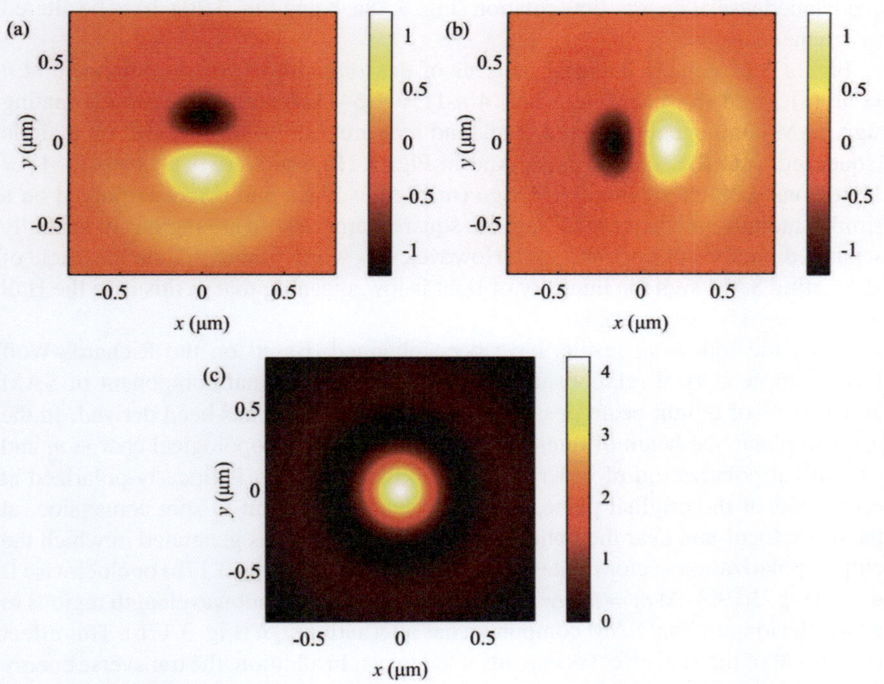

Fig. 3.18 Patterns of distributions of transverse Poynting vector components **a** P_x and **b** P_y and **c** longitudinal Poynting vector component P_z

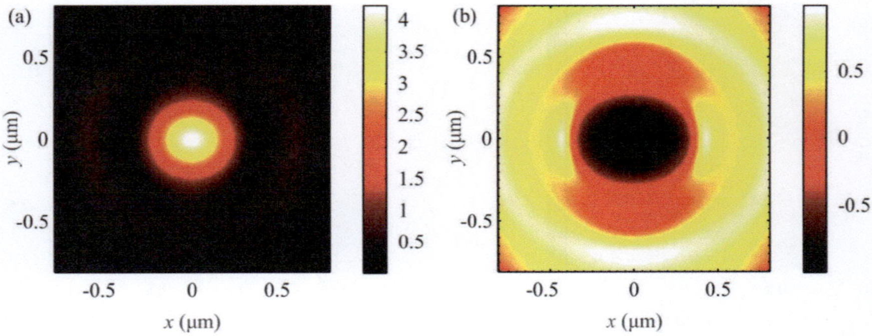

Fig. 3.19 Patterns of **a** intensity $I = I_x + I_y + I_z$ and **b** distribution of the longitudinal SAM component. Order of phase and polarization singularity: $m = -2$, $n = 2$

Figure 3.20 depicts projections of the Poynting vector at the focus of field (3.54) at $n = 2$, $m =$ -2. From Fig. 3.20a, b, the transverse energy flow at the focus is seen to rotate clockwise, unlike Fig. 3.18a, b, which depicts the anticlockwise rotation of the transverse energy flow. Thus, changing the sign of topological charge of vortex field (3.54) enables both spin orientation at the focus center (Figs. 3.17b and 3.19b) and handedness of energy flow rotation (Fig. 3.18a, b and Fig. 3.20a, b) to be altered simultaneously.

Figure 3.21 depicts intensity patterns of the longitudinal SAM component at $n = m$: (a) 3 and (b) 4. In Fig. 3.21a, 4(n-1) = 4(3–1) = 8 regions with alternating sign SAM components (marked black and red) are seen to be arranged on a circle 'squeezed' into the square frame. And in Fig. 3.21b, there are 4(n-1) = 4(4–1) = 12 regions with alternating SAM sign (marked as black and bright) arranged on a similar-radius circle 'squeezed' into the square frame. These are regions of spatially separated photons of opposite spin. However, it is worth noting that on the circle of alternating SAM sign the intensity of light is low, meaning that in this case the Hall effect is weakly pronounced.

Thus, the following results have been obtained. Based on the Richards-Wolf formalism, analytical relationship (3.61) for the longitudinal component of SAM at the focus of a light beam possessing double singularity has been derived. In the original plane, the beam of interest (3.54) has an integer topological charge m and cylindrical polarization of order n, meaning that the beam is linearly polarized at each point of the original plane. However, thanks to orbital-to-spin conversion, at the beam focus and near the optical axis, a circular region is generated in which the elliptic polarization vector rotates anticlockwise if $m > 0$ (Fig. 3.17b) or clockwise if $m < 0$ (Fig. 3.19b). At $m = n$, one finds at the focus $4(n-1)$ subwavelength regions in which the longitudinal SAM component has alternating sign (Fig. 3.17b). This effect is a variant of the Hall effect occurring at the focus. In addition, the transverse energy flow at the focus has been revealed to rotate about the optical axis anticlockwise at $m > 0$ (Fig. 3.14a, b and Fig. 3.18a, b) or clockwise at $m < 0$ (Fig. 3.20a, b). Remarkably, at $n = 2$ and $m = 1$, the transverse energy flow at the focus has been found to rotate

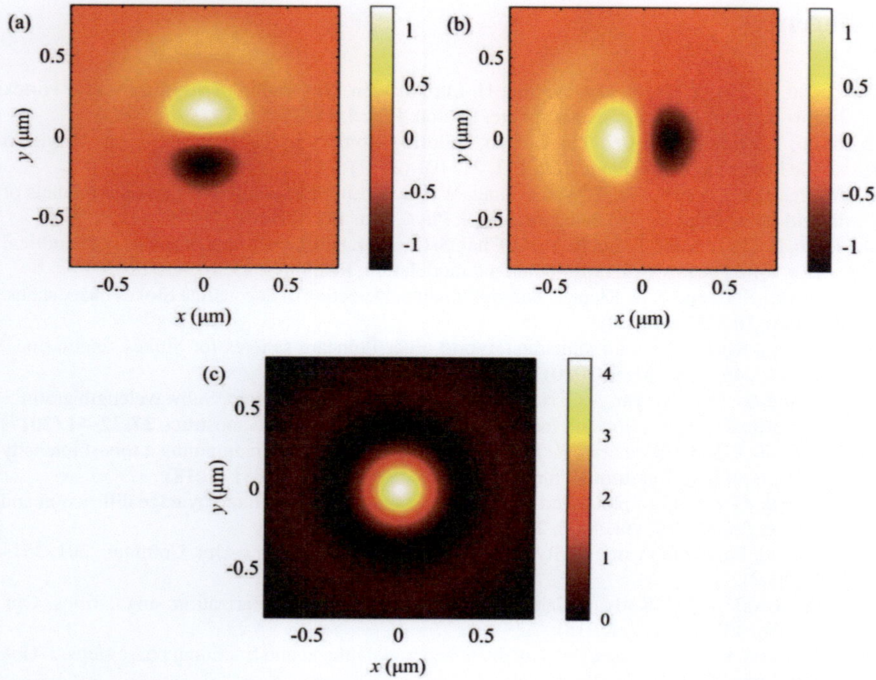

Fig. 3.20 Patterns of distributions of projections of the Poyntings **a** P_x and **b** P_y and **c** longitudinal component of the Poynting vector P_z order of phase and polarization singularity: $m = -2, n = 2$

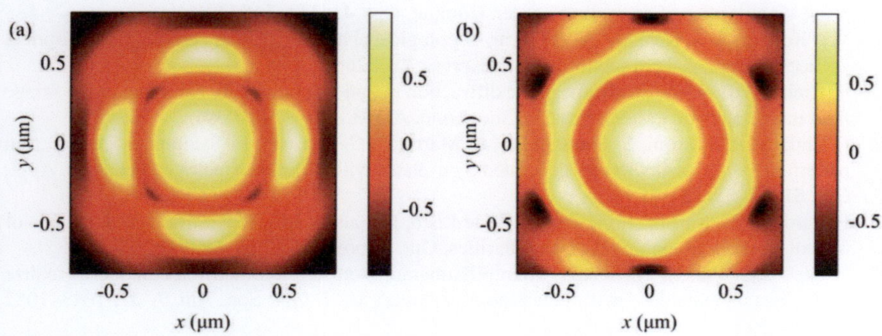

Fig. 3.21 The longitudinal SAM component at **a** $m = n = 3$ and **b** $m = n = 4$

on two optical-axis-centered rings, with the flow rotating clockwise on the smaller ring and anticlockwise on the larger one (Fig. 3.16a, b). This effect is a variant of the radial orbital Hall effect at the focus [75].

References

1. Z. Liu, Y. Liu, Y. Ke, Y. Liu, W. Shu, H. Luo, S. Wen, Generation of arbitrary vector vortex beams on hybrid-order Poincare sphere. Photon. Res. **5**, 15–21 (2017)
2. S. Fu, Y. Zhai, T. Wang, C. Yin, C. Gao, Tailoring arbitrary hybrid Poincare beams through a single hologram. Appl. Phys. Lett. **111**, 211101 (2017)
3. Y. Zhang, P. Chen, S. Ge, T. Wei, J. Tang, W. Hu, Y. Lu, Spin-controlled massive channels of hybrid-order Poincare sphere beams. Appl. Phys. Lett. **117**, 081101 (2020)
4. J. Liu, X. Chen, Y. He, L. Lu, H. Ye, G. Chai, S. Chen, D. Fan, Generation of arbitrary cylindrical vector vortex beams with cross-polarized modulation. Res. Phys. **19**, 103455 (2020)
5. G. Arora, S. Deepa, S.N. Khan, P. Senthilkumaran, Detection of degenerate Stokes index states. Sci. Rep. **10**, 20759 (2020)
6. G. Arora, Ruchi, P. Senthilkumaran, Hybrid order Poincare spheres for Stokes singularities. Opt. Lett., **45**, 5136–5139 (2020)
7. S.S. Stafeev, V.V. Kotlyar, A.G. Nalimov, M.V. Kotlyar, L. O'Faolain, Subwavelength gratings for polarization conversion and focusing of laser light. Photonic. Nanostruct. **27**, 32–41 (2017)
8. P. Lochab, P. Senthilkumaran, K. Khare, Designer vector beams maintaining a robust intensity profile on propagation through turbulence. Phys. Rev. A **98**, 023831 (2018)
9. M. Berry, Geometry of phase and polarization singularities illustrated by edge diffraction and the fides. *Proc. SPIE*, vol. 4403, 2001.
10. I. Freund, Polarization singularity indices in Gaussian laser beams. Opt. Commun. **201**, 251–270 (2002)
11. A.A. Kovalev, V.V. Kotlyar, Gaussian beams with multiple polarization singularities. Opt. Commun. **423**, 111–120 (2018)
12. M.V. Berry, Optical vortices evolving from helicoidal integer and fractional phase steps. J. Opt. A Pure Appl. Opt. **6**, 259–268 (2004)
13. M. Born, E. Wolf, *Principles of Optics* (Pergamon Press, Oxford, 1968)
14. Q. Zhan, Cylindrical vector beams: from mathematical concepts to applications. Adv. Opt. Photon. **1**, 1–57 (2009)
15. V.V. Kotlyar, S.S. Stafeev, A.A. Kovalev, Sharp focusing of a light field with polarization and phase singularities of an arbitrary order. Comput. Opt. **43**, 337–346 (2019)
16. V.V. Kotlyar, A.A. Kovalev, A.V. Volyar, Topological charge of a linear combination of optical vortices: topological competition. Opt. Express, **28**, 8266
17. B. Richards, E. Wolf, Electromagnetic diffraction in optical systems. II. Structure of the image field in an aplanatic system. Proc. R. Soc. Lond. A **253**, 358–379 (1959)
18. F. Cardano, E. Karimi, S. Slussarenko, L. Marrucci, C. de Lisio, E. Santamato, Polarization pattern of vector vortex beams generated by q-plates with different topological charges. Appl. Opt. **51**, C1–C6 (2012)
19. F. Cardano, E. Karimi, L. Marrucci, C. de Lisio, E. Santamato, Generation and dynamics of optical beams with polarization singularities. Opt. Express **21**, 8815–8820 (2013)
20. P. Kumar, S. Pal, N. Nishchal, P. Senthilkumaran, Non-interferometric technique to realize vector beams embedded with polarization singularities. J. Opt. Soc. Am. A **37**, 1043–1052 (2020)
21. B. Khajavi, E.J. Galvez, High-order disclinations in space-variant polarization. J. Opt. **18**, 084003 (2016)
22. L. Lu, Z. Wang, Y. Cai, Propagation properties of phase-locked radially-polarized vector fields array in turbulent atmosphere. Opt. Express **29**, 16833–16844 (2021)
23. M. Meier, V. Romano, T. Feurer, Material processing with pulsed radially and azimuthally polarized laser radiation. Appl. Phys. A **86**, 329–334 (2007)
24. P. Meng, S. Pereira, P. Urbach, Confocal microscopy with a radially polarized focused beam. Opt. Express **26**, 29600–29613 (2018)
25. L. Carretero, P. Acebal, S. Blaya, Three-dimensional analysis of optical forces generated by an active tractor beam using radial polarization. Opt. Express **22**, 3284–3295 (2014)

26. V.V. Kotlyar, A.A. Kovalev, S.S. Stafeev, A.G. Nalimov, S. Rasouli, Tightly focusing vector beams containing V-point polarization singularities. Opt. Las. Tech. **145**, 107479 (2022)
27. Q. Zhan, J.R. Leger, Focus shaping using cylindrical vector beams. Opt. Express **10**, 324–331 (2002)
28. S.C. Tidwell, D.H. Ford, W.D. Kimura, Generating radially polarized beams interferometrically. Appl. Opt. **29**, 2234–2239 (1990)
29. K.J. Moh, X.-C. Yuan, J. Bu, R.E. Burge, B.Z. Gao, Generating radial or azimuthal polarization by axial sampling of circularly polarized vortex beams. Appl. Opt. **46**, 7544–7551 (2007)
30. N. Davidson, N. Bokor, High-numerical-aperture focusing of radially polarized doughnut beams with a parabolic mirror and a flat diffractive lens. Opt. Lett. **29**, 1318–1320 (2004)
31. M. Rashid, O.M. Maragò, P.H. Jones, Focusing of high order cylindrical vector beams. J. Opt. A Pure Appl. Opt. **11**, 065204 (2009)
32. S.S. Stafeev, A.G. Nalimov, V.D. Zaitsev, V.V. Kotlyar, Tight focusing cylindrical vector beams with fractional order. J. Opt. Soc. Am. B **38**, 1090–1096 (2021)
33. K.S. Youngworth, T.G. Brown, Focusing of high numerical aperture cylindrical-vector beams. Opt. Express **7**(2), 77 (2000)
34. V.V. Kotlyar, S.S. Stafeev, A.A. Kovalev, Reverse and toroidal flux of light fields with both phase and polarization higher-order singularities in the sharp focus area. Opt. Express **27**, 16689–16702 (2019)
35. K.Y. Bliokh, E.A. Ostrovskaya, M.A. Alonso, O.G. Rodriguez-Herrera, D. Lara, C. Dainty, Spin-to-orbital angular momentum conversion in focusing, scattering, and imaging systems. Opt. Express **19**, 26132–26149 (2011)
36. M. Born, and E. Wolf, *Principles of optics*. Nauka: Moscow, Russia, 1973 (720 pp).
37. S.S. Stafeev, A.G. Nalimov, A.A. Kovalev, V.D. Zaitsev, V.V. Kotlyar, Circular polarization near the tight focus of linearly polarized light. Photonics **9**, 196 (2022)
38. J. Humblet, Sur le moment d'impulsion d'une onde électromagnétique. Physica **10**, 585–603 (1943)
39. K. Bliokh, A. Bekshaev, F. Nori, Extraordinary momentum and spin in evanescent waves. Nat. Commun. **5**, 3300 (2014)
40. K.Y. Bliokh, A.Y. Bekshaev, F. Nori, Dual electromagnetism: helicity, spin, momentum, and angular momentum. New J. Phys. **15**, 033026 (2013)
41. D. Wen, F. Yue, S. Kumar, Y. Ma, M. Chen, X. Ren, P.E. Kremer, B.D. Gerardot, M.R. Taghizadeh, G.S. Buller, X. Chen, Metasurface for characterization of the polarization state of light. Opt. Express **23**, 10272–10281 (2015)
42. D. Wen, F. Yue, G. Li, G. Zheng, K. Chan, S. Chen, M. Chen, K.F. Li, P.W.H. Wong, K.W. Cheah, E.Y.B. Pun, S. Zhang, X. Chen, Helicity multiplexed broadband metasurface holograms. Nat. Commun. **6**, 8241 (2015)
43. M. Khorasaninejad, W.T. Chen, A.Y. Zhu, J. Oh, R.C. Devlin, D. Rousso, F. Capasso, Multispectral chiral imaging with a metalens. Nano Lett. **16**, 4595–4600 (2016)
44. A. Ma, Y. Intaravanne, J. Han, R. Wang, X. Chen, Polarization detection using light's orbital angular momentum. Adv. Opt. Mater. **8**, 2000484 (2020)
45. V.V. Kotlyar, S.S. Stafeev, A.A. Kovalev, V.D. Zaitsev, Spin hall effect before and after the focus of a high-order cylindrical vector beam. Appl. Sci. **12**(23), 12218 (2022)
46. J.H. Poynting, The wave motion of a revolving shaft, and a suggestion as to the angular momentum in a beam of circularly polarised light. Proc. R. Soc. Lond. A **82**(557), 560–567 (1909)
47. R.P. Cameron, S.M. Barnett, A.M. Yao, Optical helicity, optical spin and related quantities in electromagnetic theory. New J. Phys. **14**, 053050 (2012)
48. G. Milione, H.I. Sztul, D.A. Nolan, R.R. Alfano, Higher-order Poincaré sphere, Stokes parameters, and the angular momentum of light. Phys. Rev. Lett. **107**, 053601 (2011)
49. L. He, H. Li, M. Li, Optomechanical measurement of photon spin angular momentum and optical torque in integrated photonic devices. Sci. Adv. **2**, e1600485 (2016)
50. D.R. Abujetas, J.A. Sánchez-Gil, Spin angular momentum of guided light induced by transverse confinement and intrinsic helicity. ACS Photonics **7**, 534–545 (2020)

51. M. Neugebauer, T. Bauer, A. Aiello, P. Banzer, Measuring the transverse spin density of light. Phys. Rev. Lett. **114**, 063901 (2015)
52. N. Bokor, Y. Iketaki, T. Watanabe, M. Fujii, Investigation of polarization effects for high-numerical-aperture first-order Laguerre-Gaussian beams by 2D scanning with a single fluorescent microbead. Opt. Express **13**, 10440–10447 (2005)
53. K. Bliokh, F. Rodríguez-Fortuño, F. Nori, A.V. Zayats, Spin–orbit interactions of light. Nat. Photon. **9**, 796–808 (2015)
54. A. Aiello, P. Banzer, M. Neugebauer, G. Leuchs, From transverse angular momentum to photonic wheels. Nat. Photonics **9**, 789–795 (2015)
55. T. Bauer, M. Neugebauer, G. Leuchs, P. Banzer, Optical polarization Möbius strips and points of purely transverse spin density. Phys. Rev. Lett. **117**, 013601 (2016)
56. W. Zhu, V. Shvedov, W. She, W. Krolikowski, Transverse spin angular momentum of tightly focused full Poincaré beams. Opt. Express **23**, 34029–34041 (2015)
57. K.Y. Bliokh, A.Y. Bekshaev, F. Nori, Dual electromagnetism: Helicity, spin, momentum and angular momentum. New J. Phys. **15**, 033026 (2013)
58. P. Meng, Z. Man, A.P. Konijnenberg, H.P. Urbach, Angular momentum properties of hybrid cylindrical vector vortex beams in tightly focused optical systems. Opt. Express **27**(24), 35336–35348 (2019)
59. M. Li, S. Yan, B. Yao, Y. Liang, P. Zhang, Spinning and orbiting motion of particles in vortex beams with circular or radial polarizations. Opt. Express **24**(18), 20604–20612 (2016)
60. R. Chen, K. Chew, C. Dai, G. Zhou, Optical spin-to-orbital angular momentum conversion in the near field of a highly nonparaxial optical field with hybrid states of polarization. Phys. Rev. A **96**(5), 053862 (2017)
61. K. Hu, Z. Chen, J. Pu, Tight focusing properties of hybridly polarized vector beams. J. Opt. Soc. Am. A **29**(6), 1099–1104 (2012)
62. M. Li, S. Yan, Y. Liang, P. Zhang, B. Yao, Spinning of particles in optical double-vortex beams. J. Opt. **20**, 025401 (2018)
63. W. Zhu, W. She, Tightly focusing vector circular airy beam through a hard aperture. Opt. Commun. **334**, 303–307 (2015)
64. J. Zhuang, L. Zhang, D. Deng, Tight-focusing properties of linearly polarized circular Airy Gaussian vortex beam. Opt. Lett. **45**, 296 (2020)
65. L. Ren, Z. Zhong, B. Zhang, Transversely polarized ultra-long optical needles generated by cylindrical polarized circular airy Gaussian vortex beams. Opt. Commun. **483**, 126618 (2021)
66. M. Chen, S. Huang, W. Shao, Tight focusing of radially polarized circular Airy vortex beams. Opt. Commun. **402**, 672–677 (2017)
67. L. Zhang, S. He, X. Peng, L. Huang, X. Yang, G. Wang, H. Liu, Y. He, D. Deng, Tightly focusing evolution of the auto-focusing linear polarized circular Pearcey Gaussian vortex beams. Chaos Solitons Fractals **143**, 110608 (2021)
68. C. Yu, Z. He, C. Huang, F. Chen, J. Zeng, Y. Li, Y. Zhang, J. Pu, H. Lin, Tight Focusing Properties of Ring Pearcey Beams with a Cross Phase. Photonics **9**(12), 964 (2022)
69. S.F. Pereira, A.S. van de Nes, Superresolution by means of polarisation, phase and amplitude pupil masks. Opt. Commun. **234**, 119–124 (2004)
70. S.M. Barnett, L. Allen, Orbital angular momentum and nonparaxial light beams. Opt. Commun. **110**, 670–678 (1994)
71. V.V. Kotlyar, A.A. Kovalev, E.S. Kozlova, A.M. Telegin, Hall effect at the focus of an optical vortex with linear polarization. Micromachines **14**, 788 (2023)
72. L. Han, S. Liu, P. Li, Y. Zhang, H. Cheng, J. Zhao, Catalystlike effect of orbital angular momentum on the conversion of transverse to three-dimensional spin states within tightly focused radially polarized beams. Phys. Rev. A **97**(5), 053802 (2018)
73. A.A. Kovalev, V.V. Kotlyar, Spin hall effect of double-index cylindrical vector beams in a tight focus. Micromachines **14**, 494 (2023)
74. V.V. Kotlyar, A.A. Kovalev, A.G. Nalimov, Energy density and energy flux in the focus of an optical vortex: reverse flux of light energy. Opt. Lett. **43**, 2921–2924 (2018)

75. V.V. Kotlyar, S.S. Stafeev, A.M. Telegin, Spin angular momentum at the tight focus of a cylindrical vector beam with an imbedded optical vortex. Optik **287**, 171103 (2023)

Chapter 4
Cylindrical Fractional-Order and Double-Index Vector Laser Beams

4.1 Tight Focusing Cylindrical Vector Beams with Fractional Order

Cylindrical vector beams (CVBs) [1] are now widely used because of the unique properties they exhibit in tight focusing. In particular, using cylindrical vector beams, it is possible to obtain focal spots with sizes smaller than the scalar diffraction limit [2–4], optical needles [5, 6], light tunnels [7, 8], chains of focal spots [9–11], focus shaping [12], surface plasmon-polaritons nanofocus [13], and pure longitudinal magnetization [14]. The polarization at each point of the CVB is linear, but its direction changes continuously, making one or several rotations when the azimuthal angle changes from 0 to 2π. Most of the works (for example, previously noted [1–14]) are devoted to the study of CVBs, in which the polarization makes only one rotation; these are radially polarized beams, in which the polarization is directed along the radius, or azimuthally polarized beams, in which the polarization is directed perpendicular to the radius. However, there are works that investigate the behavior of high-order CVBs, i.e., beams in which the direction of polarization makes several rotations [15–20]. Earlier, we showed [21–23] that for high-order CVBs in the focal region there are regions in which the projection of the Poynting vector onto the optical axis becomes negative, and if the beam order is two, then such a region is located on the optical axis. One of the ways to obtain cylindrical vector beams is to obtain sector CVBs using half-wave plates [3, 24–26], nonlinear optical crystals [27], polarizers [28], subwavelength gratings [29–31], dielectric metasurfaces [32], and spatial light modulator (SLM) [33, 34]. Earlier, we studied the effect of the number of sectors on the focusing results [35] and showed that even with the number of sectors equal to six, the differences with the focusing of the beam, in which the polarization changes continuously, become small. However, the question of what the focusing results will be if the direction of polarization in various sectors differs from the planned (radial or azimuthal) polarization—the polarization will be "twisted" or

V. V. Kotlyar et al., *Optical Hall Effect in the Sharp Focus of Laser Light*,
https://doi.org/10.1007/978-3-031-64683-6_4

"under-twisted" to a whole number of rotations, for example, due to technological errors in the manufacture of a sector element.

In this section, using the Richards-Wolf equations, the focusing of a cylindrical vector beam with a wavelength of 532 nm and fractional order was simulated by a flat diffractive lens with a numerical aperture NA = 0.95. Two series of simulations were carried out: in the first, beams with fractional orders from zero to one were investigated. The transition from linear to azimuthal polarization was carried out. In the second simulation, the influence of the deviation of the beam order from $m = 2$ (i.e., the case when the backflow is observed at the center of the focal spot) was investigated. It was shown that for integer values of the beam order, the transverse components of the Poynting vector are equal to zero, but for fractional values, they are not. At fractional values of the order of the beam, varying from zero to one, two pronounced centers with coordinates $x = 0$ and $y = \pm 0.1$ μm are observed, around which the transverse flow in the focal region is twisted. The flow rotates clockwise around the upper center, and anticlockwise around the lower center. With an increase in the order of the beam, more than one, such pronounced centers of rotation of the transverse flow become larger. It was also shown that the reverse flow remains in the center of the spot with a significant deviation of the order of the beam from $m = 2$—it appears on the axis already at $m = 1.55$. Experimentally cylindrical vector beams of fractional order could be generated using the same optical arrangement that used for generation of cylindrical vector beams of integer order [32–34]. However, in this case, an optical vortex phase $\exp\{im\varphi\}$ with a fractional topological charge m should be formed on the spatial light modulator.

4.1.1 The Richards-Wolf Formulas

In this section, the studies were carried out using the Richards-Wolf formulas [36]

$$\mathbf{U}(\rho, \psi, z) = -\frac{if}{\lambda} \int_0^{\alpha_{\max}} \int_0^{2\pi} B(\theta, \varphi) T(\theta) \mathbf{P}(\theta, \varphi) \times \exp\{ik[\rho \sin\theta \cos(\varphi - \psi) + z\cos\theta]\} \sin\theta d\theta d\varphi\,, \tag{4.1}$$

where $\mathbf{U}$ (ρ, ψ, z) is the strength of the electric or magnetic field, $B(\theta, \phi)$ is the electric or magnetic field at the input of the wide-aperture system in coordinates of the exit pupil (θ is the polar angle, φ is the azimuthal angle), $T(\theta)$ is the lens apodization function, f is the focal length, $k = 2\pi / \lambda$ is the wavenumber, λ is the wavelength (in the simulation it was considered equal to 532 nm), $\alpha_{\max}$ is the maximum polar angle determined by the numerical aperture of the lens (NA = $\sin\alpha_{\max}$), $\mathbf{P}(\theta, \phi)$ is the polarization vector, for the strength of the electric and magnetic fields having the form:

$$\mathbf{P}(\theta,\varphi)=\begin{bmatrix}1+\cos^2\varphi(\cos\theta-1)\\ \sin\varphi\cos\varphi(\cos\theta-1)\\ -\sin\theta\cos\varphi\end{bmatrix}a(\theta,\varphi)+\begin{bmatrix}\sin\varphi\cos\varphi(\cos\theta-1)\\ 1+\sin^2\varphi(\cos\theta-1)\\ -\sin\theta\sin\varphi\end{bmatrix}b(\theta,\varphi), \tag{4.2}$$

where a (θ, ϕ) and b (θ, ϕ) are functions describing the state of polarization of the x- and y-components of the intensities of the focused beam.

For a light field with cylindrical polarization of the m-th order, the Jones vectors will have the form:

$$\mathbf{E}(\theta,\phi)=\begin{pmatrix}a(\theta,\phi)\\ b(\theta,\phi)\end{pmatrix}=\begin{pmatrix}-\sin(m\phi)\\ \cos(m\phi)\end{pmatrix} \tag{4.3}$$

for the electric field strength and

$$\mathbf{H}(\theta,\phi)=\begin{pmatrix}a(\theta,\phi)\\ b(\theta,\phi)\end{pmatrix}=\begin{pmatrix}-\cos(m\phi)\\ -\sin(m\phi)\end{pmatrix} \tag{4.4}$$

for the magnetic field strength. If we put in formulas (4.3), (4.4) $m = 1$, then they will describe ordinary azimuthally polarized light, and for $m = 0$ - linearly polarized light directed along the y-axis.

4.1.2 Focusing Cylindrical Vector Beams with an Order from Zero to One

Consider focusing beams have fractional order varying from zero (linearly polarized light) to one (azimuthal polarization) with increment $\Delta m = 0.25$. Figure 4.1 shows the direction of polarization for fractional values of m is equal to 0.25, 0.5 and 0.75. Figure 4.2 shows the change in the total intensity and individual intensity components for this case, and in Fig. 4.3, similarly, the component of the Poynting vector, calculated as $\mathbf{S} = \mathrm{Re}(\mathbf{E} \times \mathbf{H}^*)$.

Figures 4.2 and 4.3 show that when the beam of order m changes from zero to unity, the intensity distribution changes from elliptical to annular, while intermediate states have the form of inclined ellipses. In this case, the inclination and broadening of the spot in one of the directions at beam of orders m close to zero is provided by the longitudinal component of the intensity, and at m close to unity, by the transverse components. For the longitudinal component S_z, the shape of the distribution changes from circular to annular (for $m = 3/4$, the distribution already looks like an asymmetric ring), but the intermediate ellipses are not inclined, but oriented along

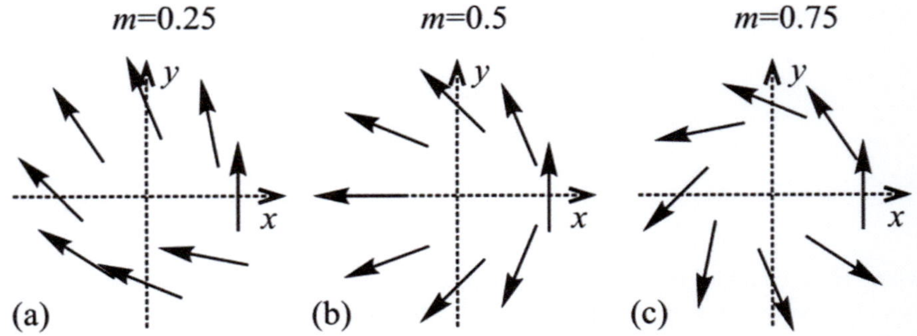

Fig. 4.1 Direction of polarization in the fractional order CVB: $m = 0.25$ (**a**), $m = 0.5$ (**b**), $m = 0.75$ (**c**)

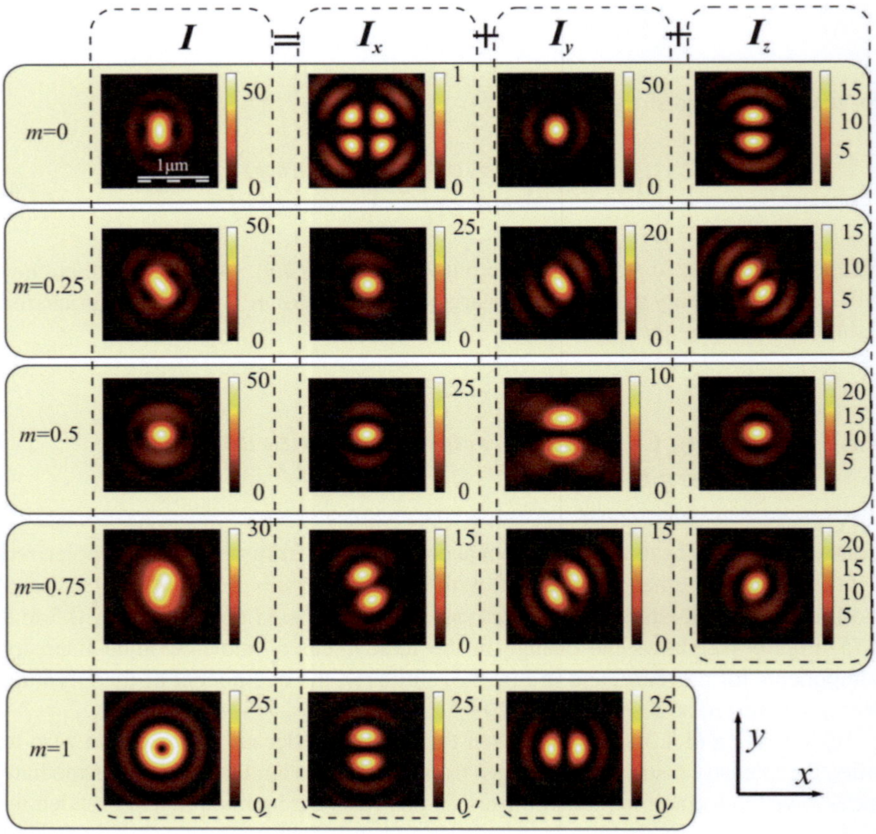

Fig. 4.2 Distribution of the total intensity I and intensity components I_z, I_x, and I_y when focusing the CVB of various orders: $m = 0$ (linear polarization), $m = 0.25$, $m = 0.5$, $m = 0.75$, and $m = 1$ (azimuthal polarization)

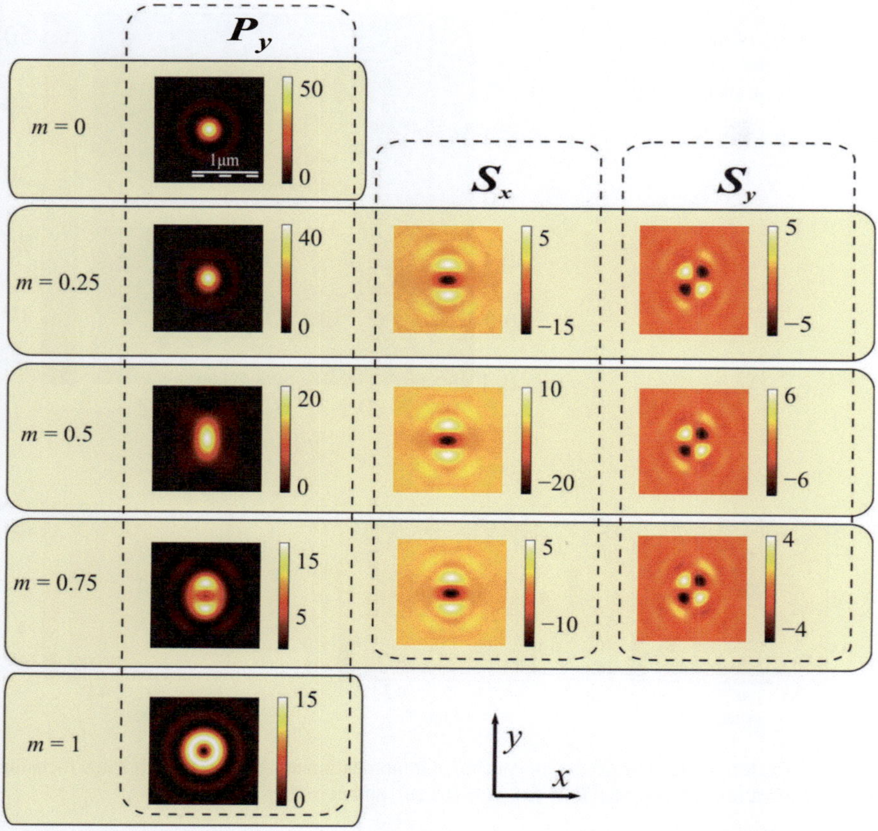

Fig. 4.3 Distribution of components of the Poynting vector P_y, S_x, and S_y when focusing the CVB of various orders: $m = 0$ (linear polarization), $m = 0.25$, $m = 0.5$, $m = 0.75$, and $m = 1$ (azimuthal polarization)

the *y*-axis (polarization directions at $m = 0$). For integer values of the beam order, the transverse components of the Poynting vector are equal to zero, but for fractional values they are not. Two centers with coordinates $x = 0$ and $y = \pm 0.1$ μm are observed, around which the transverse flow is rotated in the focal region (Fig. 4.4). Around the upper center, the flow rotates clockwise, and around the lower center, anticlockwise. We can say that the trajectory of the transverse energy flow has a shape of the figure "eight". To analyze the polarization state at the focal spot, we also calculate spin angular momentum (SAM) in the focal spot: $\mathbf{SAM} = \mathrm{Im}(\mathbf{E}^* \times \mathbf{E})$[37]. Figure 4.5 shows the longitudinal projection of SAM for CVB of different fractional orders: $m = 0.25$, $m = 0.5$, and $m = 0.75$.

Figure 4.5 shows that CVB of fractional order form areas with elliptical polarization and different directions of rotation. Moreover, the axial projection of the SAM vector is positive in areas with the energy flow rotating counterclockwise (lower part

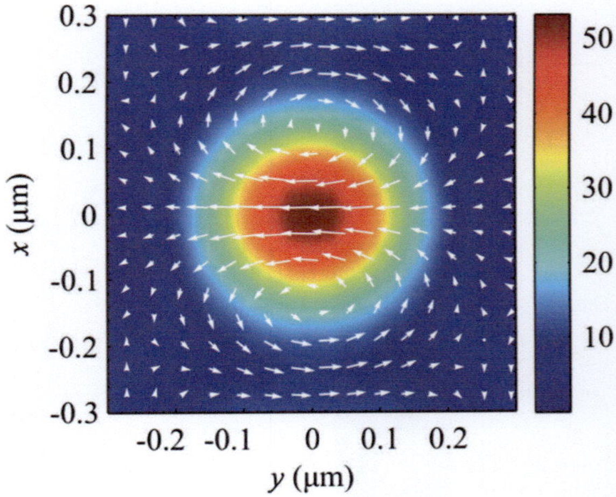

Fig. 4.4 Intensity pattern I and directions of the Poynting vector $S_x\mathbf{x} + S_y\mathbf{y}$ in the focal plane ($z = 0$, $m = 0.5$)

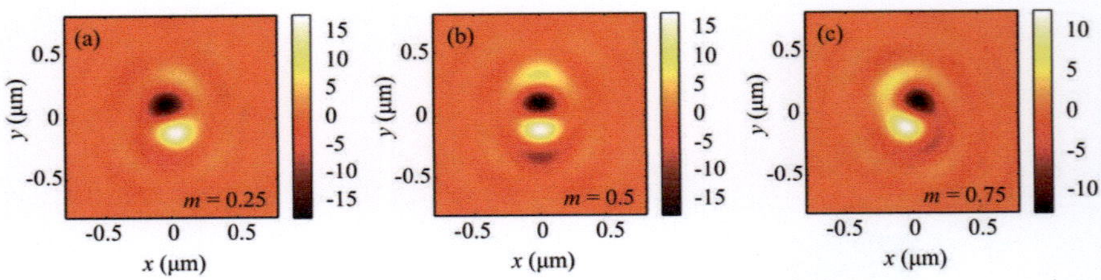

Fig. 4.5 The longitudinal projection of spin angular momentum at the focus ($z = 0$) when focusing the CVB of various orders: $m = 0.25$ (**a**), $m = 0.5$ (**b**), and $m = 0.75$ (**c**)

of Fig. 4.4). In contrast the axial projection of the SAM vector is negative in areas with the transverse energy flow rotating clockwise (upper part of Fig. 4.4).

It is also interesting that there is a change in shape: for linear polarization, the intensity is elliptical, and the longitudinal energy flux is round, while for a beam with order $m = 0.5$, on the contrary, the intensity is almost round, and the longitudinal flux is elliptical. Note that when focusing an optical vortex with azimuthal polarization (beam order $m = 1$, topological vortex charge $n = 1$), a round spot is also observed. The focal spot sizes for these three beams are shown in Table 4.1.

Comparison of the intensity values in Table 1 shows that the smallest focal spot is observed when focusing an azimuthally polarized optical vortex. But when comparing the values of the longitudinal projection S_z, it should be noted that the smallest value is observed for a focused linearly polarized beam.

Experimentally focusing of cylindrical vector beams of fractional order could be investigated using the same optical arrangement that used for generation of cylindrical vector beams of integer order [32–34]. Another way is a generation of a converging vector beam using metalens, which was previously used, for example, in

Table 4.1 Sizes of the focal spot by the full width at half maximum of the intensity and the longitudinal projection of the Poynting vector when focusing the CVB of various orders

	The order of the CVB, m	Optical vortex order, n	Full width at half maximum $I = Ix + Iy + Iz$, a.u		Spot width at half of the longitudinal projection of the Poynting vector Sz, a.u	
			FWHMx, λ	FWHMy, λ	FWHMx, λ	FWHMy, λ
Azimuthally polarized optical vortex	1	1	0.460	0.460	0.475	0.475
CVB of order $m = 0.5$	0.5	0	0.517	0.484	0.517	0.936
Linearly polarized light	0	0	0.421	0.751	0.456	0.456

[38]. In particular, to generate a beam with an order of 0.5, it could be used a subwavelength grating metalens, shown in Fig. 4.6 The metalens combines subwavelength grating based polarizer and Fresnel zone plate with focal length $f = \lambda = 633$ nm, it converts linearly polarized plane wave to CVB of order $m = 0.5$ and focuses it. The lens consists of 16 radial sectors, each of which rotates the polarization of the incident light to produce the CVB. The index of refraction of the metalens was $n = 4.352 + 0.486i$ (amorphous silicon). Figure 4.7 shows the distribution of Poynting components when focusing by this metalens (Fig. 4.6) calculated by FDTD method implemented in FullWave software (the simulation mesh step was λ/30 along all three axes).

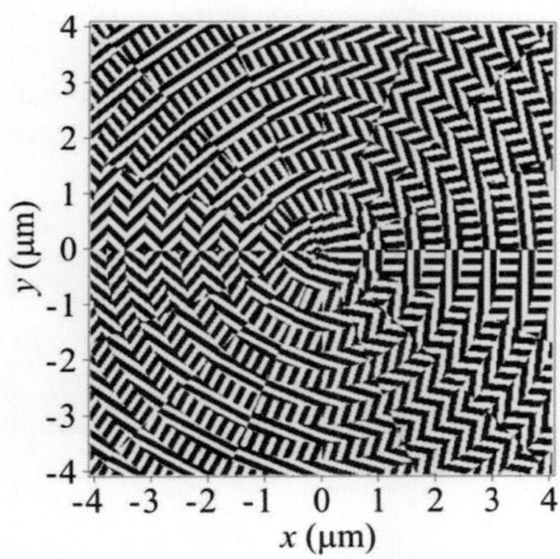

Fig. 4.6 Subwavelength grating based metalens, which generates CVB of order $m = 0.5$ and focuses it

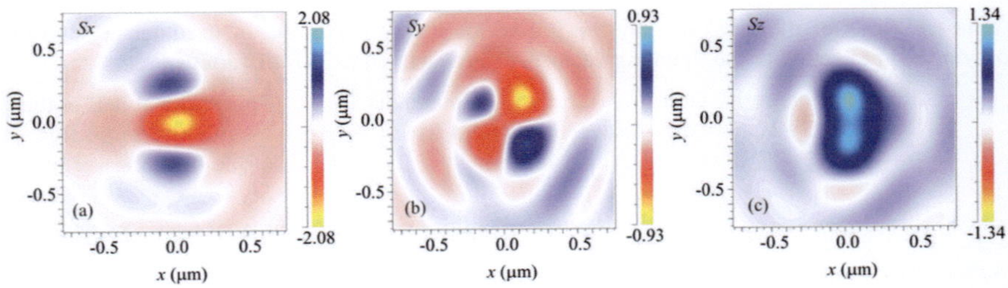

Fig. 4.7 Distribution of components of the Poynting vector S_x (**a**), S_y (**b**), and S_z (**c**) when focusing by the proposed metalens (Fig. 4.6)

Figure 4.7 shows that the results of numerical simulation of the metalens are in agreement with the results obtained using the Richards-Wolf formulas (Fig. 4.3 for $m = 0.5$).

4.1.3 *Focusing Cylindrical Vector Beams with an Order from One to Two*

Figure 4.8 shows the distribution of the longitudinal projection of the Poynting vector when the beam order changes from 1.5 to 1.9 (i.e., the polarization is "under-twisted" up to two full turns of its direction in the beam cross section).

Figure 4.8 shows that a decrease in the order of distribution m, the longitudinal component of the Poynting vector S_z becomes asymmetric view – the ring is compressed and the negative flux vanishes in the center. Figure 4.8 also shows that small deviations of the order of the beam from $m = 2$ do not immediately lead to the disappearance of the reverse energy flux on the axis. Negative values of the longitudinal component in the center of the focal spot are observed already at $m =$ 1.55. Figure 4.9 shows the change in *Sz* at the center of the focal spot.

When the beam order m deviates from an integer value, the transverse components of the Poynting vector S_x and S_y also become nonzero, although they were absent at

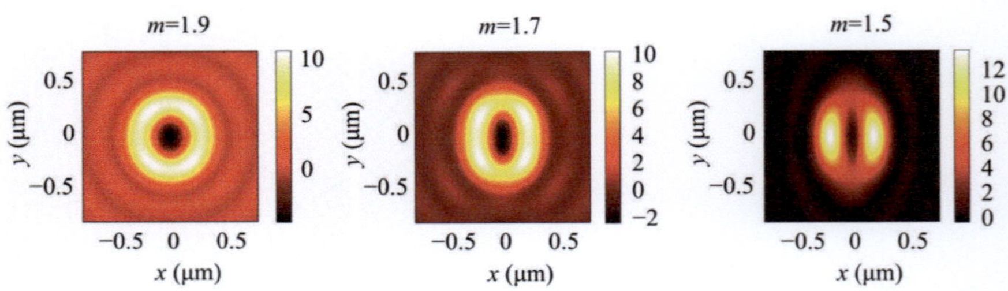

Fig. 4.8 Distribution of the longitudinal component of the Poynting vector when focusing the CVB with orders varying from 1.9 to 1.5

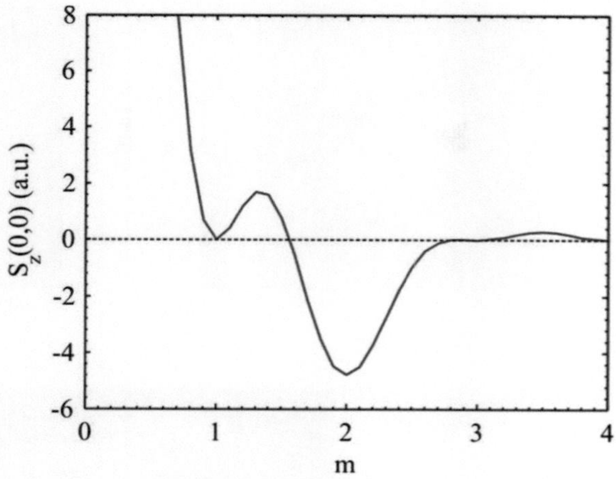

Fig. 4.9 The value of the longitudinal projection of the Poynting vector $S_z(0,0)$ at the center of the focal spot when the order of the focused beam m is changed

$m = 1$ and $m = 2$. The energy flow in the focal plane turns from laminar to turbulent. Unlike the fractional expressed centers of rotation, the cross-flow becomes larger (Fig. 4.10).

Using the Richards-Wolf formulas, the focusing of a cylindrical vector beam with a wavelength of 532 nm and a fractional topological charge was simulated by a flat diffractive lens with a numerical aperture NA = 0.95 [39, 40]. Two series of simulations were performed: the influence of the deviation of the order of the beam from $m = 2$ (i.e., the case when the backflow is observed in the center of the focal spot) was investigated and beams with fractional orders less than one were investigated: $m = 1/4$, $m = 1/2$ and $m = 3/4$. It has been shown that: when the order of the beam changes from zero to one, the intensity distribution changes from elliptical to annular, while the intermediate states have the form of inclined ellipses. For the longitudinal component of the Poynting vector, the shape of the distribution changes from round to circular, but the intermediate ellipses are not inclined. For integer values of the beam order, the transverse components of the Poynting vector are equal to zero, with deviations of the beam order from an integer value, the energy flux becomes turbulent. For the order of the beam from zero to unity, two centers with coordinates $x = 0$ and $y = \pm 0.1$ μm are observed, around which the transverse flow is rotated in the focal region. With an increase in the order of the beam above unity, such centers of rotation become larger. Small deviations of the order of the beam from $m = 2$ do not lead to the disappearance of the backflow of energy on the axis. Negative values on the axis start at $m = 1.5$. The transverse energy flows at the focus for CVB of orders $m = 0.5$ (Fig. 4.4) and $m = 1.9$ (Fig. 4.10) are connected with the spin angular momentum (SAM). Moreover, the axial projection of the SAM vector is positive in the areas with the energy flow rotating counterclockwise. In contrast the axial projection of the SAM vector is negative in the areas with the transverse energy flow rotating clockwise. In other words the flow rotates counterclockwise in the area where the polarization vector rotates counterclockwise, and the energy flow

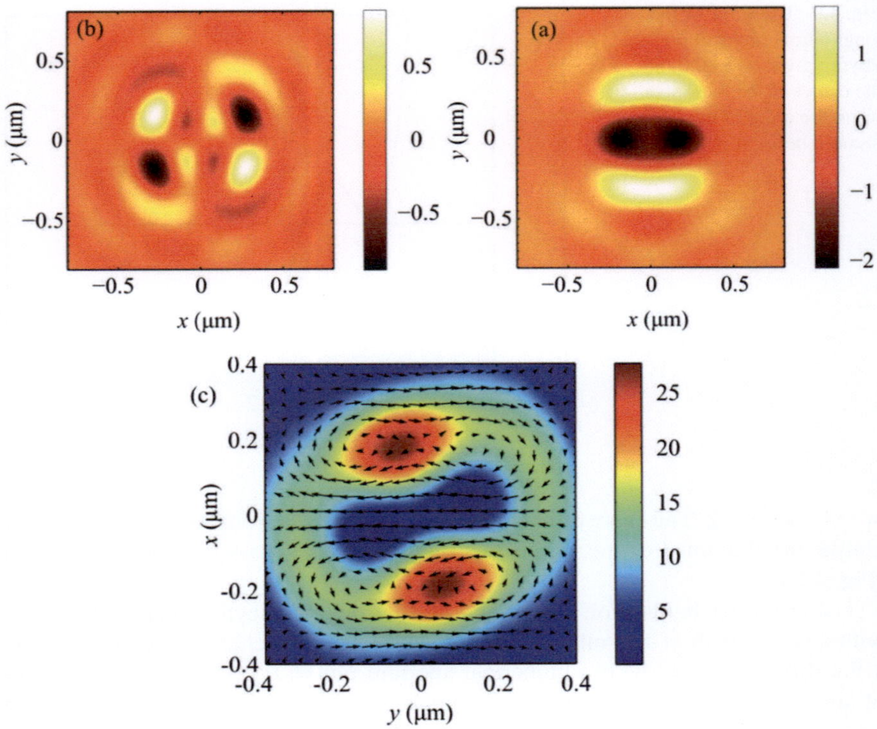

Fig. 4.10 Components of the Poynting vector S_x (**a**) and S_y (**b**) and intensity pattern I with directions of the Poynting vector $S_x\mathbf{x} + S_y\mathbf{y}$ (**c**) in the focal region when focusing the beam with the order of polarization $m = 1.9$

rotates clockwise in the area where the polarization vector rotates clockwise. This spatial separation at the focus of regions with left and right circular polarization is a demonstration of the optical Hall effect [41].

4.2 Spin Hall Effect of Double-Index Cylindrical Vector Beams in a Tight Focus

In micromachines, elements can be driven by light [42, 43]. This requires designing optical tweezers appropriate for driven elements, depending on their shape, material, and motion trajectory. The work [44] discusses how to control the mechanical motions of various particles in optical tweezers under complicated actuation of optical forces and torques by tightly focused laser beams. Typically, a light beam comes out from a laser with a Gaussian shape. Then, for certain applications, not only for optical trapping, but also for optical data transmission, laser welding, the beam should be converted to attain an on-demand shape. For this purpose, a huge branch of modern optics, laser beam shaping, is developed [45]. The beam shaping techniques are developed both within a resonator and outside it [46], by using refractional or diffractional optical elements. External beam shaping can be done to shape a beam that maintains this shape on propagation [47], or in some specific area, for instance, in the focal plane [48]. However, in optical trapping, there can also be a need not only to trap a particle in a certain point, but also to make her do some movements. For instance, the particle can be forced to travel along some trajectory, or rotate around its own center. Such a rotation occurs, when the light possesses the spin angular momentum (SAM) [49], or nonlinear polarization. Thus, in addition to the task of shaping the beam intensity distribution, there can be a task of shaping the SAM distribution. In addition to the optical trapping, the SAM can be used as information in optical data transmission [50].

In paraxial approximation, the intensity shaping can be done for a single transverse field component of a homogeneously polarized light. However, for shaping the SAM, both transverse components should be tailored with a controlled phase delay between them. The problem becomes more difficult at tight focusing conditions. Recently, it has been observed that when a linearly polarized light beam is tightly focused, then, near the focus, areas occur with elliptic polarization [51]. Since the areas with negative and positive SAM are spatially separated, this phenomenon is a manifestation of the optical spin Hall effect. Later, the same effect was discovered for tightly focused high-order cylindrical vector beam [51]. In [52], the SAM is distributed mostly on a ring and consists of alternating areas with positive and negative values. The cylindrical vector beams have the Jones vector $\mathbf{J} = [\cos m\,\varphi, \sin m\,\varphi]$ with φ being the angular polar coordinate and m being the polarization order (for $m = 1$, radial polarization). A further generalization is a two-index polarization singularity with the Jones vector $\mathbf{J} = [\cos m\,\varphi, \sin n\,\varphi]$, where $m \neq n$ [53], i.e., such a generalized vector field has different orders on the different axes. Recently, we studied such fields with V-points and for many values m and n we obtained the Poincare-Hopf index analytically [54].

In this section, based on the Richards-Wolf approximation [55], we study what happens with the SAM of a light field with the double-index polarization singularity in the tight focus. We obtain an expression for the complex amplitude near the focus. Then, based on this expression, we derive the formula for the SAM and found that it can be nonzero only for the orders m and n of different parity. For analytical prediction of the SAM distribution, we decomposed the in-focus light field into the orbital angular momentum (OAM) spectrum and estimated the contribution of each angular harmonic. It turns out that if a light field being focused is not of ring shape and has a homogeneous or decaying Gaussian shape, than the OAM spectrum consists mainly of mth and nth angular harmonics which exceed the other harmonics by an order of magnitude. It allows to estimate the polar angles with zero SAM and thus to predict the SAM distribution. As an example, we obtained SAM distributions on a ring where the areas with positive and negative SAM occur in pairs.

4.2.1 A Light Field with a Double-Index Polarization Singularity Near the Tight Focus

In [54], we investigated a generalization of cylindrical vector beams, when the polarization indices of the E_x and E_y field components were different. The amplitude of the electric vector of such a field is given by.

$$\mathbf{E}(\theta, \varphi) = A(\theta)\begin{pmatrix} \cos m\varphi \\ \sin n\varphi \end{pmatrix} \tag{4.5}$$

where $\mathbf{E}$ is the strength vector of the electric field, φ is the azimuthal angle in the source plane, (m, n) is the two-index polarization order, θ is the polar angle, describing the tilt of the light rays to the optical axis, $A(\theta)$ is the amplitude of the source field as a function of the axis tilt angle. Directions of the electric vectors are illustrated in Fig. 4.11

In work [36], we have obtained expressions for the Cartesian components of a linearly polarized optical vortex, focused by an aplanatic system. If an optical vortex with a topological charge m is linearly polarized along the axis x, then, in the input plane, the electric field is given by the following:

$$\mathbf{E}(\theta, \varphi) = A(\theta)\exp(im\varphi)\begin{pmatrix} 1 \\ 0 \end{pmatrix} \tag{4.6}$$

whereas near the tight focus, the complex amplitude reads as

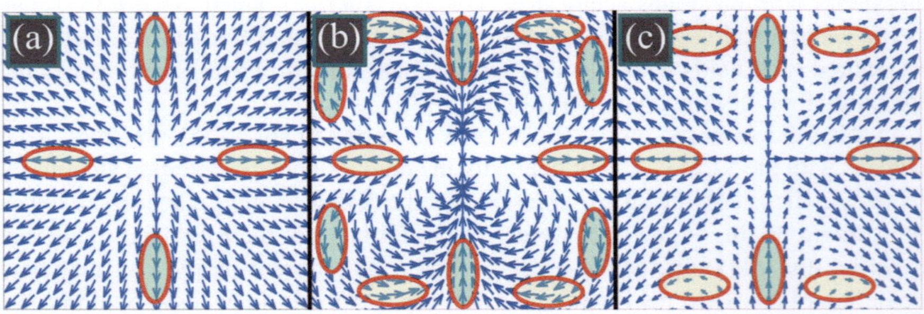

Fig. 4.11 Conventional radial polarization ($m = n = 1$) (**a**), third-order radial polarization ($m = n = 3$) (**b**), double-index polarization singularity ($m = 1, n = 3$) (**c**). For radial polarization (**a**), there are two angles (0 and π) with horizontal electric vector and two angles (π /2 and 3 π /2) with vertical electric vector. For 3rd-order radial polarization (**b**), there are six angles with horizontal electric vector and six angles with vertical electric vector. For the field with double-index polarization singularity of the order (1, 3), there are six angles with horizontal electric vector and two angles with vertical electric vector

$$\begin{cases} E_x(\rho,\psi,z) = -\frac{1}{2} i^{m+1} e^{im\psi}\left(2I_{0,m} + e^{2i\psi} I_{2,m+2} + e^{-2i\psi} I_{2,m-2}\right), \\ E_y(\rho,\psi,z) = \frac{1}{2} i^{m} e^{im\psi}\left(e^{-2i\psi} I_{2,m-2} - e^{2i\psi} I_{2,m+2}\right), \\ E_z(\rho,\psi,z) = i^{m} e^{im\psi}\left(e^{-i\psi} I_{1,m-1} - e^{i\psi} I_{1,m+1}\right), \end{cases} \quad (4.7)$$

where (ρ, ψ, z) are the cylindrical coordinates with the origin in the focus. In Eq. (4.7), the functions $I_{\nu,\mu}$ are defined as follows:

$$I_{\nu,\mu} = 2kf \int_0^\alpha \sin^{\nu+1}\left(\frac{\theta}{2}\right) \cos^{3-\nu}\left(\frac{\theta}{2}\right) \cos^{1/2}(\theta) A(\theta) e^{ikz\cos\theta} J_\mu(kr\sin\theta) d\theta, \quad (4.8)$$

where $k = 2/\lambda$ is the wavenumber of monochromatic light with the wavelength λ, f is the focal length of the focusing lens, α is the maximal tilt angle of rays to the optical axis, defining the numerical aperture of the aplanatic lens NA $= \sin(\alpha)$, $J_\nu()$ is the νth-order Bessel function of the first kind.

The same way, if such an input field is linearly polarized along the axis y, the field components near the tight focus are equal to

$$\begin{cases} E_x(\rho,\psi,z) = -\frac{1}{2} i^{m} e^{im\psi}\left(e^{2i\psi} I_{2,m+2} - e^{-2i\psi} I_{2,m-2}\right), \\ E_y(\rho,\psi,z) = -\frac{1}{2} i^{m+1} e^{im\psi}\left(2I_{0,m} - e^{2i\psi} I_{2,m+2} - e^{-2i\psi} I_{2,m-2}\right), \\ E_z(\rho,\psi,z) = i^{m+1} e^{im\psi}\left(e^{i\psi} I_{1,m+1} + e^{-i\psi} I_{1,m-1}\right). \end{cases} \quad (4.9)$$

The field with circularly symmetric amplitude distribution $A(\theta)$ and with polarization (4.5) can be decomposed into a superposition of four linearly polarized optical vortices:

$$\begin{aligned} \mathbf{E} = A(\theta)\begin{pmatrix} \cos m\varphi \\ \sin n\varphi \end{pmatrix} &= \frac{1}{2} A(\theta) e^{im\varphi}\begin{pmatrix} 1 \\ 0 \end{pmatrix} + \frac{1}{2} A(\theta) e^{-im\varphi}\begin{pmatrix} 1 \\ 0 \end{pmatrix} \\ &+ \frac{-i}{2} A(\theta) e^{in\varphi}\begin{pmatrix} 0 \\ 1 \end{pmatrix} + \frac{i}{2} A(\theta) e^{-in\varphi}\begin{pmatrix} 0 \\ 1 \end{pmatrix}. \end{aligned} \quad (4.10)$$

Using this decomposition, we get the field components of the field with polarization (4.5) near the tight focus:

$$\begin{cases} E_x(\rho,\psi,z) = -\frac{1}{2}i^{m+1}\{2I_{0,m}\cos m\psi + I_{2,m+2}\cos[(m+2)\psi] + I_{2,m-2}\cos[(m-2)\psi]\} \\ +\frac{1}{2}i^{n+1}\{I_{2,n+2}\cos[(n+2)\psi] - I_{2,n-2}\cos[(n-2)\psi]\}, \\ E_y(\rho,\psi,z) = -\frac{1}{2}i^{n+1}\{2I_{0,n}\sin n\psi - I_{2,n+2}\sin[(n+2)\psi] - I_{2,n-2}\sin[(n-2)\psi]\} \\ +\frac{1}{2}i^{m+1}\{I_{2,m-2}\sin[(m-2)\psi] - I_{2,m+2}\sin[(m+2)\psi]\}, \\ E_z(\rho,\psi,z) = i^m\{I_{1,m-1}\cos[(m-1)\psi] - I_{1,m+1}\cos[(m+1)\psi]\} \\ +i^n\{I_{1,n+1}\cos[(n+1)\psi] + I_{1,n-1}\cos[(n-1)\psi]\}. \end{cases} \quad (4.11)$$

The longitudinal component of the spin angular momentum is defined as [56]:

$$S_z = 2\mathrm{Im}\{E_x^* E_y\} \quad (4.12)$$

For simplicity, we suppose that the functions $A(\theta)$ are real valued. Thus, all the integrals $I_{\nu,\mu}$ are also real valued in the focal plane ($z = 0$). Then, substituting the transverse field components from Eq. (4.11), we get the following expression:

$$\begin{aligned} S_z = &\frac{1}{2}\mathrm{Im}\{i^{n-m}\}\{2I_{0,m}\cos m\psi + I_{2,m+2}\cos[(m+2)\psi] + I_{2,m-2}\cos[(m-2)\psi]\} \\ &\times\{2I_{0,n}\sin n\psi - I_{2,n+2}\sin[(n+2)\psi] - I_{2,n-2}\sin[(n-2)\psi]\} \\ &+\frac{1}{2}\mathrm{Im}(i^{m-n})\{I_{2,n+2}\cos[(n+2)\psi] - I_{2,n-2}\cos[(n-2)\psi]\} \\ &\times\{I_{2,m-2}\sin[(m-2)\psi] - I_{2,m+2}\sin[(m+2)\psi]\}. \end{aligned} \quad (4.13)$$

This expression is cumbersome, but it reveals that the SAM is zero unless the polarization orders n and m are of different parity. When the parity is the same for both orders, then polarization is linear in the focus, since, according to Eq. (4.11), both E_x and E_y are proportional to i^{m+1} (or i^{n+1}), multiplied by some real-valued function. Near the center ($r << \lambda$), if $n > m \geq 2$, the transverse components E_x and E_y are approximately proportional to the vector $\mathbf{J} = [\cos(m-2)\varphi, -\sin(m-2)\varphi]$. If $m > n \geq 2$, they are proportional to the vector $\mathbf{J} = [\cos(n-2)\varphi, -\sin(n-2)\varphi]$. Thus, a saddle-type polarization singularity is generated in the center [57].

For n and m of different parity, simplifications yield

$$S_z = \frac{1}{2}\mathrm{Im}\{i^{n-m}\}\left\{\begin{array}{l} 4I_{0,m}I_{0,n}\cos m\psi\sin n\psi \\ -2I_{0,m}I_{2,n+2}\cos m\psi\sin[(n+2)\psi] \\ -2I_{0,m}I_{2,n-2}\cos m\psi\sin[(n-2)\psi] \\ +2I_{0,n}I_{2,m+2}\sin n\psi\cos[(m+2)\psi] \\ +2I_{0,n}I_{2,m-2}\sin n\psi\cos[(m-2)\psi] \\ +(I_{2,n+2}I_{2,m+2} + I_{2,n-2}I_{2,m-2})\sin[(m-n)\psi] \\ -(I_{2,m+2}I_{2,n-2} + I_{2,m-2}I_{2,n+2})\sin[(m+n)\psi] \end{array}\right\}. \quad (4.14)$$

This equation is hard to analyze without knowing which terms contribute the most. Thus, we need to decompose the near-focus field into the angular harmonics and study its OAM spectrum.

4.2.2 Balance of Light Field Energy Near the Tight Focus

At first, we study the OAM spectrum of linearly polarized light after tight focusing. According to Eq. (4.7), if the input field is polarized along the axis x, then the x-component of the electric vector near the focus consists of three angular harmonics, whose topological charges are m, $m-2$, $m+2$. The y-component consists of only two angular harmonics with the topological charges are $m-2$ and $m+2$, whereas the z-component also consists of two angular harmonics with the topological charges are $m-1$ and $m+1$. Each harmonic is proportional to the function $I_{\nu,\mu}$ from Eq. (4.8). Therefore, such a harmonic has the following energy $W_{\nu,\mu}$:

$$W_{\nu,\mu} = 4\pi f^2 \int_0^\alpha \sin^{2\nu+1}\left(\frac{\theta}{2}\right)\cos^{5-2\nu}\left(\frac{\theta}{2}\right)\left|A^*(\theta)\right|^2 d\theta \tag{4.15}$$

This expression indicates that the energy of the angular harmonic is independent of the distance z from the focal plane and of the topological charge of the optical vortex that determines the index μ. As seen from the above expressions for the field near the tight focus, each angular harmonic is proportional to the function $I_{\nu,\mu}$ from Eq. (4.8). Below we derive the energy $W_{\nu,\mu}$ of such a separate angular harmonic:

$$W_{\nu,\mu} = \int_0^\infty \int_0^{2\pi} \left|I_{\nu,\mu}(r,z)\right|^2 r dr d\varphi = 2\pi \int_0^\infty \left|I_{\nu,\mu}(r,z)\right|^2 r dr \tag{4.16}$$

Substituting here the function $I_{\nu,\mu}$ from Eq. (4.8), we get

$$\begin{aligned} W_{\nu,\mu} = 8\pi k^2 f^2 \int_0^\infty & \left[\int_0^\alpha \sin^{\nu+1}\left(\frac{\theta}{2}\right)\cos^{3-\nu}\left(\frac{\theta}{2}\right)\cos^{1/2}(\theta) A^*(\theta) e^{-ikz\cos\theta} J_\mu(kr\sin\theta) d\theta\right] \\ & \times \left[\int_0^\alpha \sin^{\nu+1}\left(\frac{\theta'}{2}\right)\cos^{3-\nu}\left(\frac{\theta'}{2}\right)\cos^{1/2}(\theta') A(\theta') e^{ikz\cos\theta'} J_\mu(kr\sin\theta') d\theta'\right] r dr. \end{aligned} \tag{4.17}$$

Changing the integration variables $\eta = \sin\theta$, $\eta' = \sin\theta'$ and changing the integration order, we obtain

$$
\begin{aligned}
W_{\nu,\mu} &= 8\pi k^2 f^2 \int\limits_0^{\sin\alpha} \left(\frac{1-\sqrt{1-\eta^2}}{2}\right)^{\frac{\nu+1}{2}} \left(\frac{1+\sqrt{1-\eta^2}}{2}\right)^{\frac{3-\nu}{2}} \\
&\times \left(1-\eta^2\right)^{1/4} A^*(\arcsin\eta) \exp\left(-ikz\sqrt{1-\eta^2}\right) \frac{d\eta}{\sqrt{1-\eta^2}} \\
&\times \int\limits_0^{\sin\alpha} \left(\frac{1-\sqrt{1-\eta'^2}}{2}\right)^{\frac{\nu+1}{2}} \left(\frac{1+\sqrt{1-\eta'^2}}{2}\right)^{\frac{3-\nu}{2}} \\
&\times \left(1-\eta'^2\right)^{1/4} A(\arcsin\eta') \exp\left(ikz\sqrt{1-\eta'^2}\right) \frac{d\eta'}{\sqrt{1-\eta'^2}} \\
&\times \left[\int\limits_0^{\infty} J_\mu(kr\eta) J_\mu(kr\eta') r dr\right].
\end{aligned}
\tag{4.18}
$$

Due to the orthogonality of the Bessel functions [58], the inner integral over r reduces to the Dirac delta function (if $\mu \geq -1$):

$$
\int\limits_0^{\infty} J_\mu(k\eta r) J_\mu(k\eta' r) r dr = \frac{\delta(k\eta - k\eta')}{k\eta} = \frac{1}{k^2} \frac{\delta(\eta-\eta')}{\eta}
\tag{4.19}
$$

Thus, Eq. (4.19) is simplified and only one integral remains:

$$
W_{\nu,\mu} = 8\pi f^2 \int\limits_0^{\sin\alpha} \left(\frac{1-\sqrt{1-\eta^2}}{2}\right)^{\nu+1} \left(\frac{1+\sqrt{1-\eta^2}}{2}\right)^{3-\nu} \left|A^*(\arcsin\eta)\right|^2 \frac{d\eta}{\eta\sqrt{1-\eta^2}}
\tag{4.20}
$$

Now we return back to the trigonometric functions $\eta = \sin\theta$ and get (4.15):

$$
\begin{aligned}
W_{\nu,\mu} &= 8\pi f^2 \int\limits_0^{\alpha} \left(\frac{1-\cos\theta}{2}\right)^{\nu+1} \left(\frac{1+\cos\theta}{2}\right)^{3-\nu} \left|A^*(\theta)\right|^2 \frac{d\theta}{\sin\theta} \\
&= 4\pi f^2 \int\limits_0^{\alpha} \sin^{2\nu+1}\left(\frac{\theta}{2}\right) \cos^{5-2\nu}\left(\frac{\theta}{2}\right) \left|A^*(\theta)\right|^2 d\theta.
\end{aligned}
\tag{4.21}
$$

The integrals (4.15) can be evaluated analytically only in simple cases, but, nevertheless, contribution of each angular harmonic can be estimated. For example, if the field being focused is a uniform field with a constant amplitude $A(\theta) \equiv 1$, then.

$$W_{0,\mu} = 4\pi f^2 \int\limits_0^\alpha \sin\left(\frac{\theta}{2}\right)\cos^5\left(\frac{\theta}{2}\right)d\theta = \frac{4}{3}\pi f^2\left[1 - \cos^6\left(\frac{\alpha}{2}\right)\right] \quad (4.22)$$

$$W_{1,\mu} = 4\pi f^2 \int\limits_0^\alpha \sin^3\left(\frac{\theta}{2}\right)\cos^3\left(\frac{\theta}{2}\right)d\theta = \frac{2}{3}\pi f^2 \sin^4\left(\frac{\alpha}{2}\right)\left[1 + 2\cos^2\left(\frac{\alpha}{2}\right)\right] \quad (4.23)$$

$$W_{2,\mu} = 4\pi f^2 \int\limits_0^\alpha \sin^5\left(\frac{\theta}{2}\right)\cos\left(\frac{\theta}{2}\right)d\theta = \frac{4}{3}\pi f^2 \sin^6\left(\frac{\alpha}{2}\right) \quad (4.24)$$

According to Eq. (4.7), the x-component E_x consists of three angular harmonics with their amplitude being proportional to the functions $I_{0,m}$, $I_{2,m+2}/2$, $I_{2,m-2}/2$. The component E_y is a superposition of harmonics described by the functions $I_{2,m+2}/2$ and $I_{2,m-2}/2$. And, finally, the component E_z is a superposition of harmonics described by the functions $I_{1,m+1}$ and $I_{1,m-1}$. Therefore, the total energy of the focal field is

$$W = \left(W_{0,m} + \frac{1}{4}W_{2,m+2} + \frac{1}{4}W_{2,m-2}\right) + \left(\frac{1}{4}W_{2,m+2} + \frac{1}{4}W_{2,m-2}\right) + \left(W_{1,m+1} + W_{1,m-1}\right) \quad (4.25)$$

The brackets in Eq. (4.25) illustrate, respectively, the energies of angular harmonics in the E_x, E_y, and E_z field components. Since $I_{2,m+2} = I_{2,m-2}$ and $I_{1,m+1} = I_{1,m-1}$, we get

$$W = W_{0,m} + W_{2,m+2} + 2W_{1,m+1} \quad (4.26)$$

Substitution of Eqs. (4.22)–(4.24) into Eq. (4.26) yields

$$\begin{aligned} W = {} & \frac{4}{3}\pi f^2\left[1 - \cos^6\left(\frac{\alpha}{2}\right)\right] + \frac{4}{3}\pi f^2 \sin^6\left(\frac{\alpha}{2}\right) + \\ & + \frac{4}{3}\pi f^2 \sin^4\left(\frac{\alpha}{2}\right)\left[1 + 2\cos^2\left(\frac{\alpha}{2}\right)\right] \end{aligned} \quad (4.27)$$

After simplifications, we get.

$$W = 2\pi f^2(1 - \cos\alpha) \quad (4.28)$$

This expression is exactly the size of a part of a sphere bounded by the polar angle α. Thus, if a light field with unit amplitude is converging from a spherical surface with a numerical aperture sin(α), this field has exactly the energy given by Eq. (4.28). This means that we found the balance when the energy of the input field is equal to the sum of energies of all angular harmonics of all three Cartesian components of the electric field in the focus.

In an extreme case, when the numerical aperture is close to unit, i.e., $\alpha \approx \pi/2$, we get $W_{0,\mu} = (7/6)\pi f^2$, $W_{1,\mu} = (1/3)\pi f^2$, $W_{2,\mu} = (1/6)\pi f^2$. The whole energy is coinciding with the square of a hemisphere: $W = W_{0,m} + W_{2,m+2} + 2W_{1,m+1} = 2\pi f^2$. Thus, the total energy W of the input field is distributed in the focal field in the proportions shown in Fig. 4.12 One third $W/3$ goes into the longitudinal component E_z, and $2W/3$ goes into the transverse components E_x ($5W/8$) and E_y ($W/24$). The energy of the component E_x is distributed into the mth-order angular harmonic ($7W/12$) and into the harmonics of the orders $m - 2$ and $m + 2$, each of the energy $W/48$. The energy of the component E_y is distributed equally in the angular harmonics of the orders $m - 2$ and $m + 2$, each of the energy $W/48$. The energy of the component E_z is distributed equally in the angular harmonics of the orders $m - 1$ and $m + 1$, each of the energy $W/6$.

Obviously, if the input field is linearly polarized along the axis y, the energy distribution is the same, but the main portion ($7W/12$, or 58%) goes into the mth-order of the angular harmonic of the y-component E_y.

The above energy proportions can change if the field intensity is not homogeneous, i.e., if $A(\theta) \neq 1$. However, if the amplitude function $A(\theta)$ decays from the center to the periphery, then the contribution of the mth-order angular harmonic becomes even greater. Indeed, for instance, if the aperture is bounded by an angle α, then the energy of the side angular harmonics of the orders $m + 2$ and $m - 2$ relative to the energy of the central mth-order harmonic is.

$$\frac{W_{2,m}}{W_{0,m}} = \frac{\sin^6(\alpha/2)}{1 - \cos^6(\alpha/2)} = \frac{\sin^4(\alpha/2)}{1 + \cos^2(\alpha/2) + \cos^4(\alpha/2)} \tag{4.29}$$

When α decreases from $\pi/2$ to 0, the numerator also decreases, while the denominator increases. Thus, this portion decays. For instance, if $\alpha = \pi/2$, then $W_{2,m}/W_{0,m} = 1/7 \approx 0.143$, but even if $\sin\alpha = 0.95$, then $W_{2,m}/W_{0,m} \approx 0.057$, i.e., almost all energy goes into the central mth-order harmonic.

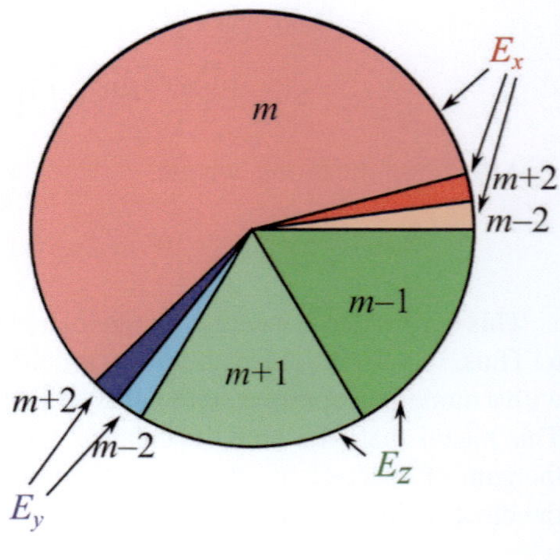

Fig. 4.12 Energy distribution of a tightly focused linearly polarized optical vortex by the field components and by the angular harmonics

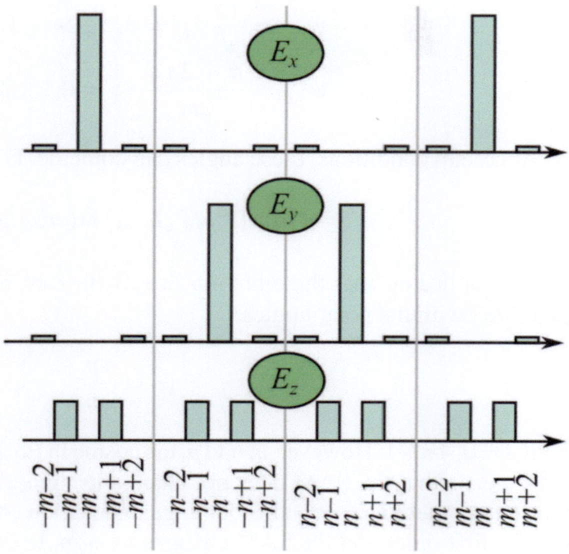

Fig. 4.13 Energy distribution of a tightly focused light field with double-index polarization singularity by the field components and by the angular harmonics

Now we apply the above introduced technique for obtaining the OAM spectrum of a light field with the double-index polarization singularity (4.5). Actually, the input field consists of four optical vortices of the orders *m*, –*m*, *n*, and –*n*. In the tight focus, each of these vortices splits into the several harmonics with the above derived energies. We suppose here for simplicity that $\alpha = \pi/2$ and that these harmonics do not coincide with each other. Thus, we have the energy distribution (OAM-spectra of the components E_x, E_y, and E_z), illustrated in Fig. 4.13

As seen in Fig. 4.13, side angular harmonics of the orders $m \pm 2$ and $n \pm 2$ have relative low energy (28 times lower than that of the orders *m* and *n*), and this energy becomes even lower when the aperture angle starts to decrease from $\alpha = \pi/2$.

4.2.3 Spin Angular Momentum of Double-Index Polarization Vortices in a Tight Focus

Since it was found that almost all of the energy of the transverse field components goes into the *m*th-order and *n*th-order angular harmonics, we can suppose approximately that the SAM from Eq. (4.14) reduces simply to

$$S_z \approx 2\mathrm{Im}\left\{i^{n-m}\right\} I_{0,m} I_{0,n} \cos(m\psi)\sin(n\psi) \tag{4.30}$$

Thus, it is seen that the SAM is equal to zero at the following polar angles:

$$\begin{aligned}\psi_{1,p} &= \frac{\pi p}{n},\ p = 0, ..., 2n - 1,\\ \psi_{2,q} &= \frac{\pi + 2\pi q}{2m},\ q = 0, ..., 2m - 1.\end{aligned} \tag{4.31}$$

At certain conditions, these angles can coincide. For example, if $n = 2m$, we get.

$$S_z \approx 4\mathrm{Im}\{i^m\}I_{0,m}I_{0,2m}\sin(m\psi)\cos^2(m\psi) \tag{4.32}$$

This indicates, that there are $4m$ lines with zero SAM, starting from the origin and tilted with the polar angles

$$\psi = \frac{\pi p}{2m} \tag{4.33}$$

with $p = 0, 4m - 1$. However, at odd p, the cosine in Eq. (4.32) is zero, but it is squared. This means that the SAM does not change its sign at these angles. Instead, due to the square, there is a second-order edge dislocation. At even p, the edge dislocation has the first order and the SAM changes its sign. In comparison with the first-order dislocations, the dislocations of the second order look as wider dark areas between the maxima. Thus, the SAM distribution should look as a set of pairs of the spots with positive and negative SAM.

Another case occurs when $m = 1$. For the SAM to be nonzero, n should be even. Thus, the angles $\alpha = \pm\pi/2$ are again the lines of second-order edge dislocation, where the SAM does not changes sign. At other angles, the SAM changes the sign.

If m and n are relatively large and close to each other, the roots of the sine and cosine do not coincide but are close to each other. Therefore, the SAM changes its sign at each such angle, but, due to the pairs of close zeros, the SAM between them is insignificant.

4.2.4 Simulation

Numerical simulation was done by the Richards-Wolf equations. At first, we studied the case when $n = 2m$. Figure 4.14 illustrates the intensity $|E_x|^2 + |E_y|^2 + |E_z|^2$ and the longitudinal SAM density of a tightly focused light field with double-index polarization singularity of three different orders (m, n): (1, 2) (Fig. 4.14a, d), (3, 6) (Fig. 4.14b, e), (7, 14) (Fig. 4.14c, f) at the following parameters: wavelength $\lambda = 532$ nm, focal length of the lens $f = 10$ μm, numerical aperture $\sin\alpha = 0.95$, amplitude apodization function is homogeneous, i.e., $A(\theta) \equiv 1$.

As seen in Fig. 4.14, indeed, the SAM distribution consists of alternating pairs of spots with positive or negative SAM. This is different from the patterns we obtained earlier near the focus, when the spots with positive and negative SAM were alternating in singles rather than in pairs [52].

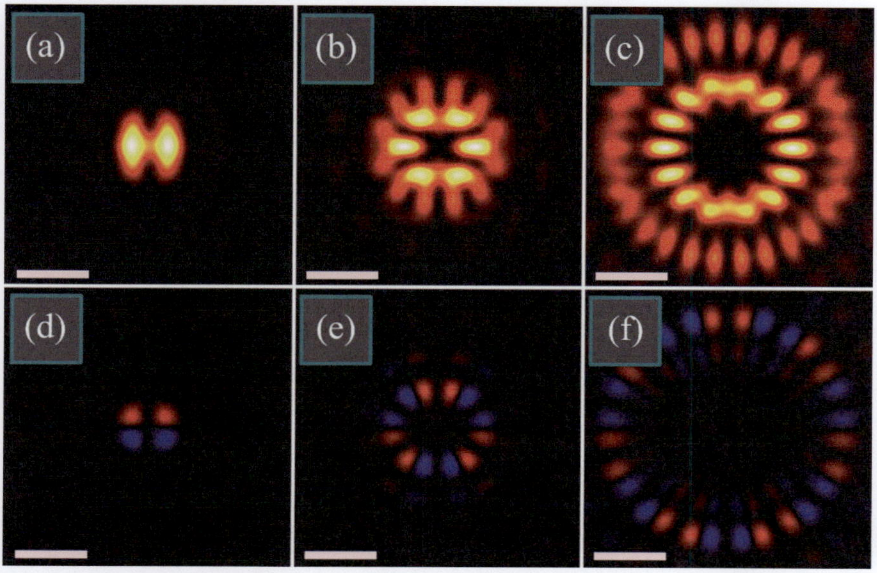

Fig. 4.14 Distributions of intensity $|E_x|^2 + |E_y|^2 + |E_z|^2$ (**a–c**) and of longitudinal SAM density $2\text{Im}\{E_x{}^* E_y\}$ (d–f) of tightly focused light field with double-index polarization singularity of the orders $(m, n) = (1, 2)$ (**a, d**), $(m, n) = (3, 6)$ (**b, e**), $(m, n) = (7, 14)$ (**c, f**) at the following parameters: wavelength $\lambda = 532$ nm, focal length of the lens $f = 10$ μm, numerical aperture $\sin \alpha = 0.95$, amplitude apodization function is homogeneous, i.e., $A(\theta) \equiv 1$. All the figures have the size 4×4 μm^2 (scale mark shows 1 μm). Red and blue colors (**d–e**) mean, respectively, positive and negative SAM

Figure 4.15 illustrates the intensity and the longitudinal SAM density of tightly focused light field with double-index polarization singularity of two different orders (m, n): (6, 7) (Fig. 4.15a, c) and (16, 17) (Fig. 4.15b, d) at the following parameters: wavelength $\lambda = 532$ nm, focal length of the lens $f = 10$ μm, numerical aperture $\sin \alpha = 0.95$, amplitude apodization function is homogeneous, i.e., $A(\theta) \equiv 1$.

According to theoretical predictions, polar angles with zero SAM should occur by pairs of close angles. Figure 4.15 confirms it. It is seen that the positive SAM is mostly in the upper side while the negative SAM is mostly in the bottom side. Actually, the SAM is alternating, but between each spot with the positive or negative SAM, there is a weak spot of the opposite SAM, which is almost invisible in Fig. 4.15.

The above theory predicts that the SAM is zero for the orders m and n of the same parity. Computation confirms this and polarization of the focal field is thus linear. Figure 4.16 depicts the intensity distributions and the polarization directions of tightly focused light fields with double-index polarization singularity of the orders $(m, n) = (3, 7)$ (Fig. 4.16a) and $(m, n) = (5, 3)$ (Fig. 4.16b) with the rest parameters being the same as in Figs. 4.14 and 4.15: wavelength $\lambda = 532$ nm, focal length of the lens $f = 10$ μm, numerical aperture $\sin \alpha = 0.95$, radial apodization function $A(\theta) \equiv 1$.

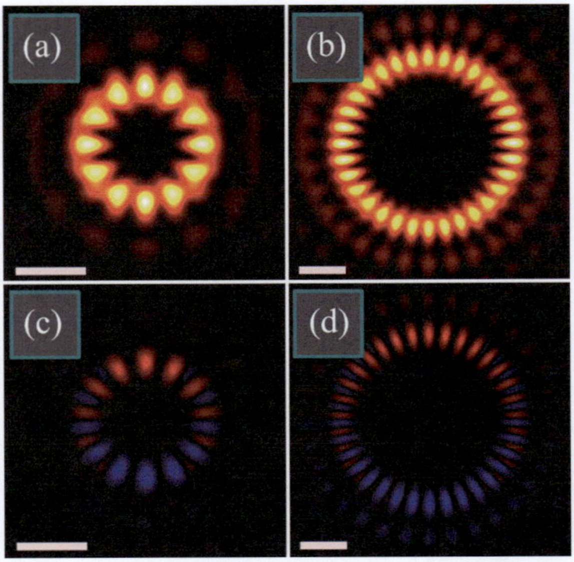

Fig. 4.15 Distributions of intensity $|E_x|^2 + |E_y|^2 + |E_z|^2$ (**a–c**) and of longitudinal SAM density $2\text{Im}\{E_x^* E_y\}$ (**d–f**) of tightly focused light field with double-index polarization singularity of the orders $(m, n) = (6, 7)$ (**a, c**) and $(m, n) = (16, 17)$ (**b, d**) at the following parameters: wavelength $\lambda = 532$ nm, focal length of the lens $f = 10$ μm, numerical aperture $\sin\alpha = 0.95$, amplitude apodization function is homogeneous, i.e., $A(\theta) \equiv 1$. Figures have the size 4×4 μm^2 (**a, c**) and 6×6 μm^2 (**b, d**) (scale mark shows 1 μm). Red and blue colors (**c, d**) mean, respectively, positive and negative SAM

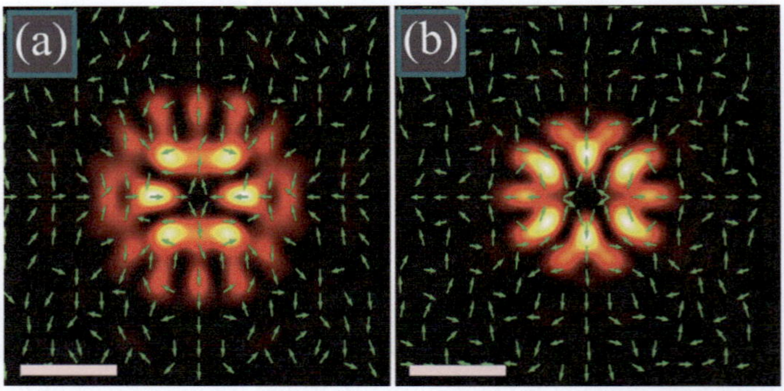

Fig. 4.16 Intensity distribution and polarization directions of a tightly focused light field with double-index polarization singularity of the orders $(m, n) = (3, 7)$ (**a**) and $(m, n) = (5, 3)$ (**b**) at the following parameters: wavelength $\lambda = 532$ nm, focal length of the lens $f = 10$ μm, numerical aperture $\sin\alpha = 0.95$, radial apodization function is $A(\theta) \equiv 1$. Scale mark shows 1 μm

It is seen in Fig. 4.16 that $E_y = 0$ on the horizontal axis ($\varphi = 0$ and $\varphi = \pi$) and $E_x = 0$ on the vertical axis ($\varphi = \pm\pi/2$), which is consistent with Eq. (4.11) for the complex amplitudes of the light field. It is also seen that in both cases a saddle-type polarization [57] singularity is generated in the center.

Based on the Richards-Wolf approximation, we have investigated here the spin angular momentum of double-index cylindrical vector beams in the tight focus. Such set of beams is a generalization of the conventional cylindrical vector beams since the polarization order is different for the different transverse field components. Thus, in the beam periphery, the number of areas with horizontal polarization is not equal to the number of areas with vertical polarization. It turns out that if the polarization orders are of different parity, then the spin Hall effect occurs in the tight focus that is there are alternating areas with positive and negative spin angular momentum, despite linear polarization of the initial light field. On the contrary, if the polarization orders are of same parity, then polarization in the tight focus remains linear (but inhomogeneous). For analytical description of the spin angular momentum distribution, we also analyzed the orbital angular momentum (OAM) spectrum of a linearly polarized mth-order vortex field in the tight focus. It turns out that if the initial light field is not of ring shape and has a homogeneous or decaying Gaussian shape, then the energy of the angular harmonics with the orders $m \pm 2$ in the transverse field components are at least 28 times lower than the energy of the mth-order angular harmonic.

This decomposition of the focused field into the OAM spectrum allowed us to predict the spin angular momentum distribution and, as an example, we demonstrated the ability to generate the focal distribution where the areas with the positive and negative spin angular momentum reside on a ring and are alternating in pairs, or separated in different semicircles.

Application areas of the results obtained are designing micromachines for optical driving biological objects [59, 60] or microtools in a lab-on-a-chip [61]. Cylindrical vector laser beams with a double index were studied in [58].

References

1. Q. Zhan, Cylindrical vector beams: from mathematical concepts to applications. Adv. Opt. Photon. **1**, 1–57 (2009)
2. R. Dorn, S. Quabis, G. Leuchs, Sharper focus for a radially polarized light beam. Phys. Rev. Lett. **91**, 233901 (2003)
3. E. Bor, M. Turduev, H. Kurt, Differential evolution algorithm based photonic structure design: numerical and experimental verification of subwavelength λ/5 focusing of light. Sci. Rep. **6**, 1–10 (2016)
4. S. Liu, Z. Guo, P. Li, B. Wei, J. Zhao, Tightly autofocusing beams: an effective enhancement of longitudinally polarized fields. Opt. Lett. **45**, 575–578 (2020)
5. T. Grosjean, I. Gauthier, Longitudinally polarized electric and magnetic optical nano-needles of ultra high lengths. Opt. Commun. Commun. **294**, 333–337 (2013)
6. J. Guan, J. Lin, C. Chen, Y. Ma, J. Tan, P. Jin, Transversely polarized sub-diffraction optical needle with ultra-long depth of focus. Opt. Commun. Commun. **404**, 118–123 (2017)

7. Y. Yu, H. Huang, M. Zhou, Q. Zhan, Engineering of multi-segmented light tunnel and flattop focus with designed axial lengths and gaps. Opt. Commun. Commun. **407**, 398–401 (2018)
8. C. Zheng, S. Su, H. Zang, Z. Ji, Y. Tian, S. Chen, K. Mu, L. Wei, Q. Fan, C. Wang, X. Zhu, C. Xie, L. Cao, E. Liang, Characterization of the focusing performance of axial line-focused spiral zone plates. Appl. Opt. **57**, 3802–3807 (2018)
9. J. Lin, R. Chen, P. Jin, M. Cada, Y. Ma, Generation of longitudinally polarized optical chain by 4π focusing system. Opt. Commun. Commun. **340**, 69–73 (2015)
10. Y. Yu, Q. Zhan, Generation of uniform three-dimensional optical chain with controllable characteristics. J. Opt. **17**, 105606 (2015)
11. Z. Xiaoqiang, C. Ruishan, W. Anting, Focusing properties of cylindrical vector vortex beams. Opt. Commun. Commun. **414**, 10–15 (2018)
12. Q. Zhan, J.R. Leger, Focus shaping using cylindrical vector beams. Opt. Express **10**, 324–331 (2002)
13. G.M. Lerman, A. Yanai, U. Levy, Demonstration of Nanofocusing by the use of Plasmonic Lens Illuminated with Radially Polarized Light. Nano Lett. **9**, 2139–2143 (2009)
14. Y. Jiang, X. Li, M. Gu, Generation of sub-diffraction-limited pure longitudinal magnetization by the inverse Faraday effect by tightly focusing an azimuthally polarized vortex beam. Opt. Lett. **38**, 2957–2960 (2013)
15. M. Rashid, O.M. Maragò, P.H. Jones, Focusing of high order cylindrical vector beams. J. Opt. A Pure Appl. Opt. **11**, 065204 (2009)
16. Y. Li, Z. Zhu, X. Wang, L. Gong, M. Wang, S. Nie, Propagation evolution of an off-axis high-order cylindrical vector beam. J. Opt. Soc. Am. A **31**, 2356–2361 (2014)
17. J. Qi, W. Wang, B. Pan, H. Deng, J. Yang, B. Shi, H. Shan, L. Zhang, H. Wang, Multiple-slit diffraction of high-polarization-order cylindrical vector beams. Proc. SPIE **10339**, 1033927 (2017)
18. X.-L. Wang, J. Ding, W.-J. Ni, C.-S. Guo, H.-T. Wang, Generation of arbitrary vector beams with a spatial light modulator and a common path interferometric arrangement. Opt. Lett. **32**, 3549–3551 (2007)
19. H. Chen, J. Hao, B.-F. Zhang, J. Xu, J. Ding, H.-T. Wang, Generation of vector beam with space-variant distribution of both polarization and phase. Opt. Lett. **36**, 3179–3181 (2011)
20. Y. Liu, Y. Ke, J. Zhou, Y. Liu, H. Luo, S. Wen, D. Fan, Generation of perfect vortex and vector beams based on Pancharatnam-Berry phase elements. Sci. Rep. **7**, 44096 (2017)
21. S.S. Stafeev, A.G. Nalimov, V.V. Kotlyar, Energy backflow in the focal spot of a cylindrical vector beam. Comput. Opt. Opt. **42**, 744–750 (2018)
22. S.S. Stafeev, V.V. Kotlyar, A.G. Nalimov, E.S. Kozlova, The Non-vortex inverse propagation of energy in a tightly focused high-order cylindrical vector beam. IEEE Photonics J. **11**, 4500810 (2019)
23. V.V. Kotlyar, S.S. Stafeev, A.G. Nalimov, A.A. Kovalev, A.P. Porfirev, Mechanism of formation of an inverse energy flow in a sharp focus. Phys. Rev. A **101**, 033811 (2020)
24. G. Machavariani, Y. Lumer, I. Moshe, A. Meir, S. Jackel, Efficient extracavity generation of radially and azimuthally polarized beams. Opt. Lett. **32**, 1468–1470 (2007)
25. G. Machavariani, Y. Lumer, I. Moshe, A. Meir, S. Jackel, Spatially-variable retardation plate for efficient generation of radially- and azimuthally-polarized beam. Opt. Commun.Commun. **281**, 732–738 (2008)
26. S.V. Alferov, S.V. Karpeev, S.N. Khonina, O.Y. Moiseev, Experimental study of focusing of inhomogeneously polarized beams generated using sector polarizing plates. Comput. Opt. Opt. **38**, 57–64 (2014)
27. R. Imai, N. Kanda, N. Higuchi, Z. Zheng, K. Konishi, M. Kuwata-Gonokami, Terahertz vector beam generation using segmented nonlinear optical crystals with threefold rotational symmetry. Opt. Express **20**, 21896–21904 (2012)
28. Z. Man, C. Min, Y. Zhang, Z. Shen, X-C.Yuan, Arbitrary vector beams with selective polarization states patterned by tailored polarizing films. Laser Phys., **23**, 105001 (2013)
29. A.G. Nalimov, L. O'Faolain, S.S. Stafeev, M.I. Shanina, V.V. Kotlyar, Reflected four-zones subwavelength microoptics element for polarization conversion from linear to radial. Comput. Opt. Opt. **38**, 229–236 (2014)

30. S.S. Stafeev, A.G. Nalimov, M.V. Kotlyar, D. Gibson, S. Song, L. O'Faolain, V.V. Kotlyar, Microlens-aided focusing of linearly and azimuthally polarized laser light. Opt. Express **24**, 29800–29813 (2016)
31. V.V. Kotlyar, S.S. Stafeev, M.V. Kotlyar, A.G. Nalimov, L. O'Faolain, Subwavelength micropolarizer in a gold film for visible light. Appl. Opt. **55**, 5025–5032 (2016)
32. A. Arbabi, Y. Horie, M. Bagheri, A. Faraon, Dielectric metasurfaces for complete control of phase and polarization with subwavelength spatial resolution and high transmission. Nat. Nanotechnol. **10**, 937–943 (2015)
33. J. Liu, X. Chen, Y. He, L. Lu, H. Ye, G. Chai, S. Chen, D. Fan, Generating of arbitrary cylindrical vector vortex beams with cross-polarized modulation. Results in Phys. **19**, 103455 (2020)
34. J. Wang, Advances in communications using optical vortices. Photonics Res. **4**, B14–B28 (2016)
35. S.S. Stafeev, V.V. Kotlyar, Tight focusing of a quasi-cylindrical optical vortex. Opt. Commun. Commun. **403**, 277–282 (2017)
36. B. Richards, E. Wolf, Electromagnetic diffraction in optical systems. II. Structure of the image field in an aplanatic system. Proc. R. Soc. Lond. ALond. A **253**, 358–379 (1959)
37. K. Bliokh, A. Bekshaev, F. Nori, Extraordinary momentum and spin in evanescent waves. Nat. Commun. Commun. **5**, 3300 (2014)
38. V.V. Kotlyar, A.G. Nalimov, S.S. Stafeev, L. O'Faolain, Subwavelength grating-based spiral metalens for tight focusing of laser light. Appl. Phys. Lett. **114**, 141107 (2019)
39. S.S. Stafeev, A.G. Nalimov, V.D. Zaitsev, V.V. Kotlyar, Tight focusing cylindrical vector beams with fractional order. J. Opt. Soc. Am. B **38**, 1090–1096 (2021)
40. V.V. Kotlyar, S.S. Stafeev, V.D. Zaitsev, A.A. Kovalev, Multiple optical spin-orbit Hall effect at the tight focus. Phys. Lett. A **458**, 128596 (2023)
41. Y. He, Z. Xie, B. Yang, X. Chen, J. Liu, H. Ye, X. Zhou, Y. Li, S. Chen, D. Fan, Controllable photonic spin Hall effect with phase function construction. Phot. Res. **8**, 963–971 (2020)
42. M.E.J. Friese, H. Rubinsztein-Dunlop, J. Gold, P. Hagberg, D. Hanstorp, Optically driven micromachine elements. Appl. Phys. Lett. **78**, 547–549 (2001)
43. Y. Shen, D.A. Weitz, N.R. Forde, M. Shayegan, Line optical tweezers as controllable micromachines: techniques and emerging trends. Soft Matter **18**, 5359–5365 (2022)
44. J. Liu, Z. Li, Controlled mechanical motions of microparticles in optical tweezers. Micromachines **9**, 232 (2018)
45. F.M. Dickey, *Laser Beam Shaping: Theory and Techniques*, 2nd ed (CRC Press, Taylor & Francis Group: Boca Raton, FL, USA, 2014)
46. U.D. Zeitner, H. Aagedal, F. Wyrowski, Comparison of resonator-originated and external beam shaping. Appl. Opt. **38**, 980–986 (1999)
47. E.G. Abramochkin, V.G. Volostnikov, *Modern Optics of Gaussian Beams*. Fizmatlit: Moscow, Russia, 2010 (in Russian), ISBN: 978-5-9221-1216-1.
48. V.A. Soifer (ed.), *Methods for Computer Design of Diffractive Optical Elements* (Wiley, USA, 2001)
49. O.V. Angelsky, A.Ya. Bekshaev, P.P. Maksimyak, A.P. Maksimyak, S.G. Hanson, C.Yu. Zenkova, Orbital rotation without orbital angular momentum: mechanical action of the spin part of the internal energy flow in light beams. Opt. Express **20**, 3563–3571 (2012)
50. J. Zhang, J. Yu, N. Chi, Transmission and full-band coherent detection of polarization-multiplexed all-optical Nyquist signals generated by Sinc-shaped Nyquist pulses. Sci. Rep. **5**, 13649 (2015)
51. S.S. Stafeev, A.G. Nalimov, A.A. Kovalev, V.D. Zaitsev, V.V. Kotlyar, Circular polarization near the tight focus of linearly polarized light. Photonics **9**, 196 (2022)
52. V.V. Kotlyar, S.S. Stafeev, A.A. Kovalev, V.D. Zaitsev, Spin hall effect before and after the focus of a high-order cylindrical vector beam. Appl. Sci. **12**(23), 12218 (2022)
53. I. Freund, Polarization singularity indices in Gaussian laser beams. Opt. Commun. Commun. **201**, 251–270 (2002)
54. V.V. Kotlyar, A.A. Kovalev, S.S. Stafeev, A.G. Nalimov, S. Rasouli, Tightly focusing vector beams containing V-point polarization singularities. Opt. Las. Tech. **145**, 107479 (2022)

55. V.V. Kotlyar, A.A. Kovalev, A.G. Nalimov, Energy density and energy flux in the focus of an optical vortex: reverse flux of light energy. Opt. Lett. **43**, 2921–2924 (2018)
56. S.M. Barnett, L. Allen, Orbital angular momentum and nonparaxial light beams. Opt. Commun. Commun. **110**, 670–678 (1994)
57. M.R. Dennis, *Topological Singularities in Wave Fields*. PhD thesis, (University of Bristol, Bristol, 2001)
58. A.A. Kovalev, V.V. Kotlyar, Spin hall effect of double-index cylindrical vector beams in a tight focus. Micromachines **14**, 494 (2023)
59. P.-K. Andrew, M.A.K. Williams, E. Avci, Optical Micromachines for Biological Studies. Micromachines **11**, 192 (2020)
60. I.A. Favre-Bulle, S. Zhang, A.V. Kashchuk, I.C.D. Lenton, L.J. Gibson, A.B. Stilgoe, T.A. Nieminen, H. Rubinsztein-Dunlop, Optical Tweezers Bring Micromachines to Biology. Opt. Photonics News **29**, 40–47 (2018)
61. Y.-J. Liu, Y.-H. Lee, Y.-S. Lin, C. Tsou, P.L. Baldeck, C.-L. Lin, Optically driven mobile integrated micro-tools for a lab-on-a-chip. Actuators **2**, 19–26 (2013)

Chapter 5
Sharp Focusing of Modified Cylindrical Vector Laser Beams

5.1 Spin-Orbital Conversion of a Strongly Focused Light Wave with High-Order Cylindrical-Circular Polarization

The rigorous description of a linearly polarized electromagnetic field in the strong focus was proposed in a classical work by Richards and Wolf [1]. Numerous follow-up publications relied on the Richards-Wolf formalism to look into the behavior of more general electromagnetic fields with various polarization states. Topics studied included characteristics of a radially polarized electromagnetic field in the strong focus [2] and spin-orbital conversion in the strong focus of a circularly polarized wave [3, 4]. Tightly focusing an elliptically polarized optical vortex has been studied [5, 6] and a concept of cylindrical vector beams has been proposed [7], which include radially and azimuthally polarized beams. Focusing promising beams with hybrid polarization has also been studied [8–10]. For this type of polarization, the transposed Jones vector takes the form $\boldsymbol{E} = (\exp(i\delta), \exp(-i\delta))$, where $\delta = \alpha r + \beta$, r is a real variable, α, β are constant, and $\mathbf{E}$ is the initial light field. This type of hybrid polarization is linear along some radii and circular on the others, however, being independent of the polar angle φ. A more general type of hybrid polarization was discussed in Ref. [11], where tightly focusing was analyzed for an incident field described by the polar angle dependent Jones vector, $\mathbf{E} = (\cos\varphi\cos\gamma - i\cos(2\psi - \varphi)\sin\gamma, \sin\varphi\cos\gamma - i\sin(2\psi - \varphi)\sin\gamma)$, where φ is the polar angle and γ, ψ are constant. This field was found to be polarized either linearly or circularly, depending on the specific value of the polar angle. However, no analytical relations to describe the hybrid field and projections of the Poynting vector were proposed in Ref. [11]. A field with hybrid polarization described by $\boldsymbol{E} = (\exp(i\delta)\sin\varphi, \ \cos\varphi)$, where φ is a polar angle and δ is constant, has also been studied [12]. We note that in this work, we discuss a more general polarization of which the above-said polarization is a particular case (at $m = 1$). We also note that in Ref. [12], projections of the Poynting vector were not defined analytically. Beams with arbitrary polarization represented

V. V. Kotlyar et al., *Optical Hall Effect in the Sharp Focus of Laser Light*,
https://doi.org/10.1007/978-3-031-64683-6_5

on a Poincare unit sphere have also been analyzed [13, 14]. In such beams, the pre-focusing polarization vector can be represented as $\mathbf{E} = (\exp(-il\varphi + i\alpha)\cos\beta + \exp(il\varphi + i\alpha)\sin\beta, i\exp(-il\varphi + i\alpha)\cos\beta + i\exp(il\varphi + i\alpha)\sin\beta)$, where l is topological charge, φ is the polar angle, (α, β) are (constant) angles on the Poincare unit sphere. It is worth noting that Refs. [13, 14] studied these beams experimentally, not offering a theoretical substantiation and expressions for the field intensity and projections of the Poynting vector in the tight focus. Tightly focusing higher-order cylindrically polarized light was studied in Refs. [15, 16], with the incident field being represented as $\boldsymbol{E} = (\cos(p\varphi + \alpha),\ \sin(p\varphi + \alpha))$, where p is the order of cylindrical polarization and α is constant. Vortex beams with arbitrary topological charge m and nth order cylindrical polarization were theoretically studied in Ref. [17]. The incident field was described by $\boldsymbol{E} = \exp(im\varphi)(\cos(n\varphi),\ \sin(n\varphi))$.

A distinctive feature of this work is that for the first time we analyze a new type of hybrid polarization of light never studied previously, with polarization of the incident field of interest being represented by $\boldsymbol{E} = (-i\sin(m\varphi),\ \cos(m\varphi))$. In this case, with changing polar angle of the initial field, polarization changes from circular, to elliptical to linear, alternating in this manner m times per full circle of the polar angle. We also propose analytical relationships for projections of the electric and magnetic field strength in the strong focus, for intensity distributions, and for projections of the Poynting vector and spin angular momentum (SAM) vector.

5.1.1 *Intensity of Light with Hybrid Polarization in the Focus*

Let the amplitudes of the original magnetic and electric field vectors for mth-order hybrid polarization be given by

$$\boldsymbol{E} = A(\theta)\begin{pmatrix} -i\sin m\varphi \\ \cos m\varphi \end{pmatrix}, \quad \boldsymbol{H} = A(\theta)\begin{pmatrix} -\cos m\varphi \\ -i\sin m\varphi \end{pmatrix}, \tag{5.1}$$

where $\mathbf{E}$ and $\mathbf{H}$ are the electric and magnetic field vectors, m is a positive integer number defining the order of cylindrical polarization, and $A(\theta)$ is the amplitude of the original light field as a function of the field angle with the optical axis. Polarization of the field in Eq. (5.1) is called hybrid because it combines properties of mth-order cylindrical polarization and circular polarization. At different polar angles φ, polarization of field (5.1) will be either circular (at $\varphi = \pi n/(4m)$, $n = 1, 3, 5...$, elliptical, or linear (at $\varphi = \pi n/(2m)$, $n = 0, 1, 2...$). From (5.1), it also follows that at $m = 0$, the field will be homogeneously linearly polarized. Relying on the Richards-Wolf formalism [1], we can derive projections of the electric field vector in the strong focus of an aplanatic optical system from the original field (5.1):

$$
\begin{aligned}
E_x &= \frac{-i^m}{2}\left[I_{0,m}\sin(m\varphi) + \frac{(1+i)}{2}I_{2,m+2}\sin((m+2)\varphi)\right.\\
&\quad\left. + \frac{(1-i)}{2}I_{2,m-2}\sin((m-2)\varphi)\right],\\
E_y &= \frac{-i^m}{2}\left[iI_{0,m}\cos(m\varphi) - \frac{(1+i)}{2}I_{2,m+2}\cos((m+2)\varphi)\right.\\
&\quad\left. + \frac{(1-i)}{2}I_{2,m-2}\cos((m-2)\varphi)\right],\\
E_z &= \frac{i^m}{2}\left[(1+i)I_{1,m+1}\sin((m+1)\varphi)\right.\\
&\quad\left. + (1-i)I_{1,m-1}\sin((m-1)\varphi)\right]
\end{aligned}
\tag{5.2}
$$

and for the magnetic field vector:

$$
\begin{aligned}
H_x &= \frac{i^m}{2}\left[iI_{0,m}\cos(m\varphi) + \frac{(1+i)}{2}I_{2,m+2}\cos((m+2)\varphi)\right.\\
&\quad\left. - \frac{(1-i)}{2}I_{2,m-2}\cos((m-2)\varphi)\right],\\
H_y &= \frac{i^m}{2}\left[-I_{0,m}\sin(m\varphi) + \frac{(1+i)}{2}I_{2,m+2}\sin((m+2)\varphi)\right.\\
&\quad\left. + \frac{(1-i)}{2}I_{2,m-2}\sin((m-2)\varphi)\right],\\
H_z &= \frac{i^m}{2}\left[(1-i)I_{1,m+1}\cos((m+1)\varphi)\right.\\
&\quad\left. -(1+i)I_{1,m-1}\cos((m-1)\varphi)\right],
\end{aligned}
\tag{5.3}
$$

where

$$
\begin{aligned}
I_{\nu,\mu} &= \left(\frac{4\pi f}{\lambda}\right)\int\limits_0^{\theta_0}\sin^{\nu+1}\left(\frac{\theta}{2}\right)\cos^{3-\nu}\left(\frac{\theta}{2}\right)\cos^{1/2}(\theta)\\
&\quad\times A(\theta)\mathrm{e}^{ikz\cos\theta}J_\mu(x)\mathrm{d}\theta,
\end{aligned}
\tag{5.4}
$$

where λ is the incident wavelength, f is the focal length of the aplanatic system, $x = kr\sin\theta$, $J_\mu(x)$ is the Bessel function of the first kind, and NA $= \sin\theta_0$ is the numerical aperture. The original amplitude function $A(\theta)$ (here, assumed to be real) can be either constant (for a plane incident wave) or given by a Gaussian beam:

$$
A(\theta) = \exp\left(\frac{-\gamma^2\sin^2\theta}{\sin^2\theta_0}\right)
\tag{5.5}
$$

where γ is constant. By way of checking Eqs. (5.2), let us reduce them to the familiar expressions [1, 4] for a linearly polarized light field ($m = 0$):

$$\begin{aligned} E_{x0} &= \frac{-i}{2} I_{2,2} \sin(2\varphi), \\ E_{y0} &= \frac{-i}{2} \left[I_{0,0} - I_{2,2} \cos(2\varphi) \right], \\ E_{z0} &= I_{1,1} \sin(\varphi) \end{aligned} \tag{5.6}$$

The only difference is that the linear polarization vector is directed along the y-axis in Eq. (5.1), being directed along the x-axis in Refs. [1, 4]. From (5.2), the intensity distribution of the electric field in the focal plane ($z = 0$) is found as:

$$\begin{aligned} I_m = &\frac{1}{4} \left[I_{0,m}^2 + I_{2,m+2}^2 + I_{2,m-2}^2 - I_{0,m} \left(I_{2,m+2} + I_{2,m-2} \right) \right] \\ &+ \frac{1}{2} \left(I_{1,m+1}^2 + I_{0,m} I_{2,m+2} \right) \sin^2((m+1)\varphi) \\ &+ \frac{1}{2} \left(I_{1,m-1}^2 + I_{0,m} I_{2,m-2} \right) \sin^2((m-1)\varphi). \end{aligned} \tag{5.7}$$

At $m = 0$, from Eq. (5.7) follows a familiar relation for the intensity distribution in the tight focus from the incident linearly polarized light field [1, 18]:

$$I_0 = \frac{1}{4} \left(I_{0,0}^2 + 2I_{1,1}^2 + 2I_{2,2}^2 \right) - \frac{1}{4} \left(I_{1,1}^2 + I_{0,0} I_{2,2} \right) \cos(2\varphi). \tag{5.8}$$

From Eq. (5.8), the intensity is seen to have two local maxima lying on the vertical axis (at $\varphi = \pm\pi/2$) because the incident field is linearly polarized and directed along the y-axis. At $m = 1$, Eq. (5.7) suggests that the intensity in the focus of the first-order hybrid field (5.1) is given by

$$\begin{aligned} I_1 = &\frac{1}{4} \left(I_{0,1}^2 + I_{1,2}^2 + I_{2,1}^2 + I_{2,3}^2 + I_{0,1} I_{2,1} \right) \\ &- \frac{1}{4} \left(I_{1,2}^2 + I_{0,1} I_{2,3} \right) \cos(4\varphi). \end{aligned} \tag{5.9}$$

From (5.9), the intensity distribution in the focus for the first-order hybrid field (5.1), i.e., for azimuthal circular polarization, features four local maxima (at $\varphi = \pm\pi/4, \pm 3\pi/4$). In the general case of an arbitrary m, the intensity distribution in Eq. (5.7) has $2(m+1)$ maxima lying on the rays formed by the polar angles $\varphi = (\pi + 2\pi n)/2(m+1)$, $n = 0, 1, 2, \ldots 2m+1$. It follows from the fact that Eq. (5.7) contains the square of the sine, which have $2(m+1)$ maxima when the angle φ changes from 0 to 2π. Numerical simulation has confirmed the theoretical conclusions.

5.1.2 Energy Flow in the Focus of Light with Hybrid Polarization

We note that in the original field (5.1), only the longitudinal projection of the energy flow is present because the longitudinal components of the electric and magnetic field vectors are zero, as is the transverse projection of the Poynting vector. Instead, the longitudinal component of SAM vector is non-zero. Hence, due to the effect of spin-orbital conversion, a transverse energy flow may be expected to be generated in the strong focus. Below, we demonstrate that this is the case. Let us derive projections of the Poynting vector (the energy flow)

$$\boldsymbol{P} = \mathrm{Re}\left(\boldsymbol{E}^{*} \times \boldsymbol{H}\right) \tag{5.10}$$

where Re is the real part of the number, $\times$ denotes a vector product of two vectors, $\mathbf{E}^{*}$ denotes complex conjugation in the focal plane ($z = 0$) for the original field with hybrid polarization, Eq. (5.1). Substituting the projections of the electric field in Eq. (5.2) and the magnetic field in Eq. (5.3) into Eq. (5.10) yields:

$$\begin{aligned}
P_x = \frac{1}{4}\big[& I_{0,m}\left(I_{1,m+1} + I_{1,m-1}\right)\cos\varphi + \\
& + I_{1,m+1}I_{2,m-2}\cos((m+1)\varphi)\cos((m-2)\varphi) \\
& + I_{1,m-1}I_{2,m+2}\cos((m-1)\varphi)\cos((m+2)\varphi) \\
& + I_{1,m+1}I_{2,m+2}\sin((m+1)\varphi)\sin((m+2)\varphi) \\
& + I_{1,m-1}I_{2,m-2}\sin((m-1)\varphi)\sin((m-2)\varphi)\big], \\
P_y = \frac{1}{4}\big[& I_{0,m}I_{1,m+1}\sin((2m+1)\varphi) + \\
& + I_{0,m}I_{1,m-1}\sin((2m-1)\varphi) \\
& + I_{1,m+1}I_{2,m-2}\cos((m+1)\varphi)\sin((m-2)\varphi) \\
& - I_{1,m-1}I_{2,m+2}\cos((m-1)\varphi)\sin((m+2)\varphi) \\
& + I_{1,m+1}I_{2,m+2}\sin((m+1)\varphi)\cos((m+2)\varphi) \\
& - I_{1,m-1}I_{2,m-2}\sin((m-1)\varphi)\cos((m-2)\varphi)\big], \\
P_z = \frac{1}{4} & \left(I_{0,m}^2 - \frac{1}{2}I_{2,m+2}^2 - \frac{1}{2}I_{2,m-2}^2\right).
\end{aligned} \tag{5.11}$$

Although the expressions for the projections of the Poynting vector in Eq. (5.11) are quite cumbersome, they allow us to make some significant general conclusions. From Eq. (5.11), the longitudinal energy flow is seen to be radially symmetric at any *m*, being φ-independent. The on-axis energy flow will be positive and non-zero only at $m = 0$ (linear polarization): $P_z(r = z = 0) = I_{0,0}^2/4$. Besides, the on-axis projection of the Poynting vector in the focus in Eq. (5.11) will be non-zero and negative only at $m = -2$ or $m = 2$: $P_z(r = z = 0) = -I_{2,0}^2/4$. Thus, we can infer that similar to conventional 2nd-order azimuthal polarization [18, 19], for hybrid incident polarization, a reverse energy flow also occurs at $m = \pm 2$. From (5.11) at $\varphi = 0$, we can derive:

$$P_x(\varphi = 0) = \frac{1}{4}\left[I_{0,m}\left(I_{1,m+1} + I_{1,m-1}\right) + I_{1,m+1}I_{2,m-2} + I_{1,m-1}I_{2,m+2}\right] > 0 \tag{5.12}$$

From (5.11), we also find that $P_y(y = 0) = 0,\ P_x(\varphi = 0) = -P_x(\varphi = \pi) > 0$. Hence, at any m, on the horizontal axis x, the transverse energy flow is always directed along the x-axis in both directions from the center. It also follows from (5.11) that on the vertical axis y, the transverse energy flow is directed along the y-axis because $P_x(\varphi = \pi/2) = P_x(\varphi = 3\pi/2) = 0$. Equation (5.11) also suggests that when passing through the y-axis zero point, the energy flow changes sign: $P_y(\varphi = \pi/2) = -P_y(\varphi = 3\pi/2) \neq 0$. The sign of the transverse flow alternate moving along the y-axis. For instance, if at $m = 1$ the energy flow on the y-axis is directed toward the center, at $m = 2$, the flow will be directed from the center. Summing up, at $m = 1$, the transverse energy flow will be directed from the center on the x-axis and toward the center on the y-axis. This can occur if the transverse energy flow rotates anticlockwise in quadrants I and III, rotating clockwise in quadrants II and IV. Next, at $m = 2$ on the vertical axis, the transverse energy flow will change sign becoming directed from the center, while remaining being directed from the center on the horizontal x-axis. This can occur if in the four quadrants there will be four lines (at an angle of 45°) along which the energy flow is directed to the center. Thus, at $m = 2$, eight transverse energy vortices will be generated (by twos in each quadrant), characterized by alternating energy rotation handedness (clockwise and anticlockwise). Using a similar reasoning, it can further be shown that at an arbitrary m, in the focus there will be $4m$ energy flow vortices. The vortex handedness will change to the opposite in passing from one vortex to the other.

For simplicity, below, we analyze particular cases of Eq. (5.11). From (5.11), it also follows that at $m = 0$ (linear polarization), the transverse energy flow components equal zero in the focus: $P_x = P_y = 0$. This can be checked by directly substituting $m = 0$ into (5.11) and taking into account properties of the integrals in Eq. (5.4): $I_{p,-q} = (-1)^q I_{p,q}$. At $m > 0$, there is a non-zero transverse energy flow of Eq. (5.11). Let us remind that for mth-order cylindrical polarization, the transverse energy flow in the focus is always zero [18]. At $m = 1$ (circular azimuthal polarization), we can derive from Eq. (5.11) the following expressions for projections of the energy flow:

$$\begin{aligned} P_x &= \frac{1}{4}\left[I_{0,1}\left(I_{1,2} + I_{1,0}\right)\cos\varphi + I_{1,2}I_{2,3}\sin 2\varphi \sin 3\varphi \right. \\ &\quad \left. + I_{1,0}I_{2,3}\cos 3\varphi - I_{1,2}I_{2,1}\cos 2\varphi \cos\varphi\right], \\ P_y &= \frac{1}{4}\left[I_{0,1}\left(I_{1,2}\sin 3\varphi - I_{1,0}\sin\varphi\right) \right. \\ &\quad + I_{1,2}I_{2,3}\cos 2\varphi \cos 3\varphi - I_{1,0}I_{2,3}\sin 3\varphi \\ &\quad \left. + I_{1,2}I_{2,1}\cos 2\varphi \sin\varphi\right], \end{aligned}$$

$$P_z = \frac{1}{4}\left(I_{0,1}^2 - \frac{1}{2}I_{2,3}^2 - \frac{1}{2}I_{2,1}^2\right). \tag{5.13}$$

From (5.13), the longitudinal energy flow component is seen to be ring-shaped with the on-axis intensity null. The transverse energy flow components are non-zero and devoid of radial symmetry. From Eq. (5.13), the transverse components of the Poynting vector in the focus are seen to have the following structure:

$$\begin{aligned}
&\varphi = 0: \quad P_x = A + B > 0,\ P_y = 0,\\
&\varphi = \pi/2: P_x = 0, P_y = -A + B < 0,\\
&\varphi = \pi: \quad P_x = -(A + B) < 0,\ P_y = 0,\\
&\varphi = 3\pi/2: P_x = 0, P_y = A - B > 0,\\
&A = I_{0,1}\left(I_{1,2} + I_{1,0}\right)/4,\ B = I_{1,0}I_{2,3} - I_{1,2}I_{2,1}.
\end{aligned} \tag{5.14}$$

From (5.14), the energy flow in the focal plane on the horizontal x-axis is seen to be directed along the x-axis from the center, being directed toward the center on the vertical y-axis. This effect occurs if the transverse energy flow rotates anticlockwise in quadrants I and III, rotating clockwise in quadrants II and IV.

One more general conclusion can be made from Eq. (5.11) without the need to do numerical simulation. In the relationship for the projection of the Poynting vector in Eq. (5.11), the sine function P_y with the maximal spatial frequency is given by $\sin(2m + 1)\varphi$. Hence, at a given r, the integrals from (5.4) that enter Eq. (5.11) will take constant values, with the entire expression for P_y being only dependent on the angle φ, so that after one full circle of radius r in the focal plane, the value of P_y will change sign $2(2m + 1)$ times.

5.1.3 SAM in the Strong Focus of a Field with Hybrid Polarization

First, we remind that the longitudinal projection of the spin density vector or SAM vector equals zero for any mth-order cylindrical polarization of the initial field [18]. In this section, we will demonstrate that given hybrid polarization of Eq. (5.1), the longitudinal projection of the SAM vector in the focus will be non-zero. Actually, let us define the SAM vector in the form [4]:

$$\boldsymbol{S} = \mathrm{Im}\left(\boldsymbol{E}^* \times \boldsymbol{E}\right) \tag{5.15}$$

where Im is the imaginary part of the number. Substituting the E-field projections from (5.2) into (5.15) yields the longitudinal component of SAM:

$$S_z = \frac{1}{4}\left[I_{0,m}\left(I_{2,m+2} - I_{2,m-2}\right)\sin 2\varphi + \left(I_{0,m}^2 - I_{2,m-2}I_{2,m+2}\right)\sin(2m\varphi)\right]. \tag{5.16}$$

Equation (5.16) suggests that at $m = 0$ (linear polarization), $S_z = 0$. At $m = 1$, Eq. (5.16) is rearranged to

$$S_z = \frac{1}{4}\left[I_{0,1}\left(I_{2,3} + I_{2,1}\right) + \left(I_{0,1}^2 + I_{2,1}I_{2,3}\right)\right]\sin 2\varphi. \tag{5.17}$$

From (5.16), the on-axis projection of the SAM vector in the focal plane changes its sign $4m$ times, because Eq. (5.16) contains the function $\sin(2m\varphi)$. Hence, there will be $4m$ local vortices of the transverse flow and $4m$ local regions with the positive or negative longitudinal projections of the SAM vector. Notably, in the focal plane regions of anticlockwise handedness of the transverse energy flow, the polarization vector also rotates anticlockwise, meaning that the projection of the SAM vector is positive ($S_z > 0$). And vice versa, in the focal plane local regions of the clockwise handedness of the transverse energy flow, the polarization vector also rotates clockwise, meaning that the longitudinal projection of the SAM vector is negative ($S_z < 0$). Dielectric microparticles, which are slightly less in size than the local energy vortex region, placed in the focal plane will start rotating around their axis. It is interesting that particles in the adjacent regions will rotate in the opposite directions.

5.1.4 Results of the Numerical Simulation of Focusing Light with Hybrid Polarization

The numerical simulation based on the Richards-Wolf formulae [1] was conducted for focusing a 532-nm plane wave with hybrid polarization, Eq. (5.1) by means of an aplanatic objective lens with NA = 0.95. Figure 5.1 depicts (a) intensity patterns and Poynting vector components (b) P_x, (c) P_y, and (d) P_z in the focal plane when focusing a plane wave with hybrid polarization of Eq. (5.1) at $m = 1$. From Fig. 5.1a, the intensity is seen to have $2(m + 1) = 2(1 + 1) = 4$ local maxima located at the corners of a square-shaped contour. At the focal spot center there occurs an intensity null. Shown in Fig. 5.1b, c are distributions of the transverse energy flow and transverse projections (b) P_x, (c) P_y of the Poynting vector. From Fig. 5.1b, c, the energy flow is seen to change its sign $2(2m + 1) = 6$ times per one full circle around center. Figure 5.1d shows the longitudinal projection P_z of the Poynting vector, which is ring-shaped and has a zero value at the center. The patterns in Fig. 5.1 confirm the conclusions made on the basis of the theoretically derived relationships for the intensity in Eq. (5.9) and the energy flow in Eq. (5.11).

Figure 5.2 depicts patterns of (a) intensity and (b, c, d) projections P_x, P_y, and P_z of the Poynting vector in the focal plane when focusing a plane wave with hybrid polarization of Eq. (5.1) at $m = 2$. The numerically simulated patterns in Fig. 5.2

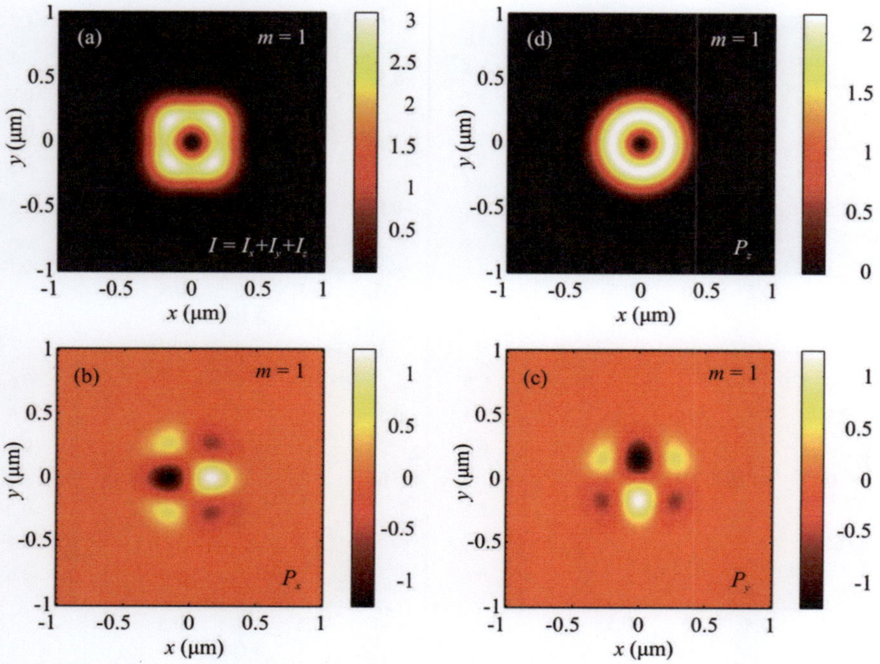

Fig. 5.1 Patterns of **a** intensity and (**b**, **c**, **d**) Poynting vector components P_x, P_y, and P_z in the focal plane when focusing a plane wave with hybrid polarization of Eq. (5.1) at $m = 1$

corroborate theoretical predictions that follow from Eqs. (5.9) and (5.11). Actually, the intensity distribution in Fig. 5.2a is seen to have $2(m + 1) = 6$ local maxima lying on a closed curve around the center. Figure 5.2b, c depicts distributions of the transverse projections P_x (b), P_y (c) of the Poynting vector, from which the energy flow is seen to change its sign $2(2m + 1) = 10$ times per full circle around the center. Figure 5.2d depicts the longitudinal projection P_z of the Poynting vector in the form of a ring. The central energy flow is negative and equal to $P_z(r = z = 0) = -I_{2,0}^2/4$, as seen from (5.11).

Figure 5.3 depicts distributions of SAM components (a, b, c) S_x, S_y, and S_z when focusing a plane wave with hybrid polarization of Eq. (5.1) at $m = 1$. From Fig. 5.3c, the longitudinal component of the SAM vector changes its sign $4m = 4$ times, which is seen from (5.17).

Figure 5.4 shows distributions of SAM vector components (a, b, c) S_x, S_y, and S_z when focusing a plane wave with hybrid polarization of Eq. (5.1) at $m = 2$. From Fig. 5.4c, the longitudinal projection of SAM vector is seen to be equal to $4m = 8$, which follows from (5.17).

Figure 5.5 depicts intensity distributions when focusing a plane wave with hybrid polarization in Eq. (5.1) at (a) $m = 2$ and (b) $m = 3$, with arrows marking the direction of the transverse Poynting vector component in the focal plane. From Fig. 5.5, the

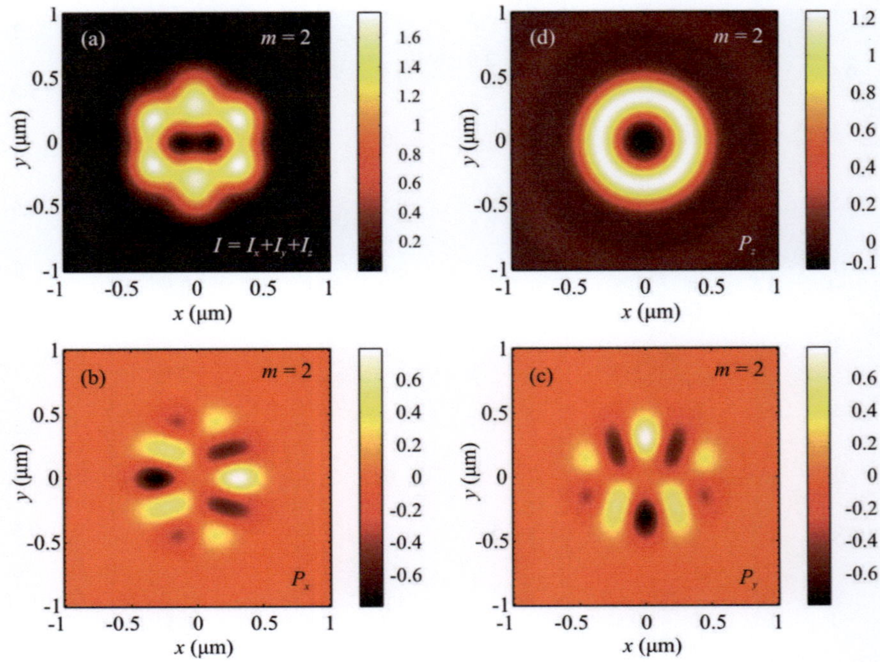

Fig. 5.2 Distributions of **a** intensity and components (**b**, **c**, **d**) P_x, P_y, and P_z of the Poynting vector in the focal plane when focusing a plane wave with hybrid polarization of Eq. (5.1) at $m = 2$

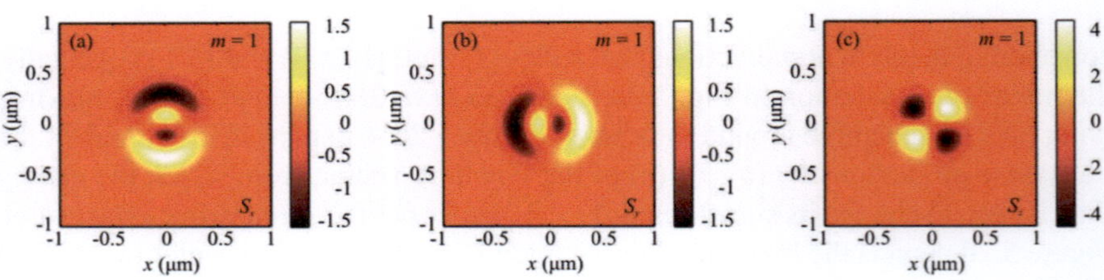

Fig. 5.3 Distribution of SAM vector components S_x (**a**) S_y (б) и S_z (**b**) when focusing a plane wave with hybrid polarization of Eq. (5.1) at $m = 1$

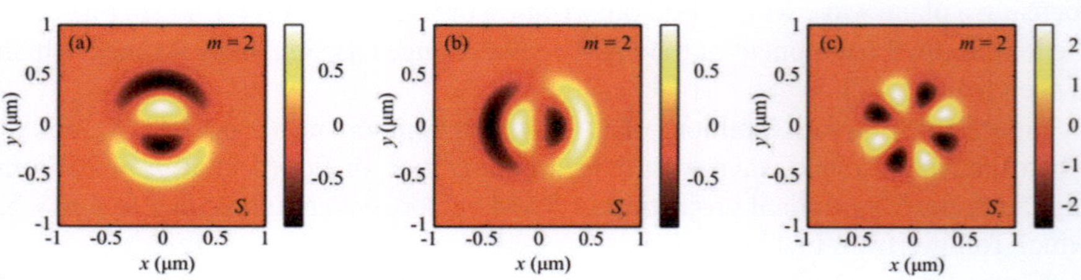

Fig. 5.4 Distribution of the SAM components (**a**, **b**, **c**) S_x, S_y, and S_z when focusing a plane wave with hybrid polarization in Eq. (5.1) at $m = 2$

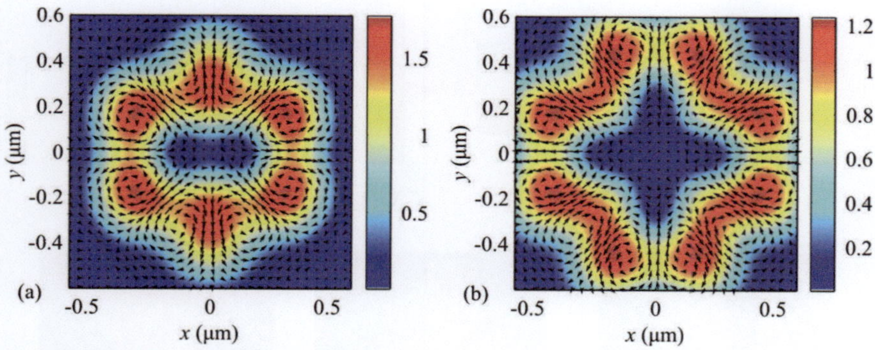

Fig. 5.5 Intensity distribution and the magnitude and direction of the Poynting vector (arrows) in the focal plane when focusing a plane wave with hybrid polarization of Eq. (5.1) for **a** $m = 2$ and **b** $m = 3$

number of the transverse flow vortices equals (a) $4m = 8$ and (b) $4m = 12$, which can be inferred from the expression (5.11) for transverse Poynting vector components. From Fig. 5.5, the centers of the transverse energy flows in the focus are also seen to not coincide with the local intensity maxima. The vortices are centered at points where the transverse energy flow is zero. From the comparison of Figs. 5.4c and 5.5a, the number of regions with positive- and negative-valued longitudinal SAM projections ($4m = 8$) is the same as the number of the transverse energy vortices ($4m = 8$). The comparison of Figs. 5.4c and 5.5a also suggests that the longitudinal SAM component is positive ($S_z > 0$) in the regions of anticlockwise handedness of the transverse energy flow. And vice versa, the longitudinal SAM component is negative ($S_z < 0$) in the regions of clockwise handedness of the transverse energy vortex. Thus, the polarization vector in the focal plane rotates anticlockwise in the regions where the transverse energy flow also has anticlockwise handedness. And vice versa, the polarization vector rotates clockwise where the transverse energy flow has clockwise handedness. This is in good agreement with the spin-orbital conversion effect. This spatial separation at the focus of left and right circular polarization is a manifestation of the optical spin Hall effect [20].

5.1.5 Experiment

Figure 5.6a shows an optical setup for generation of the beam (5.1) with $m = 2$. Figure 5.6b–d show images of obtained beam. Light from a laser Cobolt 06-MLD ($\lambda = 633$ nm, 200 mW) propagates through neutral density filter ND and Glan-Taylor polarizer GT. The resulting linearly polarized light propagates through vortex half-wave plate (Thorlabs, WPV10-633), which transform linearly polarized light into cylindrical vector beam of the second order. And finally the beam propagates

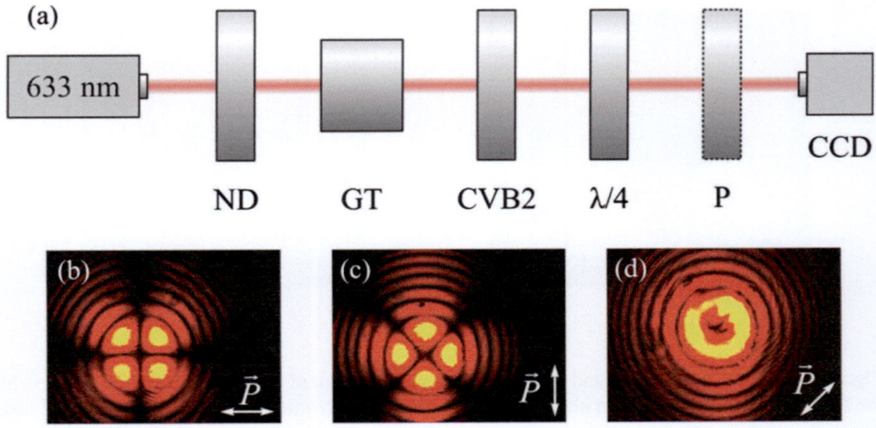

Fig. 5.6 **a** Optical setup for generation and registration of hybridly polarized beam Eq. (5.1) of the second order $m = 2$

through a quarter-wave plate. The resulting beam was registered by a CCD camera (Ucmos 10000 KPA).

ND is a neutral density filter, GT is a Glan-Taylor polarizer, CVB2 is a vortex half-wave plate (Thorlabs, WPV10-633), λ/4 is a quarter-wave plate, CCD is a camera Ucmos 10000KPA. (b)–(d) Images of the beam: a linear polarizer-analyzer P was placed before CCD camera, it was rotated by an angle θ equals to 0 (b), 90 (c), 45 (d).

To be sure that the experimentally obtained beam corresponds to the desired hybrid polarization, we simulated insertion of a linear polarizer into the beam using the Jones calculus formalism. After a linear polarizer the polarization of the beam changes in accordance with the following equation:

$$\begin{pmatrix} E_{x,\text{out}} \\ E_{y,\text{out}} \end{pmatrix} = \begin{pmatrix} \cos^2\theta & -\sin\theta\cos\theta \\ -\sin\theta\cos\theta & \sin^2\theta \end{pmatrix} \begin{pmatrix} E_{x,\text{in}} \\ E_{y,\text{in}} \end{pmatrix} \tag{5.18}$$

where $E_{x,\text{in}}$ and $E_{y,\text{in}}$ are electric field components before the polarizer (calculated by Richards-Wolf formula), $E_{x,\text{out}}$ and $E_{y,\text{out}}$ are electric field components after the polarizer-analyzer, θ is an angle between the x-axis and axis of the polarizer.

Figure 5.7 shows the intensity distribution of hybridly polarized beam Eq. (5.1) $m = 2$ propagated through a linear polarizer-analyzer, which is rotated by angle 0 (Fig. 5.7a), $\pi/2$ (Fig. 5.7b), $\pi/4$ (Fig. 5.7c). From Eq. (5.1) it follows that at the angle $\varphi = \frac{\pi}{4} + \frac{\pi n}{2}$ (along diagonal lines) there is E_x component only $\boldsymbol{E} = \begin{pmatrix} -i \\ 0 \end{pmatrix}$; however, at the angle $\varphi = \frac{\pi n}{2}$ (along Cartesian axes) there is only E_y component $\boldsymbol{E} = \begin{pmatrix} 0 \\ 1 \end{pmatrix}$.

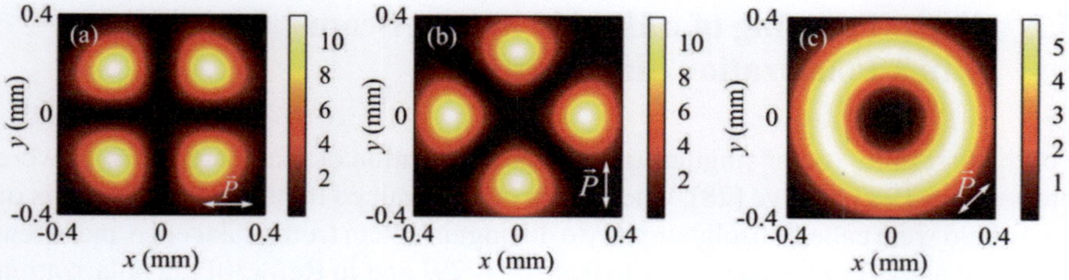

Fig. 5.7 Intensity distribution (simulation) of a hybrid beam Eq. (5.1) with $m = 2$ propagated through a linear polarizer-analyzer rotated by an angle θ equals to 0 (**a**), $\pi/2$ (**b**), $\pi/4$ (**c**)

From comparison of Figs. 5.6 and 5.7, it could be seen that the state of polarization coincides in the simulation and the experiment.

In this section, tightly focusing laser light with mth-order circular azimuthal polarization has been analyzed. This is a new type of inhomogeneous polarization which combines properties of mth-order cylindrical and circular polarizations. Using the Richards-Wolf formalism, we have deduced analytical expressions to describe projections of the E- and H-vectors, the intensity distribution, and projections of the Poynting vector and SAM vector in the tight focus of light. It has been shown theoretically and numerically that the intensity pattern in the focal spot has $2(m + 1)$ local maxima located along a closed contour centered at the on-axis zero-intensity point. We have shown that in the focus there are $4m$ vortices of the transverse energy flows with their centers located between the local intensity maxima. It has also been shown that the transverse energy flow vortex changes its handedness $2(2m + 1)$ times per full circle around the optical axis. Interestingly, the longitudinal SAM component in the focus changes its sign $4m$ times. The longitudinal SAM component has been shown to be positive in regions of anticlock handedness of the transverse energy vortex, with the polarization vector rotating anticlockwise around the optical axis. And vice versa, the polarization vector rotates clockwise and the longitudinal SAM component is negative in the regions where the energy flow rotates clockwise. The obtained results of control of intensity maxima allows the transverse mode analysis in waveguide-based sensors [21, 22]. This kind CVBs can be used in phase-sensitive surface plasmon resonance biosensors with high resolution [23] or in graphene biosensors for real-time subcellular imaging [24]. Other application include Raman spectroscopy [25] and vector magnetic field sensing [26]. Sharp focusing of modified cylindrical vector laser beams was studied in [27].

5.2 Sharp Focusing of a Hybrid Vector Beam with a Polarization Singularity

For the first time, vector singularities as a generalization of scalar singularities were proposed in 1983 by Nye [28], where lines of zero-valued transverse components of the E-field were called 'disclinations' (to distinguish them from scalar edge and screw dislocations [29]). However, both in Refs. [28, 29] and in Ref. [30] the polarization singularities were studied locally, i e., in a neighborhood of singular (critical) points. It would be of significant interest to globally investigate inhomogeneously polarized light fields characterized by different (linear, elliptical, or circular) polarization at different points of the beam cross section. That is, we aim to determine topological charges and singularity indices of the whole light field. Such studies become relevant due to a growing number of publications concerned with inhomogeneously polarized vector fields [7]. Inhomogeneously polarized beams can be generated by interferometry [31], inside a cavity [32], as well as with q-plates [33], metasurfaces [34, 35], polarization prisms [36], and spatial light modulators [37]. Points of intensity nulls at which the linear polarization vector is not defined are called V-points [30]. In a similar way, points of a light field with inhomogeneous elliptical polarization where the direction of the major axis of the polarization ellipse is undefined are called 'C-points', with the light being circularly polarized at such points. If the C-points are arranged on a line, the line is called a C-line [30]. Polarization singularities are described by singularity indices, which are calculated similar to the topological charges of scalar light fields [38]. The polarization singularity index of V-points is called a Poincare-Hopf index [30] and calculated using Stokes parameters [39–41]. Meanwhile the C-points are described by an index equal to the number of turns by π the major axis of the polarization ellipse makes around the C-point. The index of a C-point can take a fractional (half-integer) value if on a complete turn the polarization ellipse makes an odd number of turns by π. When intersecting C-lines, the polarization ellipse axis makes a jump by $\pi/2$.

In this section, we look into a hybrid nth order vector light field whose polarization varies from linear to elliptical, to circular depending on the polar angle. This field contains just C-lines with their number being equal to n. For this field, we find components of the Stokes vector and show the polarization index to be half-integer, $n/2$. Using a Richards-Wolf formalism [1], we derive analytical expressions for projections of the E-vector at the tight focus for a source hybrid nth order vector field and analytical relations for the field intensity at the focus. We find that at an even number n, the intensity has nth order symmetry and C-points at the focus. Thus, we numerically demonstrate that C-lines in the source field 'disintegrate' into C-points at the focus, which are located on the same C-lines. We also derive analytical relationships for the projection of the Stokes vector at the focus, which suggest that for an odd number n, the field at the focus is purely vector, consists of vectors of linear polarization, has n V-points, and has no C-points.

5.2.1 Source Hybrid Vector Field with Polarization Singularity Points

Let us analyze a new hybrid nth order vector field defined in the original plane by two transverse projections of the E-vector and a Jones vector in the form:

$$E_n(\varphi) = \frac{1}{\sqrt{2}}\begin{pmatrix} \cos n\varphi \\ i\alpha + \sin n\varphi \end{pmatrix} \tag{5.19}$$

where n is integer and $0 \leq |\alpha| \leq 1$. From (5.19) it follows that at $n = 0$, light field (5.19) is elliptically polarized, while at $|\alpha| = 1$ it is circularly polarized. At $\alpha = 0$, field (5.19) has inhomogeneous nth order vector polarization.

Field (5.19) has points of linear, elliptical, and circular polarization. Points of circular polarization are called C-points of polarization singularity because the direction of the major axis of the ellipse polarization is undefined at such points [30]. Topology of the polarization ellipses around a C-point is described by an index Ic, which shows how many (integer) times the major axis of the polarization ellipse changes its direction by an angle of π while making a full circle around the C-point. To find the index Ic of field (5.19), let us find all projections of the Stokes vector [40] $\boldsymbol{S} = (S_1, S_2, S_3)$, where

$$S_1 = \frac{|E_x|^2 - |E_y|^2}{|E_x|^2 + |E_y|^2},\ S_2 = \frac{2\mathrm{Re}\left(E_x^* E_y\right)}{|E_x|^2 + |E_y|^2},\ S_3 = \frac{2\mathrm{Im}\left(E_x^* E_y\right)}{|E_x|^2 + |E_y|^2}, \tag{5.20}$$

where Re and Im denote the real and imaginary parts of the number. From (5.20), the Stokes vector is seen to be of unit length: $S_1^2 + S_2^2 + S_3^2 = 1$. For field (5.19), the Stokes vector components in Eq. (5.20) take the form:

$$S_1 = 2\frac{\cos 2n\varphi - \alpha^2}{1 + \alpha^2},\quad S_2 = \frac{2\sin 2n\varphi}{1 + \alpha^2},\quad S_3 = \frac{2\alpha\cos n\varphi}{1 + \alpha^2}. \tag{5.21}$$

From (5.21) it follows that polarization of light is linear on the rays outgoing from the center at angles defined by the equation $S_3 = \cos n\varphi = 0$. At angles φ that satisfy the equation $S_3 = 1$ or $\cos n\varphi = \pm 1$ and $\alpha = +1, -1$ the light is circularly polarized. Elsewhere, the light is elliptically polarized. Thus, we can infer that field (5.19) has no isolated C-points but has C-lines, with the direction of the major axis of a polarization ellipse jumping by $\pi/2$ on crossing the line. A single C-point is equivalent to a screw dislocation and a C-line is equivalent to an edge dislocation. The number of C-lines in the source field (5.19) equals the field order n, with the lines found on $2n$ rays outgoing from the center at angles $\pi m/n$, $m = 0, 1, 2, \ldots, 2n - 1$.

In Ref. [30], a local index of hybrid vector fields for polarization singularities (C-points) was calculated and the hybrid vector field itself was locally defined near the singularity. Hereinafter, we shall calculate the topological index of the whole hybrid vector field (5.19) in a global way, in a similar way to calculating the topological charge of the whole scalar complex vortex field using Berry's formula [38]. To these ends, let us form a complex Stokes field by the rule:

$$S_c = S_1 + iS_2 \tag{5.22}$$

For the source vector field (5.19), the complex Stokes field is given by

$$S_c = 2\frac{\exp(2in\varphi) - \alpha^2}{1+\alpha^2} \tag{5.23}$$

The Stokes index σ for field (5.23) can be calculated using Berry's formula [38]:

$$\sigma = \frac{1}{2\pi}\mathrm{Im}\int_0^{2\pi} d\varphi \frac{\partial S_c(\varphi)/\partial\varphi}{S_c(\varphi)}. \tag{5.24}$$

Substituting Stokes field (5.23) into (5.24) yields:

$$\begin{aligned} \sigma &= \frac{1}{2\pi}\,\mathrm{Im}\int_0^{2\pi} d\varphi \frac{2in\exp(2in\varphi)}{\exp(2in\varphi) - \alpha^2} \\ &= \frac{n}{\pi}\int_0^{2\pi} d\varphi \frac{\left(1-\alpha^2\cos 2n\varphi\right)}{(1+\alpha^4) - 2\alpha^2\cos 2n\varphi}. \end{aligned} \tag{5.25}$$

Putting in Eq. (5.25) $\alpha^2 = 1$, we find that $\sigma = n$ and the index of the C-points and the whole field (5.19) equals $Ic = \sigma/2 = n/2$. The index Ic can be half-integer owing to the tilt of the major axis of the polarization ellipse varying from 0 to π, rather than to 2π. Putting in (5.25) $\alpha = 0$, Eq. (5.19) will describe an inhomogeneous linearly polarized field ($S_3 = 0$), containing just V-points (where the linear polarization vector is undefined), where the Stokes index of Eq. (5.25) equals $\sigma = 2n$, meanwhile the Poincare-Hopf index [30] of field (5.19) is half as large: $\eta = n$. At $0 < |\alpha| < 1$, the Stokes index in (5.25) can be calculated using a reference integral [42]:

$$\int_0^{2\pi} \frac{\cos mx}{a + b\cos x}\mathrm{d}x = \frac{2\pi}{\sqrt{a^2-b^2}}\left(\frac{\sqrt{a^2-b^2}-a}{b}\right)^m \tag{5.26}$$

In view of (5.26) and at $0 < |\alpha| < 1$, the Stokes index of field (5.19) equals $\sigma = 2n$, whereas the Poincare-Hopf index is $\eta = \sigma/2 = n$. In this case, there are no points where the light is circularly polarized.

5.2.2 Vector Field with Polarization Singularity Points in the Plane of the Tight Focus

In this subsection, using a Richards-Wolf formalism [1] we derive projections of the E-vector in the focal plane from source field (5.19). Thus, we obtain:

$$
\begin{aligned}
E_x &= -\frac{i^{n+1}}{\sqrt{2}}\left(I_{0,n}\cos n\varphi + I_{2,n-2}\cos(n-2)\varphi\right) + \frac{\alpha}{\sqrt{2}} I_{2,2}\sin 2\varphi,\\
E_y &= -\frac{i^{n+1}}{\sqrt{2}}\left(I_{0,n}\sin n\varphi - I_{2,n-2}\sin(n-2)\varphi\right) + \frac{\alpha}{\sqrt{2}}\left(I_{0,0} - I_{2,2}\cos 2\varphi\right),\\
E_z &= \sqrt{2} i^n I_{1,n-1}\cos(n-1)\varphi - i\alpha\sqrt{2} I_{1,1}\sin\varphi,
\end{aligned}
\tag{5.27}
$$

where the integrals in (5.27) take the form:

$$
\begin{aligned}
I_{\nu,\mu} = \left(\frac{4\pi f}{\lambda}\right)\int\limits_0^{\theta_0} \sin^{\nu+1}\left(\frac{\theta}{2}\right)\cos^{3-\nu}\left(\frac{\theta}{2}\right)\\
\times \cos^{1/2}(\theta) A(\theta) \mathrm{e}^{ikz\cos\theta} J_\mu(x)\mathrm{d}\theta,
\end{aligned}
\tag{5.28}
$$

where λ is the wavelength of light, f is the focal length of an aplanatic system, $x = kr\sin\theta$, $J_\mu(x)$ is a Bessel function of the first kind, and $NA = \sin\theta_0$ is the numerical aperture. The original amplitude function $A(\theta)$ (here, assumed to be real) may be constant (for a plane wave) or described by a Gaussian beam:

$$
A(\theta) = \exp\left(\frac{-\gamma^2\sin^2\theta}{\sin^2\theta_0}\right)
$$

where γ is constant. At $\alpha = 0$, the field at the focus described by Eq. (5.27) is identical (up to a constant $1/\sqrt{2}$) to the field at the focus from an nth order radially polarized wave [17]:

$$
\begin{aligned}
E_x &= -\frac{i^{n+1}}{\sqrt{2}}\left(I_{0,n}\cos n\varphi + I_{2,n-2}\cos(n-2)\varphi\right),\\
E_y &= -\frac{i^{n+1}}{\sqrt{2}}\left(I_{0,n}\sin n\varphi - I_{2,n-2}\sin(n-2)\varphi\right),\\
E_z &= \sqrt{2} i^n I_{1,n-1}\cos(n-1)\varphi.
\end{aligned}
\tag{5.29}
$$

Field (5.29) contains just V-points of polarization singularity while having neither C-points nor C-lines. At $n = 0$ and $\alpha = 1$, field (5.27) is fully identical to the field at the focus from an incident wave with right-handed circular polarization [43]:

$$
\begin{aligned}
E_x &= -\frac{i}{\sqrt{2}}\left(I_{0,0} + e^{i2\varphi} I_{2,2}\right),\\
E_y &= \frac{1}{\sqrt{2}}\left(I_{0,0} - e^{i2\varphi} I_{2,2}\right),\\
E_z &= -\sqrt{2} e^{i\varphi} I_{1,1}.
\end{aligned}
\tag{5.30}
$$

Because of this, the source field (5.19) and the field in the focus in (5.27) can be called hybrid, as at some points they have linear, elliptical, or circular polarization. For field (5.27), the intensity at the focus is given by

$$
\begin{aligned}
I = \frac{1}{2}\{ & I_{0,n}^2 + I_{2,n-2}^2 + 2I_{0,n}I_{2,n-2}\cos 2(n-1)\varphi \\
& + \alpha^2 I_{0,0}^2 + \alpha^2 I_{2,2}^2 - 2\alpha I_{0,0}I_{2,2}\cos 2\varphi \\
& + 4I_{1,n-1}^2 \cos^2(n-1)\varphi + 4\alpha^2 I_{1,1}^2 \sin^2\varphi \\
& - 2\alpha \cos\left(\frac{n+1}{2}\right)\pi \left[\sin n\varphi\left(I_{0,0}I_{0,n} + I_{2,2}I_{2,n-2}\right)\right. \\
& - \sin(n-2)\varphi\left(I_{0,0}I_{2,n-2} + I_{2,2}I_{0,n}\right) \\
& \left. - \sin\varphi \sin(n-1)\varphi I_{1,1}I_{1,n-1}\right]\}.
\end{aligned}
\tag{5.31}
$$

Equation (5.31) is rather cumbersome, but putting $n = 2p$ (even) yields $\cos(n + 1)\pi/2 = 0$, leading to a simpler relationship of the intensity:

$$
\begin{aligned}
I_{n=2p} = \frac{1}{2}\{ & I_{0,n}^2 + I_{2,n-2}^2 + 2I_{0,n}I_{2,n-2}\cos 2(n-1)\varphi \\
& + \alpha^2 I_{0,0}^2 + \alpha^2 I_{2,2}^2 - 2\alpha I_{0,0}I_{2,2}\cos 2\varphi \\
& + 4I_{1,n-1}^2 \cos^2(n-1)\varphi + 4\alpha^2 I_{1,1}^2 \sin^2\varphi\}.
\end{aligned}
\tag{5.32}
$$

From (5.32), the intensity at the center of the focal plane is seen to be non-zero because the term $\alpha^2 I_{0,0}^2$ is non-zero. The intensity pattern has central symmetry as Eq. (5.32) contains cosines of the double angle 2φ, as well as squared cosine and sine functions, meaning that replacing φ with $\varphi + \pi$ introduces no changes to the intensity pattern. From (5.32), the intensity pattern is also seen to have $2(n - 1)$ local intensity peaks (not considering central intensity maximum) because the term $\cos 2(n - 1)\varphi$ changes sign $2(n - 1)$ times per full circle. At odd numbers $n = 2p + 1$, we obtain $\cos(n + 1)\pi/2 = \pm 1$, which means that the intensity in Eq. (5.31) has no central symmetry due to different intensity values at φ and $\varphi + \pi$, but has a central intensity peak, similar to the previous case.

Let us derive formulae for projections of the Stokes vector at the focus. Since these formulae are rather cumbersome, below, we give only relationships for projections of symmetrical fields at the focus for an even number $n = 2p$. The Stokes vector can be defined in a different way using four projections, rather than three use in definition (5.20):

$$s_0 = |E_x|^2 + |E_y|^2,\ s_1 = |E_x|^2 - |E_y|^2,$$
$$s_2 = 2\mathrm{Re}(E_x^* E_y),\ s_3 = 2\mathrm{Im}(E_x^* E_y). \tag{5.33}$$

Based on (5.33), for field (5.27) (at $n = 2p$) we obtain:

$$\begin{aligned}
s_0 &= \frac{1}{2}\big(I_{0,n}^2 + I_{2,n-2}^2 + 2I_{0,n}I_{2,n-2}\cos 2(n-1)\varphi \\
&\quad + \alpha^2 I_{0,0}^2 + \alpha^2 I_{2,2}^2 - 2\alpha^2 I_{0,0}I_{2,2}\cos 2\varphi\big),\\
s_1 &= \frac{1}{2}\big(I_{0,n}^2\cos 2n\varphi + I_{2,n-2}^2\cos 2(n-2)\varphi \\
&\quad + 2I_{0,n}I_{2,n-2}\cos 2\varphi - \alpha^2 I_{0,0}^2 - \alpha^2 I_{2,2}^2\cos 4\varphi \\
&\quad + 2\alpha^2 I_{0,0}I_{2,2}\cos 2\varphi\big),\\
s_2 &= \frac{1}{2}\big(I_{0,n}^2\sin 2n\varphi + I_{2,n-2}^2\sin 2(n-2)\varphi \\
&\quad + 2I_{0,n}I_{2,n-2}\sin 2\varphi - \alpha^2 I_{2,2}^2\cos 4\varphi \\
&\quad + 2\alpha^2 I_{0,0}I_{2,2}\cos 2\varphi\big),\\
s_3 &= \alpha\sin\left(\frac{n+1}{2}\right)\pi\big[\cos n\varphi\big(I_{0,0}I_{0,n} - I_{2,2}I_{2,n-2}\big) \\
&\quad + \cos(n-2)\varphi\big(I_{0,0}I_{2,n-2} - I_{2,2}I_{0,n}\big)\big].
\end{aligned} \tag{5.34}$$

In (5.34), the relations for $s_0,\ s_1,\ s_2$ are given for even numbers $n = 2p$, except for s_3 which holds at any n. The purpose is to demonstrate that at odd $n = 2p + 1$ we have $s_3 = S_3 = 0$, hence we can infer that the field at the focus has no C-points, being purely vector and composed of linear polarization vectors.

Using two components of the field from Eq. (5.34), the complex Stokes field can be expressed as

$$\begin{aligned}
S = s_1 + is_2 &= \frac{1}{2}\big[I_{0,n}^2 e^{2in\varphi} + I_{2,n-2}^2 e^{2i(n-2)\varphi} \\
&-\alpha^2 I_{2,2}^2 e^{4i\varphi} + 2e^{2i\varphi}\big(\alpha^2 I_{0,0}I_{2,2} + I_{0,n}I_{2,n-2}\big) - \alpha^2 I_{0,0}^2\big].
\end{aligned} \tag{5.35}$$

From (5.35), it follows that the topological charge of the vortex Stokes field is undefined, varying over the entire focal plane, for at large radii r, the amplitudes by the exponents vary in magnitudes, making it impossible to determine which term in the sum (5.35) is larger in the absolute value at each particular case. For instance, at some radii, the Stokes index of field (5.35) can be $\sigma = 2n$, being $\sigma = 2(n-2)$ at other radii and taking values of 4, 2 or 0 elsewhere. What may be said definitely is that near the optical axis only the last term in (5.35) remains non-zero, which has no vortex phase. Hence, at any n, the Stokes index at the center of the focus is zero ($\sigma = 0$). Conclusions arrived at in this section are validated by the numerical modeling.

5.2.3 Numerical Modeling

Figure 5.8 depicts a distribution of polarization ellipses in the source field (5.19) at different n: 3(a), 2(b), 1(c), and 4(d). Indices for C-lines of the fields in Fig. 5.8, derived from Eq. (5.25) using a complex Stokes field, equal $Ic = \sigma/2 = n/2$: 3/2(a), 1(b), 1/2(c), and 2(d). From Fig. 5.8*a*, field (5.19) with $n = 3$ is seen to have three C-lines located at angles $\varphi = \pi m/3$, $m = 0, 1, 2$. The tilt of the major axis of the polarization ellipses changes by $\pi/2$ at each of six sectors between the adjacent C-lines. Thus, after a full circle, the tilt of the major axis changes by $6\pi/2 = 3\pi$, meaning that the index of the field in Fig. 5.8a is $Ic = 3\pi/(2\pi) = 3/2$. In a similar way, in Fig. 5.8b, field (1) with $n = 2$ has two C-lines located on the Cartesian axes. With the angle φ in a sector between C-lines changing from 0 to $\pi/2$, the tilt of the major axis of the polarization ellipse is rotated by an angle of $\pi/2$, hence after a full circle around the center, the tilt of the major axis changed by $4\pi/2$. Hence, the index of the field in Fig. 5.8b equals $Ic = 2\pi/(2\pi) = 1$. In Fig. 5.8c, the C-line is found on the horizontal Cartesian axis. With the angle φ in one of the sectors between C-lines changing from 0 to π (in the upper semi-plane), the tilt of the polarization ellipse major axis is rotated by $\pi/2$, and in the bottom semi-plane the tilt of the major axis is also rotated by $\pi/2$. Hence, after a full circle, the polarization ellipse makes a turn by π and the singularity index equals $Ic = \pi/(2\pi) = 1/2$. Finally, in Fig. 5.8d, a change in the tilt of the major axis of polarization ellipses can be analyzed in a similar way.

Shown in Fig. 5.9a is a total intensity at the focus of field (5.19) at $\alpha = 1$ and $n = 2$. The numerical modeling was conducted using a Richards-Wolf formalism [1] for a wavelength of $\lambda = 532$ nm and numerical aperture NA $= 0.95$. Shown in Fig. 5.9b, c are an amplitude and phase of the complex Stokes field $S_c = s_1 + is_2$, which was calculated with the aid of the Stokes vector components in Eq. (5.34). From Fig. 5.9a, it is seen that according to theoretical predictions in Eqs. (5.31) and (5.32), the intensity pattern at the focus remains unchanged after replacing φ by $\varphi + \pi$, with an intensity peak located at the center. From Fig. 5.9c, it is seen that there is no singular point at the center of the phase pattern for the Stokes field (5.35), because there is no isolated intensity null. Two isolated intensity nulls (singularity points) in Fig. 5.9c, each having the topological charge 1, are seen on the vertical axis (1 and 2 points in Fig. 5.9d). In Fig. 5.9d, the arrows specify a pattern of the polarization ellipses at the focus. Figure 5.9e depicts C-points at the focus, which are all located on the Cartesian axes, it is where the C-lines are located in the source plane (Fig. 5.8b). Thus, the numerical modeling has shown that as a result of tightly focusing, C-lines 'disintegrate' into a number of C-points arranged on the same lines. This effect is analogous to an effect of astigmatic edge-to-screw dislocation conversion of a wavefront in scalar paraxial optics [44]. Indices of two symmetrical and closest to the center C-points on the horizontal Cartesian axis are $Ic = \pm 1/2$ (1 and 2 points in Fig. 5.9c), with the indices of the next two neighboring C-points located farther from the center on the horizontal axis being $Ic = \mp 1/2$ (3 and 4 points in Fig. 5.9c).

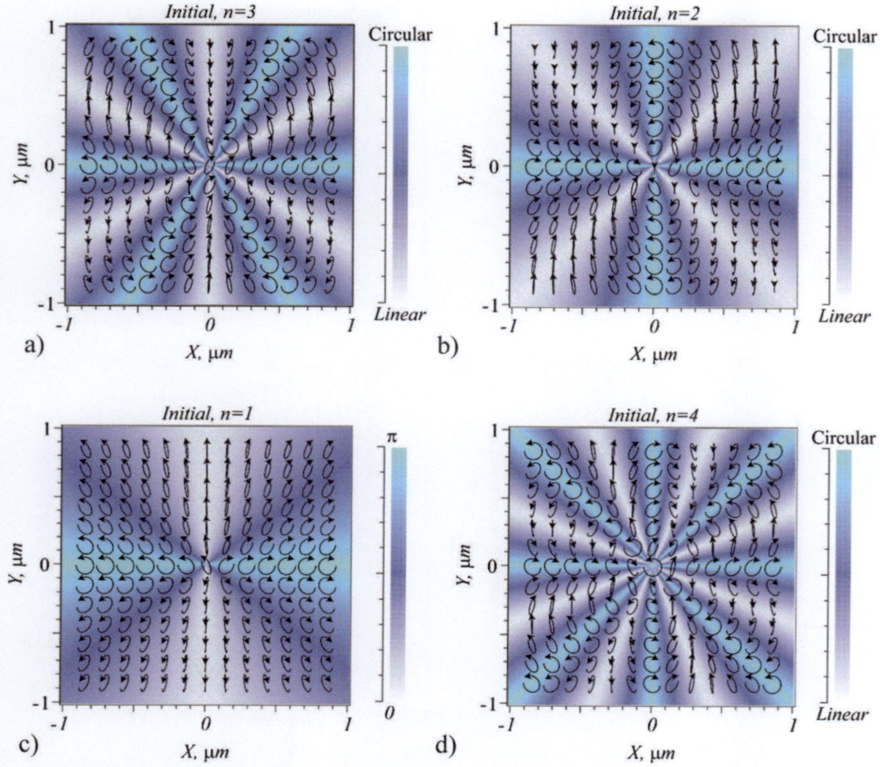

Fig. 5.8 Polarization patterns (shown in half-tones and with arrows) in the source field (5.19) at $\alpha = 1$ at different orders n: 3 (**a**), 2 (**b**), 1 (**c**), 4 (**d**). The arrows show the handedness of the E-vector; the origin of the ellipse is determined based on the phase of the field at this point

In a similar way, Fig. 5.10 depicts numerical simulation results at the focus for $n = 3$ (the rest parameters being the same as in Fig. 5.9). Figure 5.10a suggests that in agreement with the theoretical prediction in (5.31), the intensity pattern at the focus at odd $n = 3$ is asymmetric. As can be inferred from (5.34), there is a vector field with purely inhomogeneous linear polarization at the focus (Fig. 5.10d), as putting $n = 2p + 1$, we obtain $s_3 = S_3 = 0$. In a phase pattern of the complex Stokes field in Fig. 5.10c, there occur three phase singularity points with the topological charge + 2 (1, 2, and 3 points in Fig. 5.10c). In total, the Stokes index is $\sigma = 6$, and the V-points singularity index is $Ic = \sigma / 2 = 3$. The pattern in Fig. 5.10d is seen to contain three V-points (2 points of the "center" type and 1 point of the "knot" type). Thus, when field (5.19) has an odd order n, the C-lines (Fig. 5.8a) of the original plane are transformed into a number of V-points (Fig. 5.10d) and the whole field becomes vectorial (with no points of elliptical polarization).

Shown in Fig. 5.11a is an intensity pattern at the focus, which has a fourth-order symmetry relative to the Cartesian coordinates. The amplitude and phase of the

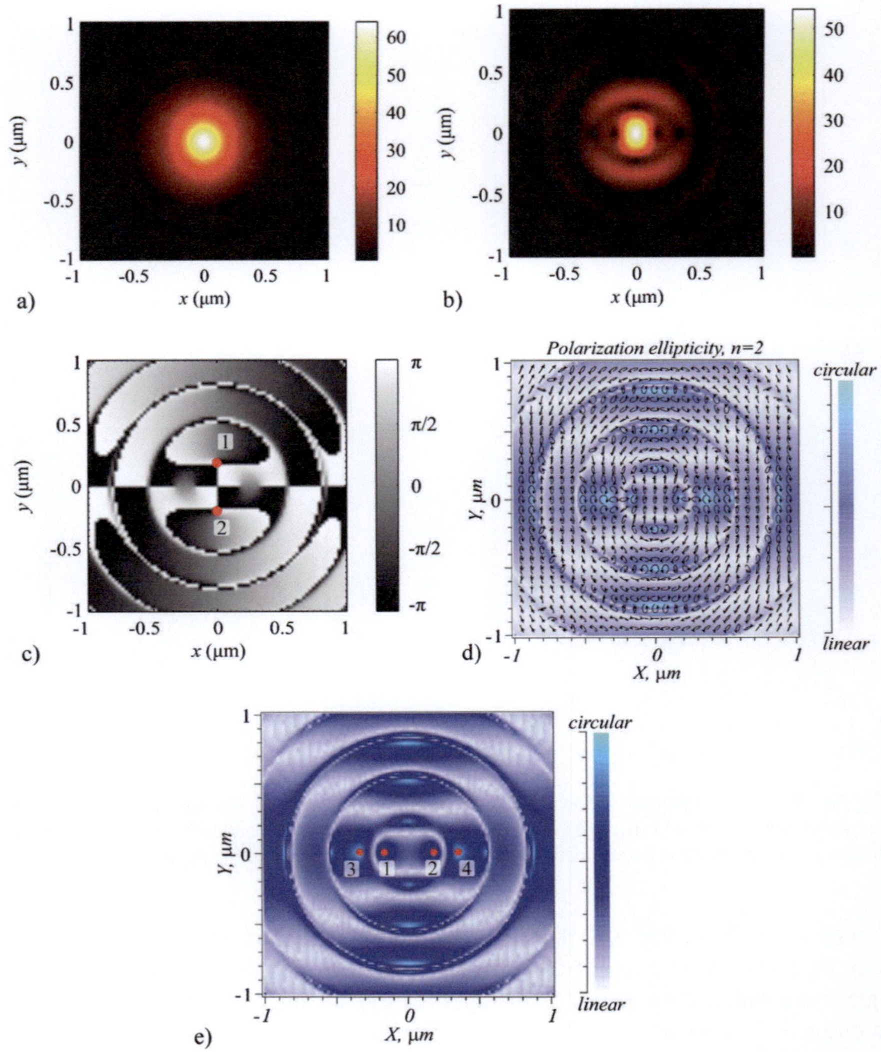

Fig. 5.9 **a** An intensity $I = Ix + Iy + Iz$, **b** amplitude, and **c** phase of a complex Stokes field (5.35) when focusing field (5.19) at $n = 2$. **d** Pattern of elliptic polarization at the focus, and **e** pattern of points with circular, elliptic, and linear polarization

complex Stokes field are depicted in Figs. 5.11b, c. Phase singularities points linked with the C-points are observed in the phase pattern of the Stokes field (Fig. 5.11e). Finally, Fig. 5.11f depicts a plot for the Stokes index against the radius R of an origin-centered circle along which the phase delay of the Stokes field (5.35) of Fig. 5.11c is calculated. From the plot it is seen that at different radii R, the Stokes index takes values of 8, 6, 2, 0, which is in good agreement with the theory in Eq. (5.35).

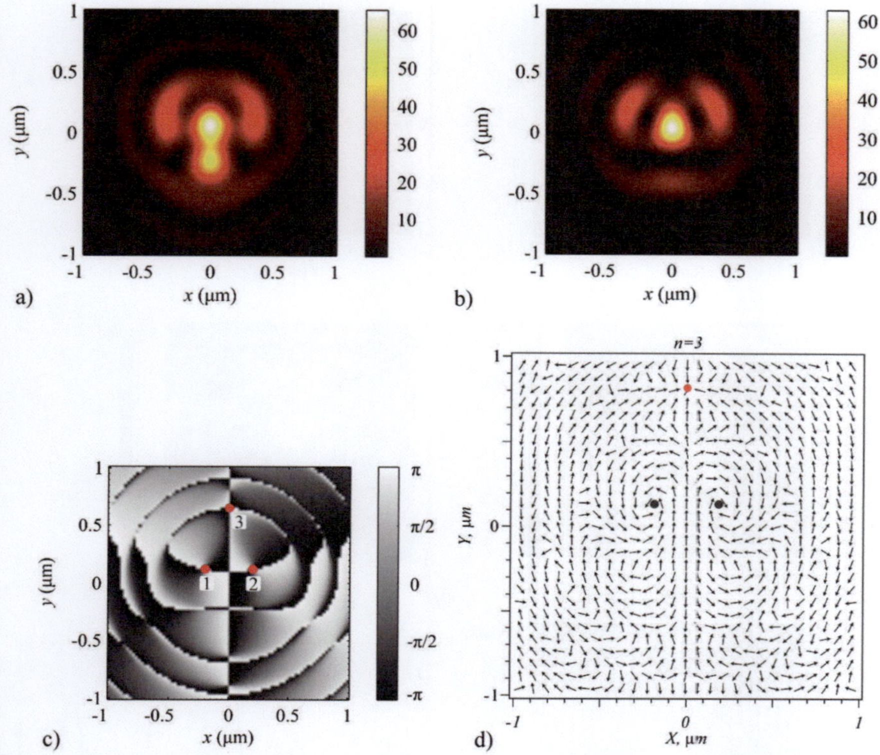

Fig. 5.10 **a** An intensity $I = Ix + Iy + Iz$, **b** amplitude, and **c** phase of the complex Stokes field (5.35) when focus in field (5.19) at $n = 3$. **d** A pattern of linear polarization vectors at the focus: (black V-points of the "center" type and red V-point of the "knot" type)

From Fig. 5.11e, d, the C-points are seen to lie on the Cartesian axes and two diagonal lines, where C-lines were located in the original plane (Fig. 5.8d). Two center-symmetrical C-points located on the horizontal axis (Fig. 5.11e, f) have a singularity index of ± 1/2, producing a 'lemon'-shaped topology. Remarkably, when making a circle around such a C-point, the surface of the polarization ellipses produces a Mobius strip in the 3D space [45–47].

Figure 5.12 depicts the neighborhood of a C-point (marked with bold black) in more detail, showing a characteristic tilt of the major axis of polarization ellipses ('lemon'-type topology) lying on a circle centered on the C-point. The axes of the polarization ellipses are seen to turn by an angle of π when making a full circle, i.e., the index of the C-point equals 1/2.

Summing up, we have theoretically and numerically studied a new type of nth order hybrid vector light field that is tightly focused with an aplanatic system [48]. The polarization of the source hybrid vector field varies in the original plane from linear, to elliptical and circular with the polar angle. The polarization pattern of the field has

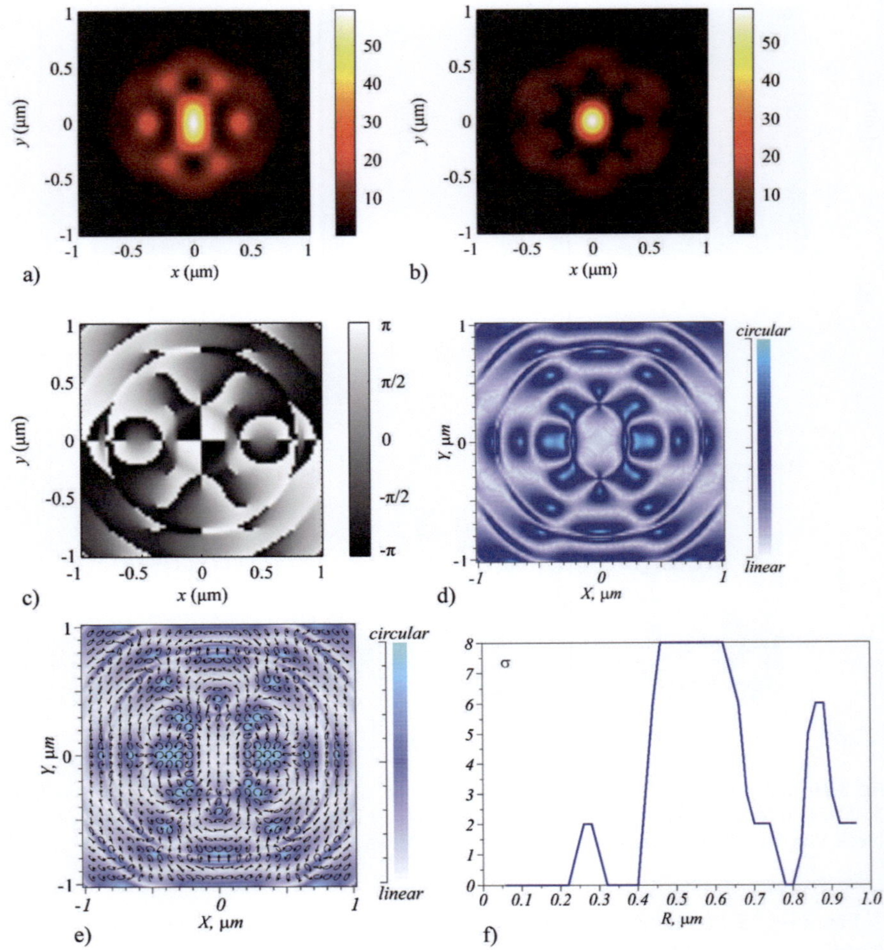

Fig. 5.11 **a** An intensity $I = Ix + Iy + Iz$, **b** amplitude, and **c** phase of the complex Stokes field (5.35) when focusing field (5.19) at $n = 4$. **d** pattern of points characterized by circular, elliptic, and linear polarization at the focus, **e** pattern of the elliptical polarization in focus, and **f** dependence of the Stokes index σ on the radius R of the circle on which it is calculated (1 μm frame size)

n C-lines of circular polarization that pass through the center. Components for the Stokes vector of such a field have been analytically derived and the field polarization singularity index has been shown to equal $n/2$. Based on a Richards-Wolf formalism, analytical relationships for projections of the E-vector and an intensity of light at the tight focus have been deduced. At even n, the intensity at the focus has been shown to possess a center symmetry and have C-points lying on lines coincident with the C-lines of the source field. Analytical relationships have been deduced to describe

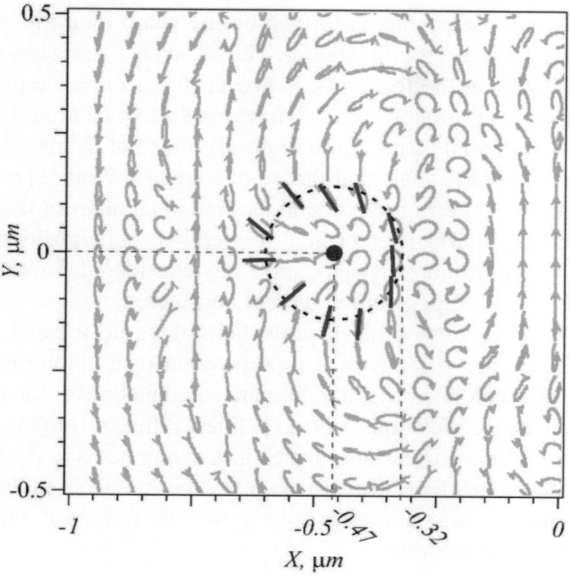

Fig. 5.12 Polarization ellipses for a field fragment at the focus depicted in Fig. 5.11e limited by (− 1 μ < x < 0; − 0.5 μm < y < 0.5 μm) and a C-point with Ic = 1/2 found at (− 0.47 μm; 0) marked with a black point around which the major axes of the polarization ellipses centered on a circle of radius 0.15 μm are depicted as blue lines

projections of the Stokes vector at the focus, which suggest that at odd n, the field at the focus is purely vector and has several V-points, while having no C-points.

5.3 Spin-Orbital Conversion in a Tight Focus of an Axial Superposition of a High-Order Cylindrical Vector Beam and a Beam with Linear Polarization

Cylindrical vector fields, including those of higher orders, are well known [7, 49]. They are an example of inhomogeneously polarized light beams, in the cross section of which the local linear polarization vector changes its direction from point to point. Cylindrical beams of low orders are called beams with radial and azimuthal polarizations [49]. Such beams can, for example, be formed using two half-wave plates rotated relative to each other [49], multisection polarizers [50], metasurfaces [51], quarter-wave plates, and a light modulator [52]. Cylindrical vector beams (CVB) are used in particle micromanipulation [53], microscopy [54], quantum informatics [55, 56].

CVBs of any order do not have a spin angular momentum (SAM) and the third component of the Stokes vector is zero. This means that both in the initial plane and in any other section of the beam during its propagation the polarization is locally linear. It has recently been shown that local subwavelength regions with circular and elliptical polarization appear at the focus of CVBs with a fractional order [57]. The optical effect of spin-orbital conversion is known in laser beams, when a transverse vortex energy flux is formed during sharp focusing of an ordinary Gaussian beam with a circular polarization. That is, such a beam has an orbital angular momentum (OAM) in the focus [4]. However, the occurrence of a local SAM in the focus of fractional-order CVBs [57] have not the explanation since such a beam has not a SAM in the initial plane. Another disadvantage of the research [57] is that it is impossible to apply the analytical theory of Richards-Wolf [1] to describe the electromagnetic field in a sharp focus due to the fractional order of the beam. In [58], a modification of a cylindrical vector field formed by a superposition of such a beam with a beam of uniform linear polarization was considered. However, in [58] the tight focusing of such a superposition did not considered.

In this section, all six projections of the electric and magnetic field strength vectors in the sharp focus of a superposition of a cylindrical vector field and a uniform field with linear polarization are calculated theoretically and numerically, using the Richards-Wolf approach [1]. Energy fluxes (projections of the Poynting vector), intensity distributions, and Stokes components in the focus are also obtained. It was shown that local transverse vortex energy fluxes and subwavelength regions with an elliptical and a circular polarization are formed in the focus of such a field with an odd integer order.

5.3.1 Projections of Vectors of Electric and Magnetic Fields in Focus

We consider an initial light field with a non-uniform polarization, the Jones vector for the electric and magnetic fields of which has the form:

$$\begin{aligned} E_m(\varphi, a) &= \begin{pmatrix} \cos(m\varphi) - a \\ \sin(m\varphi) \end{pmatrix} = \begin{pmatrix} \cos(m\varphi) \\ \sin(m\varphi) \end{pmatrix} - a\begin{pmatrix} 1 \\ 0 \end{pmatrix}, \\ H_m(\varphi, a) &= \begin{pmatrix} -\sin(m\varphi) \\ \cos(m\varphi) - a \end{pmatrix}, \end{aligned} \tag{5.36}$$

where (r, φ) are polar coordinates in the initial plane, a is a real number. This light field was considered in [58]. It was shown that the field (5.36) has the Poincaré-Hopf polarization singularity index η [30], which is $\eta = m$ for $|a| < 1$, $\eta = m/2$ for $|a| = 1$, and $\eta = 0$ for $|a| > 1$. The light field (5.36) is an axial superposition of two well-known light fields: a cylindrical vector field of the mth order and a light field with the linear polarization directed along the horizontal axis. The real number a determines the polarization singularity index of the field (5.36) and the distribution of intensity, the energy flux, and the density of SAM in the sharp focus of the field (5.36).

For $a = 0$, the field (5.36) is a well-known CVB of a high order [7, 17]. The beam (5.36) at $a = 0$ has an inhomogeneous polarization and the polarization is locally linear in each point of the beam cross section. The polarization of mth order CVB is also linear in each point of the focus. The purpose of this study is to show the presence of local areas in the focus of the field (5.36) with $a \neq 0$, where a transverse energy flow (the energy circulates in a closed loop) is formed, and the longitudinal projection of the SAM vector is different from zero, i.e., there is an elliptical and a circular polarization.

Using the Richards-Wolf formalism [1], explicit expressions for all projections of electric and magnetic field intensity vectors in the sharp focus of the light field (5.36) were obtained:

$$\begin{aligned}
E_x(r,\varphi) &= i^{m-1}\big(\cos(m\varphi)I_{0,m} + \cos((m-2)\varphi)I_{2,m-2}\big) + ia\big(I_{0,0} + \cos(2\varphi)I_{2,2}\big),\\
E_y(r,\varphi) &= i^{m-1}\big(\sin(m\varphi)I_{0,m} - \sin((m-2)\varphi)I_{2,m-2}\big) + ia\sin(2\varphi)I_{2,2},\\
E_z(r,\varphi) &= 2i^{m}\cos((m-1)\varphi)I_{1,m-1} + 2a\cos\varphi I_{1,1},\\
H_x(r,\varphi) &= -i^{m-1}\big(\sin(m\varphi)I_{0,m} + \sin((m-2)\varphi)I_{2,m-2}\big) + ia\sin(2\varphi)I_{2,2},\\
H_y(r,\varphi) &= -i^{m-1}\big(-\cos(m\varphi)I_{0,m} + \cos((m-2)\varphi)I_{2,m-2}\big) + ia\big(I_{0,0} - \cos(2\varphi)I_{2,2}\big),\\
H_z(r,\varphi) &= -2i^{m}\sin((m-1)\varphi)I_{1,m-1} + 2a\sin\varphi I_{1,1}.
\end{aligned} \tag{5.37}$$

In (5.37), functions $I_{\nu,\mu}$ depend only on the radial variable r and are equal to the expression:

$$I_{\nu,\mu} = \left(\frac{4\pi f}{\lambda}\right)\int_0^{\theta_0} \sin^{\nu+1}\left(\frac{\theta}{2}\right)\cos^{3-\nu}\left(\frac{\theta}{2}\right)\cos^{1/2}(\theta)A(\theta)\mathrm{e}^{ikz\cos\theta}J_\mu(x)\mathrm{d}\theta, \tag{5.38}$$

where k is the wave number of light, λ is the wavelength of light, f is the focal length of an ideal spherical lens that forms the focus, z is an optical axis ($z = 0$ is the focal plane), $x = kr\sin\theta$, $J_\mu(x)$ is the Bessel function of the first kind and of the μth order, $\mathrm{NA} = \sin\theta_0$ is the numerical aperture of the aplanatic optical system, $A(\theta)$ is any real function that describes the input field amplitude, which has axial symmetry (plane wave, Gaussian beam, Bessel-Gaussian beam). For the integrals $I\nu$, μ (5.38), the first index $\nu = 0, 1, 2$ describes the type of the integral, and the second index $\mu = 0, 1, 2, \ldots, m$ is equal to the order of the Bessel function.

Each projection of electric and magnetic field vectors (5.37) is the sum of two beams vectors projections: a cylindrical vector field of the mth order and a light field with linear polarization. This is easy to verify if we recall what projections an electromagnetic field with linear polarization directed along the horizontal axis has in the focus [1]:

$$
\begin{aligned}
E_{Lx}(r,\varphi) &= -i\big(I_{0,0}+\cos(2\varphi)I_{2,2}\big),\\
E_{Ly}(r,\varphi) &= -i\sin(2\varphi)I_{2,2},\\
E_{Lz}(r,\varphi) &= -2\cos\varphi I_{1,1},\\
H_{Lx}(r,\varphi) &= -i\sin(2\varphi)I_{2,2},\\
H_{Ly}(r,\varphi) &= -i\big(I_{0,0}-\cos(2\varphi)I_{2,2}\big),\\
H_{Lz}(r,\varphi) &= -2a\sin\varphi I_{1,1}.
\end{aligned}
\tag{5.39}
$$

5.3.2 The Intensity Distribution in the Focal Plane

Based on the obtained amplitudes of the electric field vector projections in the focus (5.37), it is possible to derivate expressions for the intensity and its components along the Cartesian axes. It should be noted that the expressions for the intensity with even and odd numbers m will be different. Indeed, formula (5.37) implies expressions for $I = I_x + I_y + I_z = |E_x|^2 + \left|E_y\right|^2 + |E_z|^2$:

$$
I_x = \begin{cases}
\left[a(-1)^{p+1}\big(I_{0,0}+\cos(2\varphi)I_{2,2}\big)+\cos(m\varphi)I_{0,m}+\cos((m-2)\varphi)I_{2,m-2}\right]^2,\\
m=2p,\\
\left[a\big(I_{0,0}+\cos(2\varphi)I_{2,2}\big)\right]^2+\left[\cos(m\varphi)I_{0,m}+\cos((m-2)\varphi)I_{2,m-2}\right]^2,\\
m=2p+1,
\end{cases}
\tag{5.40}
$$

$$
I_y = \begin{cases}
\left[a(-1)^{p+1}\sin(2\varphi)I_{2,2}+\sin(m\varphi)I_{0,m}-\sin((m-2)\varphi)I_{2,m-2}\right]^2,\\
m=2p,\\
\left[a\sin(2\varphi)I_{2,2}\right]^2+\left[\sin(m\varphi)I_{0,m}-\sin((m-2)\varphi)I_{2,m-2}\right]^2,\\
m=2p+1,
\end{cases}
\tag{5.41}
$$

$$
I_z = \begin{cases}
4\left[a(-1)^{p}\cos(\varphi)I_{1,1}+\cos((m-1)\varphi)I_{1,m-1}\right]^2,\\
m=2p,\\
4\left[a\cos(\varphi)I_{1,1}\right]^2+4\left[\cos((m-1)\varphi)I_{1,m-1}\right]^2,\\
m=2p+1.
\end{cases}
\tag{5.42}
$$

It can be seen from (5.42) that for $m = 2$ and $p = 1$ the intensity is equal to the simple expression:

$$
I_{2z}(r,\varphi) = 4\cos^2(\varphi)I_{1,1}^2(1-a)^2.
\tag{5.43}
$$

It can be seen from 5.43) that the longitudinal intensity is zero at $a = 1$, and it has two local maxima on the horizontal x-axis ($\varphi = 0$ and $\varphi = \pi$) at $a \neq 1$.

Next, we obtained the expressions for the total intensity in the focus at $m = 2p + 1$, because, as it will be shown later, the transverse energy fluxes and the longitudinal projection of the SAM arise in the focus only for odd numbers m:

$$\begin{aligned} I = a^2\big[I_{0,0}^2 + I_{2,2}^2 + 2I_{1,1}^2 + 2\cos 2\varphi\big(I_{0,0}I_{2,2} + I_{1,1}^2\big)\big] \\ + \big[I_{0,m}^2 + I_{2,m-2}^2 + 2I_{1,m-1}^2 + 2\cos(2(m-1)\varphi) \\ \big(I_{0,m}I_{2,m-2} + I_{1,m-1}^2\big)\big]. \end{aligned} \tag{5.44}$$

The expressions for the intensity (5.40) and (5.44) for an arbitrary m contains a term $a^2I_{0,0}^2 > 0$, in which according to the integral (5.38) the zero-order Bessel function is used as one of the factor. Therefore, the intensity will be different from zero (there will be a local maximum) on the optical axis (at $r = 0$), since $J_0(0) = 1$. The arguments of the cosines are even in the expression (5.44) for the intensity. This means that the intensity pattern, although it does not have radial symmetry, has axial symmetry, i.e., $I(r, \varphi) = I(r, \varphi + \pi)$. Also it can be seen from (5.40) that the intensity I_x will have a maximum on the optical axis due to the term $a^2I_{0,0}^2$, and it follows from (5.41) that I_y will have a zero on the optical axis. It should be also noted that that the intensity pattern I_y will have $2m$ local maxima since the expression for I_y contains the squared $\cos(m\varphi)$. The total intensity (5.44) will have $2(m - 1)$ local maxima (except the intensity maximum on the optical axis), since formula (5.44) has $\cos(2(m - 1)\varphi)$. These conclusions will be confirmed by modeling.

5.3.3 The Energy Flux Density in the Focal Plane

In this section, the expressions for three projections of the Poynting vector in the focus of the light field (5.36) are obtained. It is known [17, 59] that a cylindrical vector field of any order with $a = 0$ does not have a spin or vortex energy flows both in the initial plane $(E_x, E_y) = (\cos(m\varphi), \sin(m\varphi))$ and in the focus (5.37). It means that the longitudinal projections of the SAM and the OAM vectors in the focus are zero at each point. Below we will show that the superposition of a CVB and a light field with linear polarization (5.37) has a local spin and a vortex energy flux. The Poynting vector is given by the next formula [1]:

$$\boldsymbol{P} = \frac{c}{2\pi}\mathrm{Re}\big(\boldsymbol{E}^* \times \boldsymbol{H}\big), \tag{5.45}$$

where **E** and **H** are vectors of electric and magnetic fields, * is a complex conjugation, × is a vector multiplication, c is the light speed in vacuum. Further the constant $c/(2\pi)$ will be ignored. We substituted the expressions for the projections of the electromagnetic field in the focus (5.37) into expression (5.45) and obtained:

$$P_x(r,\varphi)=\begin{cases}2a(-1)^{(m-1)/2}\left[\cos((m-1)\varphi)\left(I_{1,1}I_{2,m-2}-I_{0,0}I_{1,m-1}\right)\right.\\ \left.+\cos((m+1)\varphi)\left(I_{2,2}I_{1,m-1}-I_{1,1}I_{0,m}\right)\right], \quad m=2p+1,\\ 0, \quad m=2p, \quad p=0,1,2,...\end{cases} \tag{5.46}$$

$$P_y(r,\varphi)=\begin{cases}2a(-1)^{(m-1)/2}\left[-\sin((m-1)\varphi)\left(I_{1,1}I_{2,m-2}-I_{0,0}I_{1,m-1}\right)\right.\\ \left.+\sin((m+1)\varphi)\left(I_{2,2}I_{1,m-1}-I_{1,1}I_{0,m}\right)\right], \quad m=2p+1,\\ 0, \quad m=2p, \quad p=0,1,2,...\end{cases} \tag{5.47}$$

$$P_z(r,\varphi)=\begin{cases}a^2\left(I_{0,0}^2-I_{2,2}^2\right)+\left(I_{0,m}^2-I_{2,m-2}^2\right)\\ +2a(-1)^p\cos(m\varphi)\left(I_{2,2}I_{2,m-2}-I_{0,0}I_{0,m}\right), \ m=2p,\\ a^2\left(I_{0,0}^2-I_{2,2}^2\right)+\left(I_{0,m}^2-I_{2,m-2}^2\right), \ m=2p+1, \ p=0,1,2,...\end{cases} \tag{5.48}$$

It can be seen from (5.48) that the distribution of the Poynting vector longitudinal component for odd numbers m has a circular symmetry in the focus. For an even number m it depends on the polar angle and has m maxima when going around the optical axis. It can be seen from (5.46) and (5.47) that the transverse energy flow takes place only for odd numbers m and is equal to zero for an even m. To characterize the transverse energy flux in the focus in more detail, we proceed to the polar projections of the transverse energy flux vector. Using the transition from the Cartesian projections of the Poynting vector to polar:

$$\begin{aligned}P_r&=P_x\cos\varphi+P_y\sin\varphi,\\ P_\varphi&=-P_x\sin\varphi+P_y\cos\varphi,\end{aligned} \tag{5.49}$$

from (5.46) and (5.47) we find the transverse components of the Poynting vector in the tight focus of field (5.36) in polar coordinates for odd numbers m (for even numbers m, the Poynting vector components are equal to zero):

$$\begin{aligned}P_r&=2a(-1)^p\cos(m\varphi)Q_1(r), \quad m=2p+1,\\ Q_1(r)&=I_{1,m-1}\left(I_{2,2}+I_{0,0}\right)-I_{1,1}\left(I_{2,m-2}+I_{0,m}\right),\\ P_\varphi&=2a(-1)^p\sin(m\varphi)Q_2(r), \quad m=2p+1,\\ Q_2(r)&=I_{1,m-1}\left(I_{2,2}-I_{0,0}\right)+I_{1,1}\left(I_{0,m}-I_{2,m-2}\right).\end{aligned} \tag{5.50}$$

It can be seen from (5.50) that the transverse energy flow rotates non-uniformly at different radii and for different p the rotation occurs counterclockwise or clockwise. The irregularity lies in the fact that the transverse vector of the energy flux rotates around the optical axis not tangentially to some circle, but at a different angle to some circle. There are an $2m$ subwavelength regions on a circle centered on the optical axis in which the transverse energy flow rotates along a closed trajectory. This follows from (5.50), in which the argument of cosine and sine is equal to $m\varphi$. From Eqs. (5.46) and (5.47) it follows that the transverse flow changes sign $2(m +$

1) times per complete revolution. This follows from the fact that in Eqs. (5.46) and (5.47) there are terms with sin($m + 1$) or with cos($m + 1$), which change sign 2(m + 1) times. Moreover, the energy flow rotates in different directions (clockwise or counterclockwise) in neighboring areas. Integration of radial and azimuthal energy fluxes in (5.50) over the angle φ gives zero. It means that the total transverse energy flux is zero in the focus.

5.3.4 The Density of the Stokes Vector in the Focal Plane

In this section, we find the projections of the Stokes vector in the focus of the initial vector field (5.36). The components of the Stokes vector **S** are calculated by the formulas [60]:

$$\begin{aligned} S_1 &= \frac{|E_x|^2 - |E_y|^2}{|E_x|^2 + |E_y|^2}, \quad S_2 = \frac{2\mathrm{Re}\left(E_x^* E_y\right)}{|E_x|^2 + |E_y|^2}, \quad S_3 = \frac{2\mathrm{Im}\left(E_x^* E_y\right)}{|E_x|^2 + |E_y|^2}, \\ s_1 &= |E_x|^2 - |E_y|^2, \quad s_2 = 2\mathrm{Re}\left(E_x^* E_y\right), \quad s_3 = 2\mathrm{Im}\left(E_x^* E_y\right), \end{aligned} \tag{5.51}$$

where Re and Im determine the real and the imaginary parts of a complex number. In (5.51), the small letters (s_1, s_2, s_3) denote the unnormalized components of the Stokes vector. The normalized Stokes vector, as it can be seen from (5.51), has a unit length $S_1^2 + S_2^2 + S_3^2 = 1$. Due to the cumbersomeness of the expressions, and in order to find out whether the circular polarization will be in focus, we obtained expressions only for the third Stokes projection without normalization, i.e., we calculated a function of the form $s_3 = 2\mathrm{Im}\left(E_x^* E_y\right)$. It should be preliminarily noted that the third component of the Stokes vector is proportional to the longitudinal projection of the SAM [59]:

$$\boldsymbol{S} = \frac{1}{16\pi\omega}\mathrm{Im}\left(\boldsymbol{E}^* \times \boldsymbol{E}\right), \tag{5.52}$$

where ω is a cyclic frequency of light. Further the constant $1/(16\pi\omega)$ will be ignored. It can be seen from (5.52) that the longitudinal component of the SAM (without taking into account the constant) coincides with the unnormalized third component of the Stokes vector:

$$s_3 = S_z = 2\mathrm{Im}\left(E_x^* E_y\right) \tag{5.53}$$

Substituting the projections of the electric field (5.37) into (5.53), we obtain:

$$s_3 = S_z(r, \varphi) = \begin{cases} 2a(-1)^{(m-1)/2}\left[\sin((m-2)\varphi)\left(I_{0,0}I_{2,m-2} - I_{2,2}I_{0,m}\right)\right. \\ \left. - \sin(m\varphi)\left(I_{0,0}I_{0,m} - I_{2,2}I_{2,m-2}\right)\right], \quad m = 2p+1, \\ 0, \quad m = 2p, \quad p = 0, 1, 2, \ldots \end{cases} \tag{5.54}$$

It can be seen from (5.54) that there are no regions with a circular (elliptical) polarization in the focus of field (5.36) for an even m. If m is odd and $a \neq 0$ than there are $2m$ local regions in the focus, in which the light has an elliptical polarization. It should be noted from (5.50) that for an odd m there are also $2m$ local regions of transverse vortex energy flows in the focus. Comparison of (5.50) and (5.54) shows that the number of regions with a transverse vortex energy flow in focus is $2m$ and is equal to the number of regions with an elliptical polarization. Moreover, the direction of the transverse energy flow rotation is different in neighboring regions, just as the direction of the polarization vector rotation alternates in neighboring regions. Since the field (5.36) does not have the transverse energy flow and the longitudinal SAM in the initial plane then both the total (over the entire focal plane) longitudinal SAM and the total transverse energy flow must be equal to zero in the focus. It should be noted that if we integrate the spin density (5.54), i.e., the longitudinal component of the SAM, over the entire beam cross section in the focus, then the integrals over the angle φ will give zero, and the total beam spin (5.36) in the focus, as in the initial plane, will be equal to zero:

$$ss_3 = \int_0^{\infty}\int_0^{2\pi} s_3(r, \varphi) r \mathrm{d}r \mathrm{d}\varphi = 0.$$

To compare the theory and simulation results, we derive an expression for s_2 only for the even number $m = 2p$, $p = 0, 1, 2, \ldots$:

$$\begin{gathered} s_2 = \sin(2\varphi)\left[2I_{0,m}I_{2,m-2} \; + a^2 I_{2,2}\left(I_{0,0} + \cos(2\varphi)I_{2,2}\right)\right] \\ \sin(2m\varphi)I_{0,m}^2 - \sin(2(m-2)\varphi)I_{2,m-2}^2\,, \; m = 2p. \end{gathered} \tag{5.55}$$

It can be seen from (5.55) that, the distribution of s_2 in the focus will be axisymmetric since all arguments of the cosines and sines are even. The maximum argument in (5.55) has $\sin(2m\varphi)$ which is equal to $2m$. Therefore, the number of sign changes for the function s_2 will be equal to $4m$.

The mechanism of occurrence of an even number of local vortices of energy fluxes at the focus for field (5.36) with an odd order number of a cylindrical vector field can be described as follows. First, only when m is odd and when $a = 1$ does the index of the polarization singularity of the field (5.36) become half-integer $\eta = m/2$. Second, for even m, there are no transverse energy fluxes at the focus (5.46), (5.47). The half-integer polarization singularity index leads to the fact that the initial light field has m polarization singularity lines emanating from the center (the direction of linear polarization is not determined on the singularity lines) and dividing the beam cross section into m parts. In each of these parts (between two adjacent singularity lines), two local regions are formed at the focus (there are $2m$ such regions at the focus), in which the polarization is circular (elliptical), but of different signs (left and right). This also follows from expression (5.54). On the other hand, the presence in the focus of regions with non-zero spin density (5.52), according to the effect of spin–orbit interaction, must simultaneously generate local regions with a vortex transverse energy flux (non-zero longitudinal projection of the orbital angular momentum). In total, such regions with a vortex energy flow should be equal to the number of regions with circular polarization, that is, equal to $2m$. For the total OAM of the beam to be equal to zero, the number of local regions with a vortex transverse energy flow must be even. And the direction of rotation of the energy flow in neighboring areas should be opposite.

The presence of local regions with a vortex energy flow in the focus can be explained in another way. Field (5.36) with $a = -1$ can be represented as

$$E_m(\varphi, a = -1) = \begin{pmatrix} \cos(m\varphi) + 1 \\ \sin(m\varphi) \end{pmatrix} = 2\cos\left(\frac{m\varphi}{2}\right)\begin{pmatrix} \cos\left(\frac{m\varphi}{2}\right) \\ \sin\left(\frac{m\varphi}{2}\right) \end{pmatrix}. \tag{5.56}$$

It can be seen from (5.56) that the initial field (5.36) for $a = -1$ and odd m is a cylindrical vector field of fractional order ($m/2$). And it was shown in [57] that such a light field has local regions with a vortex energy flow and with circular polarization at the focus.

The formation at the focus of local regions with circular and elliptical polarization for the field (5.36) immediately follows from the expressions for the projections of the electric vector at the focus (5.37). From (5.37) it can be seen that at $a = 0$ (there is no field with linear polarization), the product of the transverse projections of the electric vector will be real $\mathrm{Im}(E_x^* E_y) = 0$, since both projections will have the same factor $(-i)^{m-1}$. If $a \neq 0$, then both expressions for the transverse projections of the electric vector in (5.37) consist of two terms with factors $(-i)^{m-1}$ and i. Therefore, the product $E_x^* E_y$ will have a factor $(-i)^m$, which will be imaginary when m is odd. Therefore, the longitudinal projection of the SAM or the third Stokes component will be non-zero $\mathrm{Im}(E_x^* E_y) \neq 0$. This means that there are local regions at the focus in which the polarization is elliptical or circular.

5.3.5 Numerical Simulations Results and Discussion

In this subsection, we present the simulation results for the intensity distribution, the projections of the Stokes vectors (or the SAM longitudinal projection), as well as the Poynting vector projection in the focus of the initial light field (5.36). The calculation was carried out using the general formulas of Richards-Wolf [1], which describe the light in the focus area:

$$\begin{aligned} \boldsymbol{U}(\rho, \psi, z) = -\frac{if}{\lambda}\int_0^{\alpha}\int_0^{2\pi} & B(\theta, \varphi)T(\theta)\boldsymbol{P}(\theta, \varphi)\times \\ & \times\ \exp\{ik[\rho\sin\theta\cos(\varphi - \psi) + z\cos\theta]\}\sin\theta\, \mathrm{d}\theta\, \mathrm{d}\varphi, \end{aligned} \tag{5.57}$$

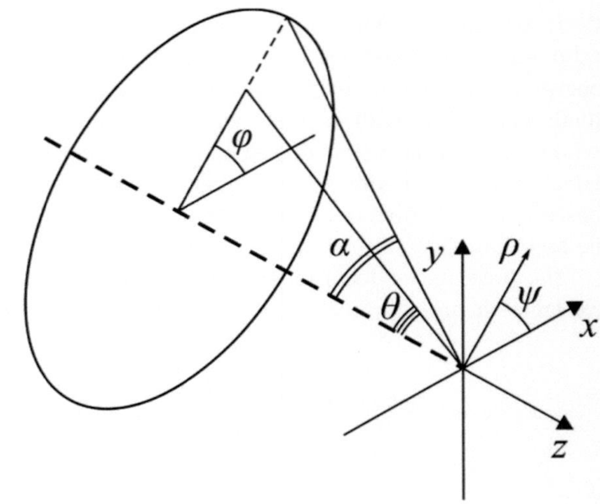

Fig. 5.13 The geometric interpretation of the problem

where $\mathbf{U}(\rho, \psi, z)$ is an electric or a magnetic field strength, $B(\theta, \varphi)$ is an electric or a magnetic field at the input of a wide-aperture optical system dependent on the exit pupil coordinates (θ is a polar angle, φ is an azimuth angle), $T(\theta)$ is a lens apodization function, f is a focal length, $k = 2\pi/\lambda$ is a wave number, λ is a wavelength, α is the maximum polar angle defined by the lens numerical aperture (NA = sinα), and $\mathbf{P}(\theta, \varphi)$ is a polarization matrix. Integral (5.57) allows to calculate the distribution of the electromagnetic field components in the exit pupil coordinates (Fig. 5.13).

The polarization matrix $\mathbf{P}(\theta, \varphi)$ for the strength of electric and magnetic fields has the form [61, 62]:

$$\boldsymbol{P}(\theta, \phi) = \begin{bmatrix} \sin^2\phi + \cos^2\phi\cos\theta & \sin\phi\cos\phi(\cos\theta - 1) & \cos\phi\sin\theta \\ \sin\phi\cos\phi(\cos\theta - 1) & \cos^2\phi + \sin^2\phi\cos\theta & \sin\phi\sin\theta \\ -\sin\theta\cos\phi & -\sin\theta\sin\phi & \cos\theta \end{bmatrix} \times \begin{bmatrix} a(\theta, \phi) \\ b(\theta, \phi) \\ c(\theta, \phi) \end{bmatrix} \tag{5.58}$$

where $a(\theta, \phi)$, $b(\theta, \phi)$, and $c(\theta, \phi)$ are polarization functions for x-, y-, and z-component of an incident field. For example, for the linearly polarized along the x-axis light, the components will be equal to $a = 1$, $b = 0$, and $c = 0$. For all the examples considered in this section, the longitudinal component of the focused field was proposed to be zero: $c = 0$ (the initial plane), then:

$$\boldsymbol{P}(\theta, \varphi) = \begin{bmatrix} 1 + \cos^2\varphi(\cos\theta - 1) \\ \sin\varphi\cos\varphi(\cos\theta - 1) \\ -\sin\theta\cos\varphi \end{bmatrix} a(\theta, \varphi) + \begin{bmatrix} \sin\varphi\cos\varphi(\cos\theta - 1) \\ 1 + \sin^2\varphi(\cos\theta - 1) \\ -\sin\theta\sin\varphi \end{bmatrix} b(\theta, \varphi), \tag{5.59}$$

For the initial field (5.36), the polarization functions will have the form:

$$\mathbf{E}(\theta,\varphi)=\begin{pmatrix}a(\theta,\varphi)\\ b(\theta,\varphi)\end{pmatrix}=\begin{pmatrix}\cos(m\varphi)-a\\ \sin(m\varphi)\end{pmatrix} \tag{5.60}$$

for the electric field strength and

$$\mathbf{H}(\theta,\varphi)=\begin{pmatrix}a(\theta,\varphi)\\ b(\theta,\varphi)\end{pmatrix}=\begin{pmatrix}-\sin(m\varphi)\\ \cos(m\varphi)-a\end{pmatrix} \tag{5.61}$$

for the magnetic field strength.

The distribution of linear vectors over the beam cross section in the initial plane depends on the number m and the parameter a. Figure 5.14 shows the distributions of linear polarization vectors in the cross section of the field (5.36) for $m = 2$ (a, c, e) and $m = 3$ (b, d, f), and $a = 1/2$ (a, b), $a = 3/2$ (c, d), $a = 1$ (e, f).

It was shown in [58] that the polarization singularity index (the number of linear polarization vector rotations by 2π while going around the closed contour around the optical axis—around the singularity point) at $a < 1$ is equal to m. Thus, the polarization singularity index is equal to 2 (a) and 3 (b) in Fig. 5.14. The polarization singularity index is equal to $m/2$ for $a = 1$, since there are only singularity lines Fig. 5.14e, f in the field (5.36) and the linear polarization vectors rotate from line to line by an angle π. Therefore, the index is equal to 1 (e) and 3/2 (f) in Fig. 5.14. For $a > 1$, the polarization singularity index of the field (5.36) is equal to zero (Fig. 5.14c, d).

The calculation of the intensity in the focus of the field (5.36) was carried out using the Richards-Wolf formulas (5.57)–(5.61) for the wavelength of 633 nm and the numerical aperture NA = 0.95. The intensity distribution and its components were calculated for the vector beam (5.36) of the second ($m = 2$) (Figs. 5.15, 5.16 and 5.17) and third ($m = 3$) (Figs. 5.18, 5.19 and 5.20) orders. The beam parameter a (5.36) was chosen to be 1 (Figs. 5.15 and 5.18), 1/2 (Figs. 5.16 and 5.19), and 3/2 (Figs. 5.17 and 5.20).

Figures 5.15a, 5.16 and 5.17a show that for any $a > 0$ an elliptical focal spot elongated along the x-axis is formed near the optical axis in the center of focus. The intensity maximum on the optical axis follows from formulas (5.42)–(5.44). The intensity distributions in Figs. 5.15a, 5.16 and 5.17a differ only by the size of the central and two side lobes. For $a > 1$, the side lobes are small and almost all the intensity is in the central elliptical spot. For $a = 1$, the energy of the side lobes increases. For $a < 1$, the energy of the two side lobes, whose intensity maxima lie on the vertical y-axis, is comparable to the intensity of the central focal spot. A feature of the intensity distribution in Figs. 5.15, 5.16, and 5.17 is that the longitudinal component is absent for case of $a = 1$. This unique case takes place only when $m = 2$ and $a = 1$, and is described by the formula (5.43). Figures 5.15d, 5.16 and 5.17d confirm formula (5.42), according to which the longitudinal intensity I_z at $a \neq 1$ has two local intensity maxima on the horizontal axis at $\varphi = 0$ and $\varphi = \pi$. There

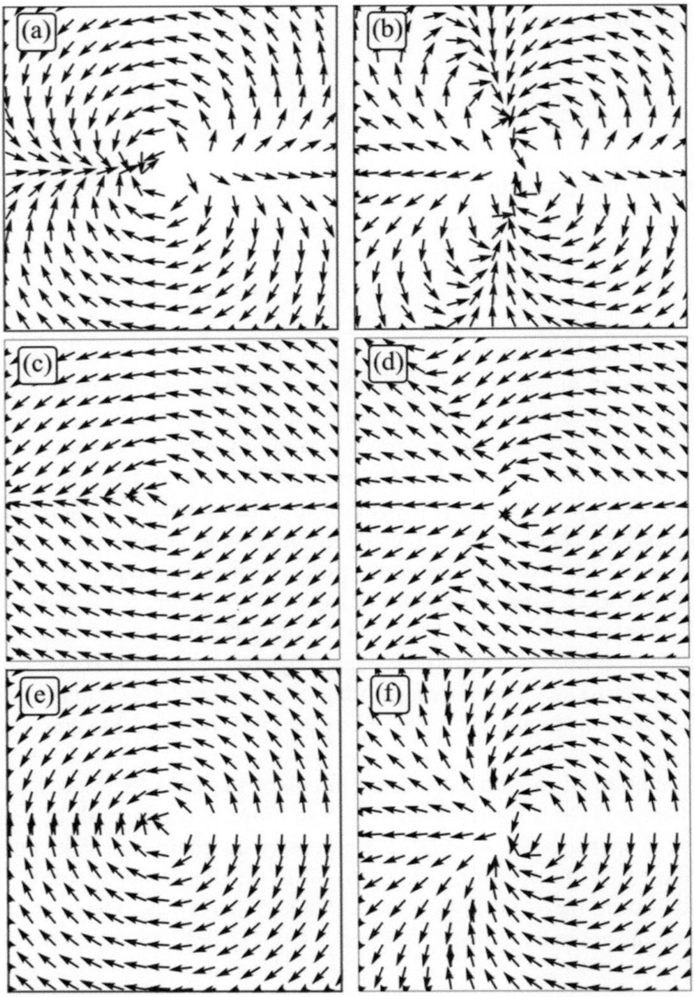

Fig. 5.14 The distributions of linear polarization vectors in the cross section of the field (5.36) for $m = 2$ (**a**, **c**, **e**) and $m = 3$ (**b**, **d**, **f**), and $a = 1/2$ (**a**, **b**), $a = 3/2$ (**c**, **d**), $a = 1$ (**e**, **f**)

are 4 local maxima for any a in Figs. 5.15c, 5.16 and 5.17c. This is consistent with formula (5.41), since I_y should have $2m$ such maxima. Figures 5.18, 5.19 and 5.20 show intensity distributions similar to those shown in Figs. 5.15, 5.16, and 5.17 but for the odd $m = 3$.

Figures 5.18a, 5.19a and 5.20a show a focal elliptical spot with side lobes that in the center of the intensity pattern near the optical axis. The number of lobes, according to the formula (5.44), is $2(m - 1) = 4$. These lobes are clearly visible in Fig. 5.19a where the parameter $a < 1$ and almost invisible in Fig. 5.20a where $a > 1$. Figures 5.18d, 5.19d and 5.20d confirm formula (5.42) according to which the

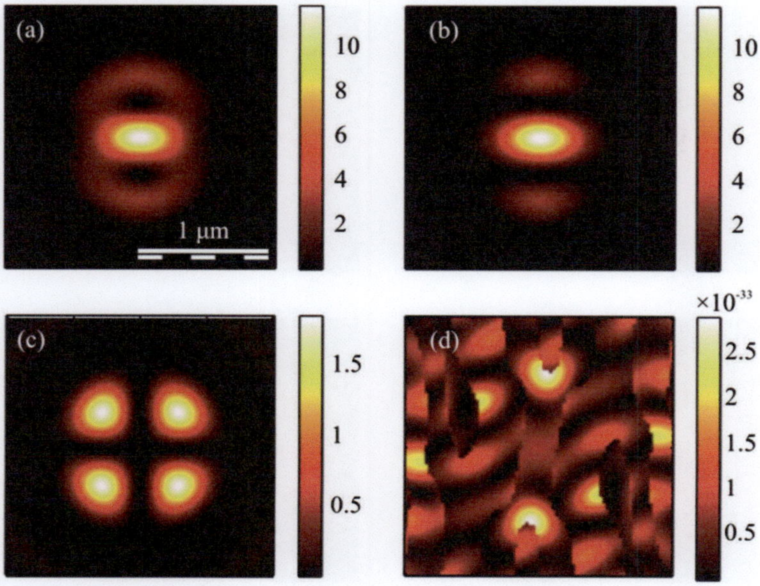

Fig. 5.15 The intensity I (**a**) and its components I_x (**b**), I_y (**c**) I_z (**d**) of the focused vector field (5.36) of the second ($m = 2$) order with $a = 1$

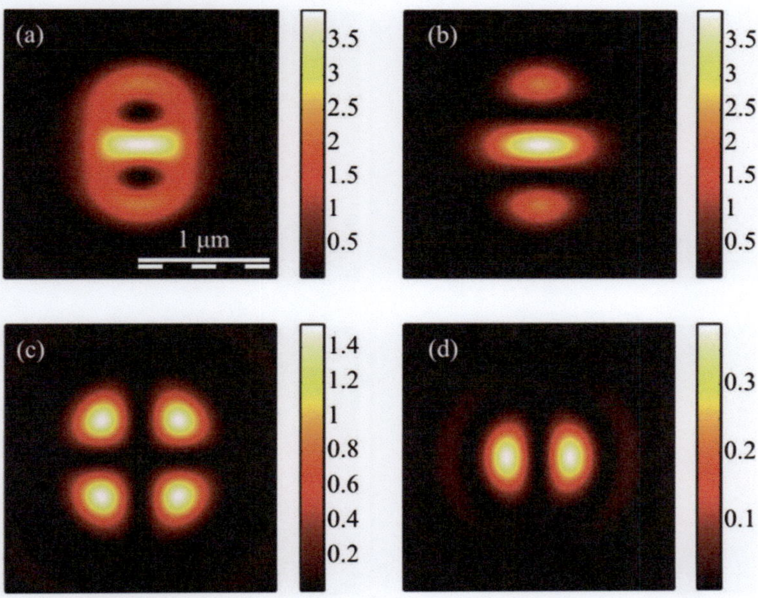

Fig. 5.16 The intensity I (**a**) and its components I_x (**b**), I_y (**c**) I_z (**d**) of the focused vector field (5.36) of the second ($m = 2$) order with $a = 1/2$

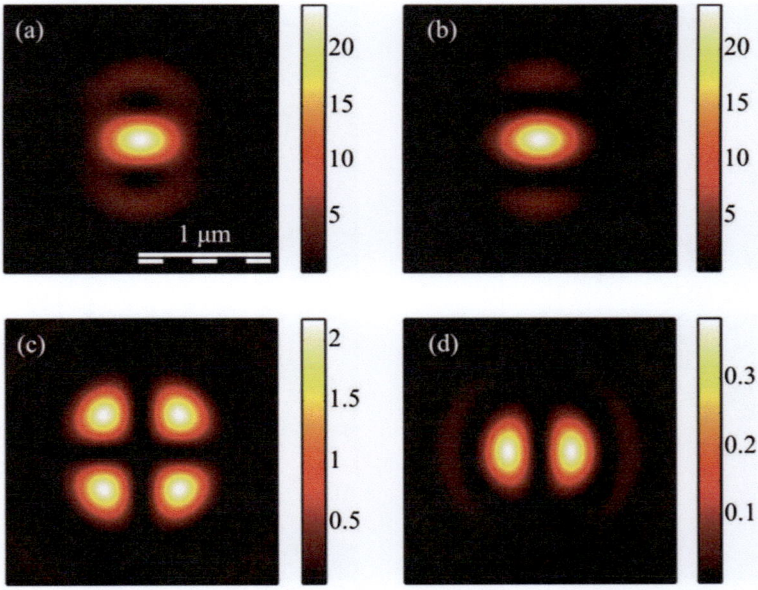

Fig. 5.17 The intensity I (**a**) and its components I_x (**b**), I_y (**c**) I_z (**d**) of the focused vector field (5.36) of the second ($m = 2$) order with $a = 3/2$

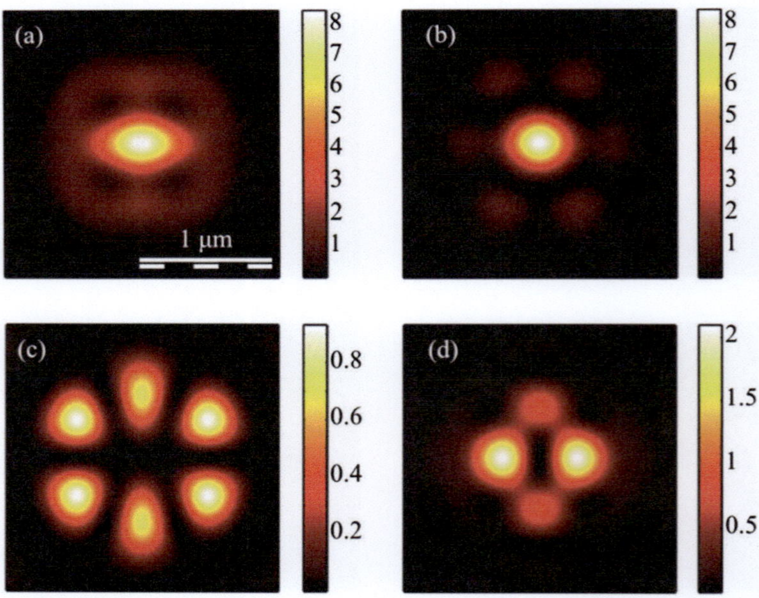

Fig. 5.18 The intensity I (**a**) and its components I_x (**b**), I_y (**c**), I_z (**d**) of the focused vector field (5.36) of the second ($m = 3$) order with $a = 1$

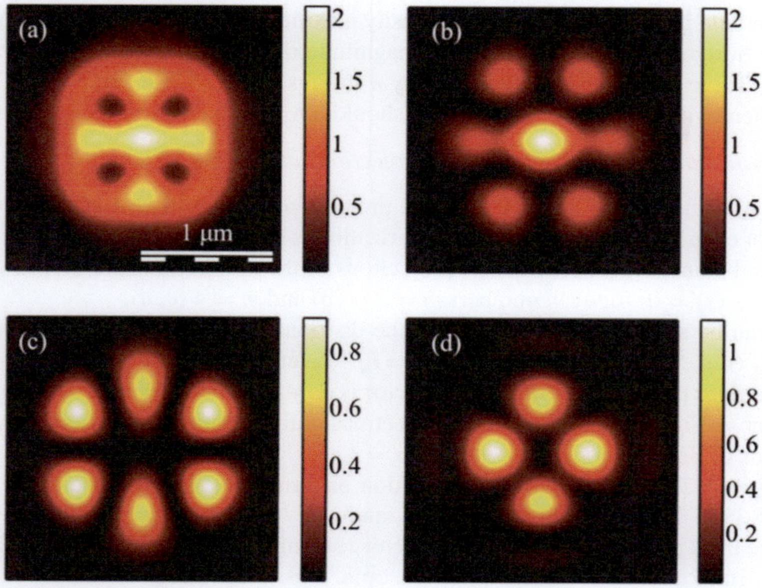

Fig. 5.19 The intensity I (**a**) and its components I_x (**b**), I_y (**c**), I_z (**d**) of the focused vector field (5.36) of the second ($m = 3$) order with $a = 1/2$

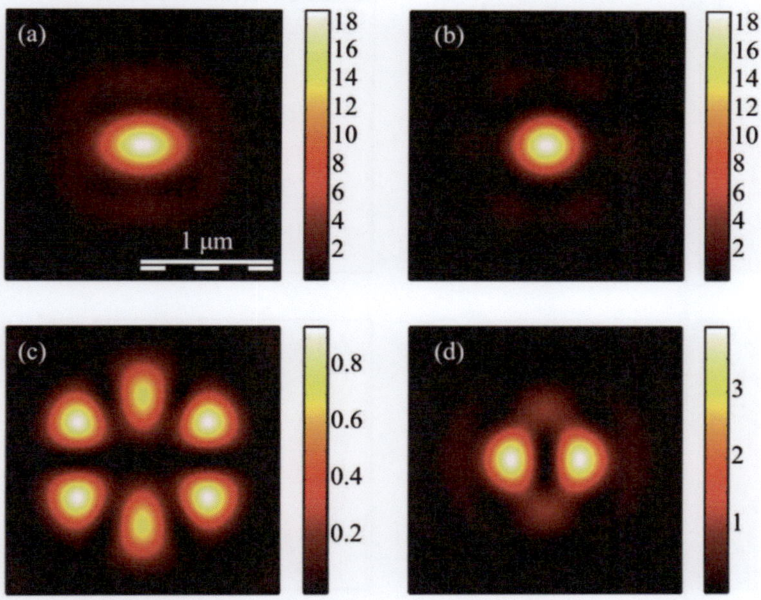

Fig. 5.20 The intensity I (**a**) and its components I_x (**b**), I_y (**c**), I_z (**d**) of the focused vector field (5.36) of the second ($m = 3$) order with $a = 3/2$

longitudinal intensity I_z has 4 local intensity maxima for any a. Two maxima located on the horizontal x-axis are larger in magnitude than two maxima on the vertical y-axis. There are 6 local maxima for any a in Figs. 5.18c, 5.19c and 5.20c. This is consistent with formula (5.41), since I_y should have $2m$ such maxima.

The Distribution of the Stokes Vector Projections in the Focal Plane

It can be seen from (5.54) that the third projection of the Stokes vector is equal to zero for even m. This means that the polarization is linear at each point of the field in the focal plane. Figure 5.21 shows projections of the non-normalized Stokes vector s_1 and s_2 ($s_3 = 0$) for even numbers $m = 2$ (a, b) and $m = 4$ (c, d).

It can be seen from Fig. 5.21a that the distribution pattern s_1 almost coincides with I_x (Fig. 5.16b). This is because $s_1 = I_x - I_y$ and $I_x > I_y$.

The third projection of the Stokes vector is non-zero in the focus only for an odd number m. Figure 5.22 shows three projections of the non-normalized Stokes vector in focus of field (5.36) with $m = 3$ and $a = 1$.

Figure 5.22a shows that the distribution pattern of s_1 almost coincides with I_x (Fig. 5.19c). This is because $s_1 = I_x - I_y$ and $I_x > I_y$. It can be seen from Fig. 5.22c that the third projection of the Stokes vector s_3 changes sign $2m = 6$ times on circles with certain radii and with center on the optic axis. This is consistent with formula (5.54), which includes $\sin(m\varphi)$. This function changes sign $2m$ times per turn. The

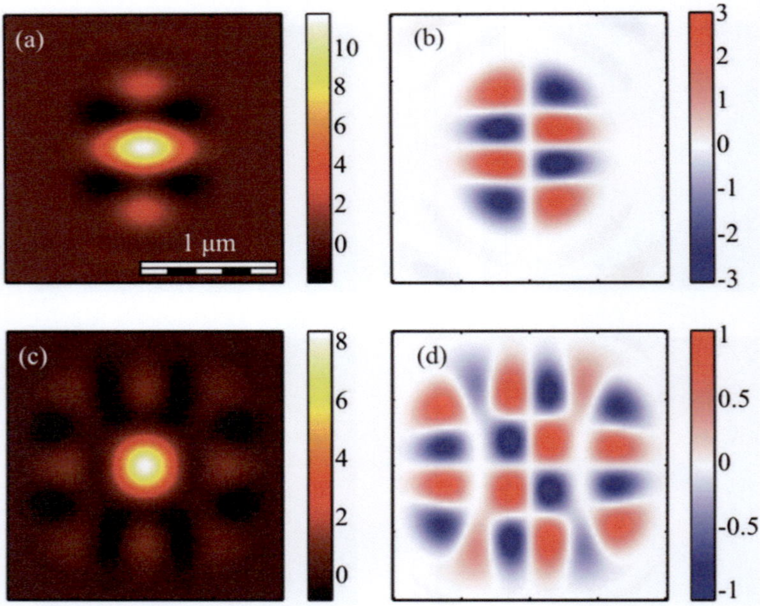

Fig. 5.21 The Stokes vector components s_1 (**a**, **c**) and s_2 (**b**, **d**) of the focused vector field (5.36) with $m = 2$ (**a**, **b**) and $m = 4$ (**c**, **d**) for $a = 1$

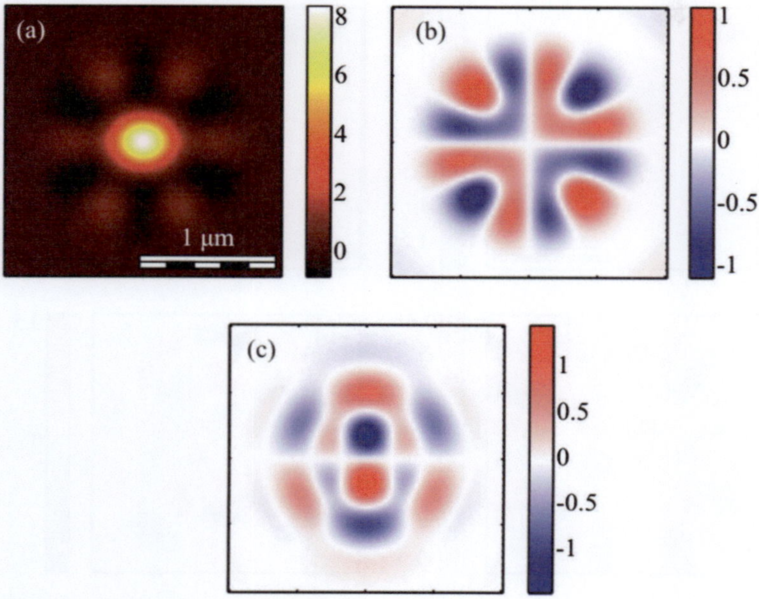

Fig. 5.22 The Stokes vector components s_1 (**a**), s_2 (**b**), and s_3 (**c**) of the focused vector field (5.36) with $m = 3$ and $a = 1$

second Stokes projection changes sign when going around a closed trajectory around the optical axis $4m$ times: 8 (Fig. 5.21b), 16 (Fig. 5.21d), and 12 (Fig. 5.22b). This is consistent with formula (5.55), in which the term with the maximum argument has the form $\sin(2m\varphi)$.

For comparison, Fig. 5.23 shows the second s_2 (a, c) and the third s_3 (b, d) components of the Stokes vector of the focused vector field (5.36) with other odd numbers m: 1 (a, b) and 5 (c, d).

It can be seen from Fig. 5.23 that the distribution s_2 changes sign $4m$ times when going around the optical axis: 4 (Fig. 5.23a) and 20 (Fig. 5.23c). This is consistent with formula (5.55). And the distribution s_3 changes sign $2m$ times: 2 (Fig. 5.23b) and 10 (Fig. 5.23d). This is consistent with formula (5.44).

For the sake of completeness, we show the distributions of the third component of the normalized Stokes vector (Fig. 5.24).

Figure 5.24 shows that the magnitude and size of the regions with an elliptical and a circular polarization, where the component S_3 is close to + 1 (the brightest) or − 1 (the darkest), decreases with decreasing parameter a. The comparison of Figs. 5.22c and 5.24 shows that the structures of normalized S_3 and non-normalized s_3 qualitatively agree.

In this subsection, it is shown by numerical examples that local regions (size is about 200–250 nm) with an elliptical or a circular polarization are formed in the focus of the vector field (1) (wavelength is 532 nm, $NA = 0.95$). The number of

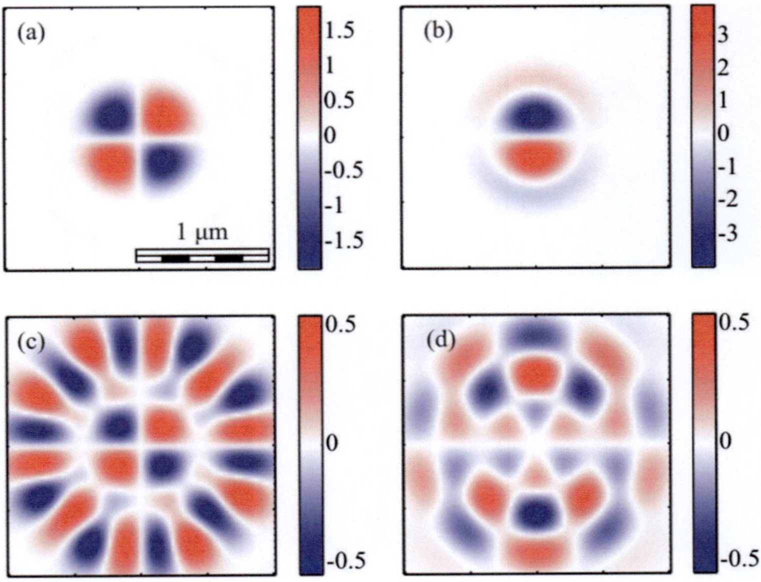

Fig. 5.23 The Stokes vector components s_2 (**a**, **c**) and s_3 (**b**, **d**) of the focused vector field (5.36) with $m = 1$ (**a**, **b**) and $m = 5$ (**c**, **d**)

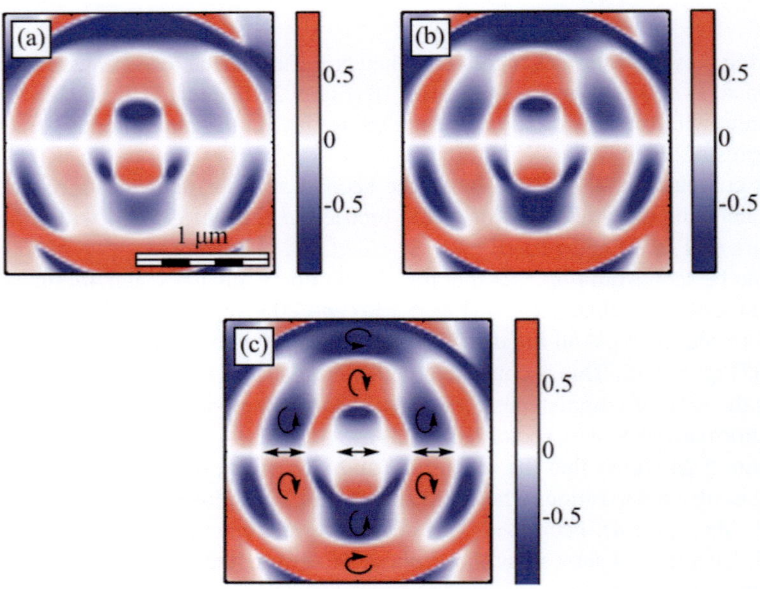

Fig. 5.24 The third component of the normalized Stokes vector (5.51) S_3 of the focused vector field (5.36) with $m = 3$ and different parameters a: 1/2 (**a**), 1 (**b**), and 3/2 (**c**)

such regions is related to the field number m. The number of such regions is $2m$ on some circle centered on the optical axis. It should be noted that the regions with local elliptical polarization appear in the focus only for an odd number m and for a non-zero parameter a. If the parameter $a = 0$, then the field (1) reduces to the well-known cylindrical vector field of order m, which has for any m only a local linear polarization in the focus, and regions with an elliptical polarization are absent.

The Distribution of the Poynting Vector Projections in the Focal Plane

This subsection presents the calculation of the energy flux vector (the Poynting vector) (5.45) in the tight focus of the field (5.36) carried out by the Richards-Wolf formulas (5.56)–(5.60).

Figure 5.25 shows the projections of the Poynting vector in the focus of the field (5.36) with the even number $m = 2$ and $a = 1$. According to the obtained for an even number m formulas (5.46)–(5.48), the transverse projections of the energy flux vector P_x and P_y are equal to zero (Fig. 5.25b, c), and the longitudinal component P_z does not have radial symmetry (Fig. 5.25a). It can be seen from Fig. 5.25a and formula (5.48) that the Poynting vector longitudinal component has a local maximum on the optical axis and two local maxima (side lobes) are located on the vertical axis at $\varphi = \pi/2$ and $\varphi = 3\pi/2$, since the function P_z (5.48) for $m = 2$ depends on the angle as $\cos(2\varphi)$. The calculation parameters in Figs. 5.25 and 5.26 are the same as in all previous figures. Figure 5.25a also shows that the longitudinal component of the Poynting vector outwardly coincides with the intensity distribution in Fig. 5.15a ($m = 2$). This is explained by the fact that the expression for the intensity (5.44) at $m = 2$, as well as (5.48), depends on the angle as $\cos(2\varphi)$.

Figure 5.26 shows the Poynting vector projections for the odd number $m = 3$ and $a = 1$ in the focus. It can be seen Fig. 5.26a that the longitudinal component P_z is radially symmetric and has a maximum value on the optical axis. This is consistent with Eq. (5.48). It can be seen from Fig. 5.26b, c that the transverse energy flow rotates in 8 local subwavelength regions: counterclockwise in 4 regions and clockwise in other 4 regions. Both transverse projections of the energy flux P_x and P_y change sign 8 times when going around the optical axis along a circle of some radius. This is consistent with formulas (5.46) and (5.47), since the dependence on the angle in these formulas is determined by the function $\cos((m + 1)\varphi)$ at $m = 3$. But local areas with a transverse vortex flow will be $2m$. This follows from Eq. (5.50), in which there are factors in the form $\cos(m\varphi)$ and $\sin(m\varphi)$. The number of local regions with a vortex energy flow must be equal to the number of regions with circular polarization, that is, $2m$. This follows from the effect of spin–orbit interaction. In Fig. 5.26d, their number is 6, and they lie along a circle of some radius. Their size is about 200 nm. Integrating in (5.50) the angular P_φ and the radial P_r projections of the Poynting vector over the angle φ we obtain that the total transverse energy flux is equal to zero in the focus.

Comparing Figs. 5.22c and 5.26d shows that there are 6 regions lying on a circle of a certain radius centered on the optical axis, in which the polarization is circular, and also 6 regions in which the energy flow rotates around the ring.

In the section, theoretically and numerically, using the Richards-Wolf approach, all six projections of electric and magnetic field strength vectors in the sharp focus

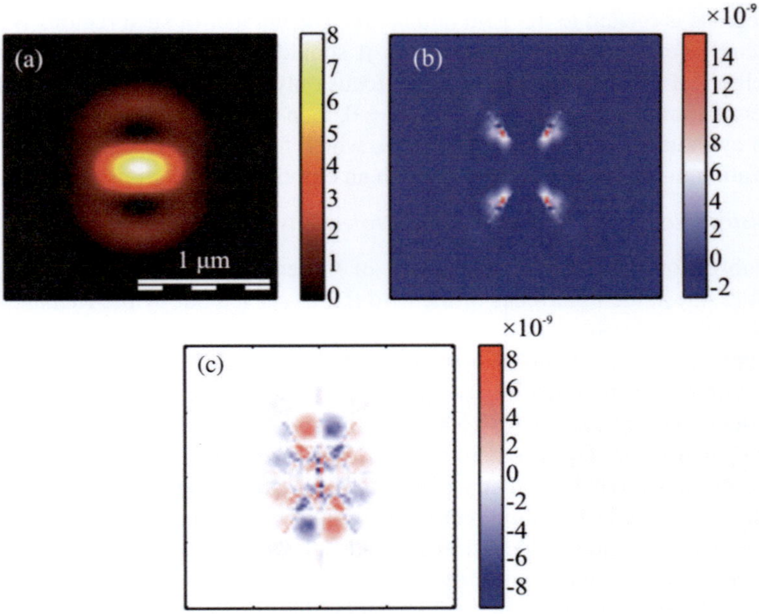

Fig. 5.25 The Poynting vector components in focus for $m = 2$ and $a = 1$: P_z (**a**), P_x (**b**), P_y (**c**)

of a superposition of a cylindrical vector field with order m and a uniform field with linear polarization are calculated [63]. Energy fluxes (projections of the Poynting vector), intensity distributions, and Stokes components are calculated in the focus plane. The study shows that local regions of about 200–250 nm in size (wavelength is 532 nm, NA = 0.95) with the an elliptical or circular polarization are formed in the focal plane of such an incident field. The number of such areas is related to the field number m. The number of such regions on some circle with center on the optical axis is $2m$ in focus. It should be noted that regions with local elliptical polarization appear in the focus only for an odd number m and for a non-zero parameter a. If the parameter a of the initial field is zero then the field reduces to the well-known cylindrical vector field of order m, which for any m has only a local linear polarization in the focus. And regions with an elliptical polarization are absent in this case. In addition to the presence of a local SAM for case of an odd number m, there are local subwavelength regions in the focus where the transverse energy flux propagates along a closed contour. The number of such regions lying on a circle of some radius is $2m$. Moreover, the energy flow in neighboring areas rotates in different directions (clockwise and counterclockwise). The total flow of the transverse energy flow is zero. The total transverse energy flow is zero. These beams can be used to create a micromachine in which two microparticles in the form of gears are captured in the focus of the beam into neighboring local areas in which the energy flow rotates in different directions, and therefore, these gears will also rotate in different directions.

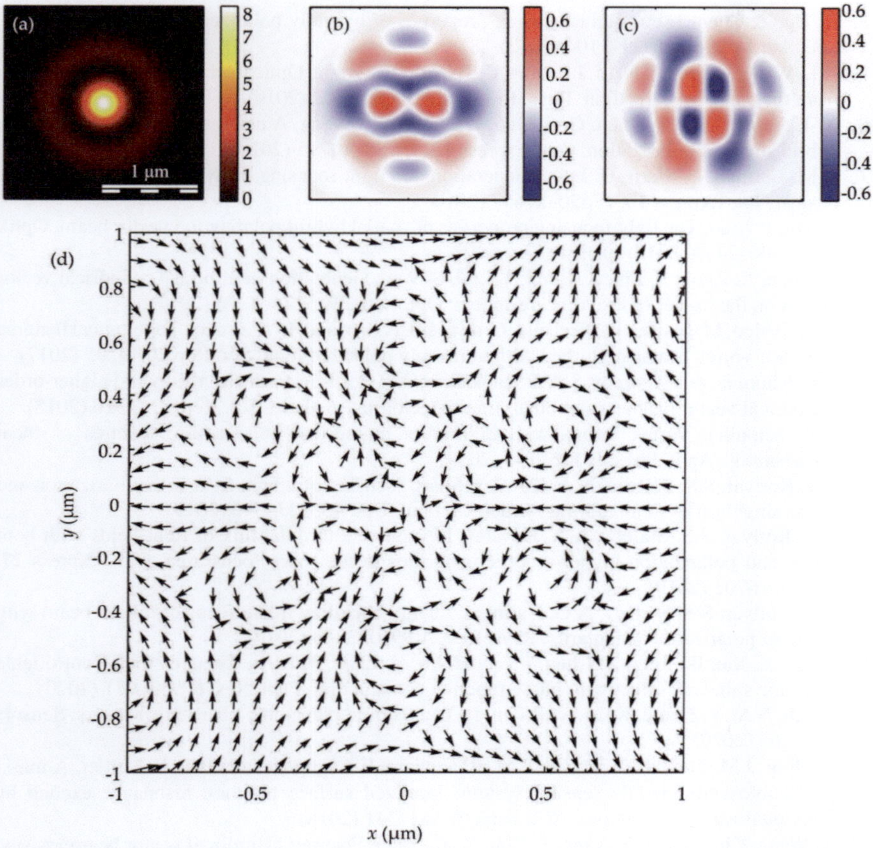

Fig. 5.26 The Poynting vector components in focus for $m = 3$ and $a = 1$: P_z (**a**), P_x (**b**), P_y (**c**). The arrows show the direction of the transverse Poynting vector at the focus (**d**)

References

1. B. Richards, E. Wolf, Electromagnetic diffraction in optical systems. II. Structure of the image field in an aplanatic system. Proc. R. Soc. Lond. A **253**, 358–379 (1959)
2. R. Dorn, S. Quabis, G. Leuchs, Sharper focus for a radially polarized light beam. Phys. Rev. Lett. **91**, 233901 (2003)
3. Y. Zhao, J.S. Edgar, G.D.M. Jeffries, D. McGloin, D.T. Chiu, Spin-to-orbital angular momentum conversion in a strongly focused optical beam. Phys. Rev. Lett. **99**, 073901 (2007)
4. V.V. Koltyar, A.G. Nalimov, S.S. Stafeev, Exploiting the circular polarization of light to obtain a spiral energy flow at the subwavelength focus. J. Opt. Soc. Am. B **36**, 2850–2855 (2019)
5. V.V. Kotlyar, A.A. Kovalev, A.G. Nalimov, Energy density and energy flux in the focus of an optical vortex: reverse flux of light energy. Opt. Lett. **43**, 2921–2924 (2018)
6. B. Chen, J. Pu, Tight focusing of elliptically polarized vortex beams. Appl. Opt. **48**, 1288–1294 (2009)
7. Q. Zhan, Cylindrical vector beams: from mathematical concepts to applications. Adv. Opt. Photon. **1**, 1–57 (2009)

8. K. Hu, Z. Chen, J. Pu, Tight focusing properties of hybridly polarized vector beams. J. Opt. Soc. Am. A **29**(6), 1099–1104 (2012)
9. X.L. Wang, J. Chen, Y.N. Li, J.P. Ding, C.S. Gou, H.T. Wang, Optical orbital angular momentum from the curl of polarization. Phys. Rev. Lett. **105**, 253602 (2010)
10. X.-L. Wang, Y. Li, J. Chen, C.-S. Guo, J. Ding, H.-T. Wang, A new type of vector fields with hybrid states of polarization. Opt. Express **18**, 10786–10795 (2010)
11. G.M. Lerman, L. Stern, U. Levy, Generation and tight focusing of hybridly polarized vector beams. Opt. Express **18**, 27650–27657 (2010)
12. H. Hu, P. Xiao, The tight focusing properties of spatial hybrid polarization vector beam. Optik **124**, 2406–2410 (2013)
13. S. Chen, X. Zhou, Y. Liu, X. Ling, H. Luo, S. Wen, Generation of arbitrary cylindrical vector beams on the higher order Poincare sphere. Opt. Lett. **39**, 5274–5276 (2014)
14. A. D'Errico, M. Maffei, B. Piccirillo, C. de Lisio, F. Cardano, L. Marrucci, Topological features of vector vortex beams perturbed with uniformly polarized light. Sci. Rep. **7**, 40195 (2017)
15. S.N. Khonina, A.V. Ustinov, S.A. Fomchenkov, A.P. Porfirev, Formation of hybrid higher-order cylindrical vector beams using binary multi-sector phase plates. Sci. Rep. **8**, 14320 (2018)
16. S.N. Khonina, Vortex beams with high-order cylindrical polarization: features of focal distributions. Appl. Phys. B **125**, 100 (2019)
17. V.V. Kotlyar, S.S. Stafeev, A.A. Kovalev, Sharp focusing of a light field with polarization and phase singularities of an arbitrary order. Comput. Opt. **43**, 337–346 (2019)
18. V.V. Kotlyar, S.S. Stafeev, A.A. Kovalev, Reverse and toroidal flux of light fields with both phase and polarization higher-order singularities in the sharp focus area. Opt. Express **27**, 16689–16702 (2019)
19. V.V. Kotlyar, S.S. Stafeev, A.G. Nalimov, Energy backflow in the focus of a light beam with phase or polarization singularity. Phys. Rev. A **99**, 033840 (2019)
20. Y. He, Z. Xie, B. Yang, X. Chen, J. Liu, H. Ye, X. Zhou, Y. Li, S. Chen, D. Fan, Controllable photonic spin Hall effect with phase function construction. Phot. Res. **8**, 963–971 (2020)
21. M. Qi, N.M.Y. Zhang, K. Li, S.C. Tjin, L. Wei, Hybrid plasmonic fiber-optic sensors. Sensors **20**, 3266 (2020)
22. Z. Ding, J.M. Stubbs, D. McRae, J.M. Blacquiere, F. Lagugné-Labarthet, S. Mittler, A mass-producible and versatile sensing system: localized surface plasmon resonance excited by individual waveguide modes. ACS Sens. **3**, 334–341 (2018)
23. R. Wang, Ch., Zhang, Y. Yang, S. Zhu, X.-C. Yuan, Focused cylindrical vector beam assisted microscopic pSPR biosensor with an ultra wide dynamic range. Opt. Lett. **37**, 2091–2093 (2012)
24. L. Sun, Yu., Zhang, Y. Wang, Y. Yang, Ch., Zhang, X. Weng, S. Zhu, X. Yuan, Real-time subcellular imaging based on graphene biosensors. Nanoscale **10**(4), 1759–1765 (2018)
25. F. Lu, T. Huang, L. Han, H. Su, H. Wang, M. Liu, W. Zhang, X. Wang, T. Mei, Tip-enhanced Raman spectroscopy with high-order fiber vector beam excitation. Sensors **18**(11), 3841 (2018)
26. H. Wu, J. Qiu, P. Wang, Y. Shang, L. Zhang, H. Wei, F. Pang, Cylindrical vector beam for vector magnetic field sensing based on magnetic fluid. IEEE Photonic. Tech. L. **33**(14), 703–706 (2021)
27. V.V. Kotlyar, S.S. Stafeev, E.S. Kozlova, A.G. Nalimov, Spin-orbital conversion of a strongly focused light wave with high-order cylindrical-circular polarization. Sensors **21**(19), 6424 (2021)
28. J.F. Nye, Polarization effects in the diffraction of electromagnetic waves: the role of disclinations. Proc. R. Soc. Lond. A **387**, 105–132 (1983)
29. J.F. Nye, M.V. Berry, Dislocations in wave trains. Proc. R. Soc. Lond. A **336**, 165–190 (1974)
30. I. Freund, Polarization singularity indices in Gaussian laser beams. Opt. Commun. **201**, 251–270 (2002)
31. X.-L. Wang, J. Ding, W.-J. Ni, C.-S. Guo, H.-T. Wang, Generation of arbitrary vector beams with a spatial light modulator and a common path interferometric arrangement. Opt. Lett. **32**, 3549–3551 (2007)

32. D. Naidoo, F.S. Roux, A. Dudley, I. Litvin, B. Piccirillo, L. Marrucci, Controlled generation of higher-order Poincaré sphere beams from a laser. Nat. Photonics **10**, 327–333 (2016)
33. L. Marrucci, C. Manzo, D. Paparo, Optical spin-to-orbital angular momentum conversion in inhomogeneous anisotropic media. Phys. Rev. Lett. **96**, 163905 (2006)
34. V.V. Kotlyar, A.G. Nalimov, S.S. Stafeev, L. O'Faolain, Subwavelength grating-based spiral metalens for tight focusing of laser light. Appl. Phys. Lett. **114**, 141107 (2019)
35. Z. Bomzon, G. Biener, V. Kleiner, E. Hasman, Radially and azimuthally polarized beams generated by space-variant dielectric subwavelength gratings. Opt. Lett. **27**(5), 285–287 (2002)
36. Z. Ren, Z. Chen, X. Wang, J. Ding, H. Wang, Polarization interferometric prism: a versatile tool for generation of vector fields, measurement of topological charges, and implementation of a spin-orbit controlled-Not gate. Appl. Phys. Lett. **118**, 011105 (2021)
37. V. Kumar, N.K. Viswanathan, Topological structures in the Poynting vector field: an experimental realization. Opt. Lett. **38**(19), 3886–3889 (2013)
38. M.V. Berry, Optical vortices evolving from helicoidal integer and fractional phase steps. J. Opt. A Pure Appl. Opt. **6**, 259–268 (2004)
39. G. Arora, S. Deepa, S.N. Khan, P. Senthilkumaran, Detection of degenerate Stokes index states. Sci. Rep. **10**, 20759 (2020)
40. M. Born, E. Wolf, The basis of optics. M. Sci. 720 (1973)
41. G. Arora, Ruchi, and P. Senthilkumaran, Hybrid order Poincare spheres for Stoks singularities, Opt. Lett., **45**, 5136 (2020)
42. A.P. Prudnikov, Y.A. Brychkov, O.I. Marichev, Integrals and series, in *Special Functions.* Gordon and Breach, New York (1986), vol. 2.
43. V.V. Kotlyar, A.G. Nalimov, S.S. Stafeev, Inversion of the axial projection of the spin angular momentum in the region of the backward energy flow in sharp focus. Opt. Express **28**(23), 33830–33839 (2020)
44. A.A. Kovalev, V.V. Kotlyar, Optical vortex beams with the infinite topological charge. J. Opt. **23**(5), 055601 (2021)
45. V.V. Kotlyar, A.G. Nalimov, A.A. Kovalev, A.P. Porfirev, S.S. Stafeev, Spin-orbit and orbit-spin conversion in the sharp focus of laser light: theory and experiment. Phys. Rev. A **102**, 033502 (2020)
46. T. Bauer, M. Neugebauer, G. Leuchs, P. Banzer, Optical polarization Möbius strips and points of purely transverse spin density. Phys. Rev. Lett. **117**, 013601 (2016)
47. T. Bauer, P. Banser, E. Karimi, S. Orlov, A. Rubano, L. Marrucci, E. Santamato, R.W. Boyd, G. Leuchs, Observation of optical polarization Möbius strips. Science **347**(6225), 964–966 (2015)
48. V.V. Kotlyar, S.S. Stafeev, A.G. Nalimov, Sharp focusing of a hybrid vector beam with a polarization singularity. Photonics **8**, 227 (2021)
49. Q. Zhan, J.R. Leger, Focus shaping using cylindrical vector beams. Opt. Express **10**, 324–331 (2002)
50. G. Machavariani, Y. Lumer, I. Moshe, A. Meir, S. Jackel, Efficient extracavity generation of radially and azimuthally polarized beams. Opt. Lett. **32**, 1468–1470 (2007)
51. Z. Liu, Y. Liu, Y. Ke, Y. Liu, W. Shu, H. Luo, S. Wen, Generation of arbitrary vector vortex beams on hybrid-order Poincare sphere. Photon. Res. **5**, 15–21 (2017)
52. J. Liu, X. Chen, Y. He, L. Lu, H. Ye, G. Chai, S. Chen, D. Fan, Generation of arbitrary cylindrical vector vortex beams with cross-polarized modulation. Res. Phys. **19**, 103455 (2020)
53. S. Yan, B. Yao, Radiation forces of a highly focused radially polarized beam on spherical particles. Phys. Rev. A **76**, 053836 (2007)
54. R. Chen, K. Agarwal, C.J. Sheppard, X. Chen, Imaging using cylindrical vector beams in a high-numerical-aperture microscopy system. Opt. Lett. **38**, 3111–3114 (2013)
55. R. Fickler, R. Lapkiewicz, S. Ramelow, A. Zeilinger, Quantum entanglement of complex photon polarization patterns in vector beams. Phys. Rev. A **89**, 4172–4183 (2014)
56. A. Hollezek, A. Aiello, C. Gabriel, C. Morquargt, G. Leuchs, Classical and quantum properties of cylindrically polarized states of light. Opt. Express **19**, 9714–9736 (2011)

57. S.S. Stafeev, A.G. Nalimov, V.D. Zaitsev, V.V. Kotlyar, Tight focusing cylindrical vector beams with fractional order. J. Opt. Soc. Am. B **38**, 1090–1096 (2021)
58. V.V. Kotlyar, A.A. Kovalev, V.D. Zaitsev, Topological charge of light fields with a polarization singularity. Photonics **9**, 298 (2022)
59. K.Y. Bliokh, E.A. Ostrovskaya, M.A. Alonso, O.G. Rodriguez-Herrera, D. Lara, C. Dainty, Spin-to-orbital angular momentum conversion in focusing, scattering, and imaging systems. Opt. Express **19**, 26132–26149 (2011)
60. M. Born, E. Wolf, *Principles of Optics* (Nauka, Moscow, 1973), 720 pp
61. S.F. Pereira, A.S. van de Nes, Superresolution by means of polarisation, phase and amplitude pupil masks. Opt. Commun. **234**, 119–124 (2004)
62. S.N. Khonina, S.G. Volotovsky, Control by contribution of components of vector electric field in focus of a high-aperture lens by means of binary phase structures. Comput. Opt. **34**, 58–68 (2010)
63. V. Kotlyar, S. Stafeev, V. Zaitsev, E. Kozlova, Spin-orbital conversion with the tight focus of an axial superposition of a high- order cylindrical vector beam and a beam with linear polarization. Micromachines **13**, 1112 (2022)

Chapter 6
Poincare Beams at the Tight Focus

6.1 Poincare Beams at the Tight Focus: Inseparability, Radial Spin Halls Effect, and Reverse Energy Flow

Among other recipients of the 2022 Nobel Prize in physics, A. Zeilinger was recognized for research in quantum entanglement between photons. By way of illustration, his work [1] has amassed 6,000 + citations. As far as photons are concerned, (hybrid) quantum entanglement (in terms of both polarization and angular momentum) implies that two photons are in superposition, which can be described by the expression [2]:

$$|\psi\rangle = a|H\rangle|l\rangle + e^{i\delta}\sqrt{1-a^2}|V\rangle|-l\rangle,$$

where H and V stand for the horizontal and vertical polarization vectors, l is the orbital angular momentum (OAM) of photon or topological charge, a is the weight coefficient (probability density of detecting the photon state), and δ is the phase delay. At $a = 0.5$, the state $|\psi\rangle$ gets fully entangled. It turns out that similar entanglement may be found in classical physics and refers to a light field described by on-axis superposition of two vortex laser beams with different polarization states and different topological charges. Physically, a classical entangled state is when the light field is impossible to describe as a direct product $|P\rangle|OAM\rangle$, with no way to simultaneously determine a particular polarization state $|P\rangle$ and a particular OAM state $|OAM\rangle$ of the light field of interest. Hence, the light field described by the aforementioned superposition is ill-defined. In this work, we discuss an example of ill-defined light fields in the form of a wide class Poincare beams [3, 4], which include linearly and circularly polarized vortex laser beams and cylindrical vector beams [5]. Below, we show that given specific parameters, remarkable optical effects occur at the tight focus of such beams, including a reverse on-axis energy flow [6, 7] and an optical Hall effect [8]. The optical Hall effect is divided into spin (SHE) [9–13], orbital [14], and spin-orbital [15] effects. An optical Hall effect occurs due to conservation of angular momentum of light and spin–orbit interaction. As a rule, the optical Hall

V. V. Kotlyar et al., *Optical Hall Effect in the Sharp Focus of Laser Light*,
https://doi.org/10.1007/978-3-031-64683-6_6

effect occurs because of the light–medium interaction, for instance when light reflects at a metasurface [16]. What we are interested in more is studying the SHE that occurs without the involvement of medium and microparticles but by tightly focusing laser beams propagating in free space (see the review [17]). For instance, an off-axis focal spot has been shown to be generated by passing a tightly focused vortex laser beam through an aperture [18]. In this work, analyzing free-space propagation of Poincare beams, we show that given specific parameters, there is a transverse energy flow at the tight focus, rotating about the optical axis clockwise or anticlockwise. We also reveal that at the tight focus, a radial spin Hall effect occurs, when on-axis SAM projections (that may be looked at as 'spin') have different signs on different-radius circles centered at the optical axis. That is, on the different-radius circles, vectors of elliptic polarization rotate oppositely.

6.1.1 Inseparability of Vector and Spatial Degrees of Freedom

In a previous work [19], we derived expressions to describe an intensity pattern from Poincare beams at the tight focus. In this work, our purpose is to derive expressions for Poynting vector projections and an on-axis projection of the spin angular momentum (SAM) at the tight focus of the Poincare beam. But beforehand it makes sense to discuss a topic of inseparability (entanglement) of polarization and spatial degrees of freedom of the Poincare beam.

The light beams whose polarization state is described by unit vectors on a Poincare sphere are described by a Jones vector given by [3, 4]:

$$\mathbf{E}_P(\varphi) = \frac{1}{\sqrt{2}} \begin{pmatrix} ae^{-in\varphi} + be^{in\varphi} \\ iae^{-in\varphi} - ibe^{in\varphi} \end{pmatrix}, \tag{6.1}$$

where $a = \cos(\theta/2)e^{-i\psi/2}$, $b = \sin(\theta/2)e^{i\psi/2}$, $a^2 + b^2 = 1$, θ and ϕ are the polar and azimuthal angles on the sphere, and n is an integer number that describes the topological charge of the optical vortex or the order of cylindrical (azimuthal or radial) polarization. Beam (6.1) can be represented as superposition of two beams with one having right-handed and the other left-handed polarization:

$$\frac{1}{\sqrt{2}} \begin{pmatrix} ae^{-in\varphi} + be^{in\varphi} \\ iae^{-in\varphi} - ibe^{in\varphi} \end{pmatrix} = \frac{ae^{-in\varphi}}{\sqrt{2}} \begin{pmatrix} 1 \\ i \end{pmatrix} + \frac{be^{in\varphi}}{\sqrt{2}} \begin{pmatrix} 1 \\ -i \end{pmatrix} = ae^{-in\varphi}|R\rangle + be^{in\varphi}|L\rangle, \tag{6.2}$$

where $|R\rangle$ and $|L\rangle$, respectively, denote Jones vectors for the right- and left-handed polarization. Considering that $a^2 + b^2 = 1$, Eq. (6.2) can be rearranged as follows:

$$E_p(\varphi) = \sqrt{\gamma}e^{-in\varphi}|R\rangle + \sqrt{1-\gamma}e^{in\varphi}|L\rangle, \ \gamma = a^2. \tag{6.3}$$

The two quantities are said to be inseparable [20–22] when it not possible to represent the vector field as the direct product of the spatial scalar function by a Jones vector that describes the polarization state of the field. In other words, the vector field of inseparable quantities is not possible to represent as $U(r, \varphi, z)|P\rangle$, where $U(r, \varphi, z)$ is a scalar function that describes the solution of a paraxial Helmholtz equation and $|P\rangle$ is the state of transverse polarization of a paraxial light beam.

The entanglement notion is a classical analog of quantum entanglement [1, 2, 20–22]. The degree of inseparability is defined based on Neumann's entropy *Ent*, which in a simple case of Eq. (6.3) is given by the following:

$$\mathrm{Ent}(E_p) = -\left[\gamma \log_2(\gamma) + (1-\gamma)\log_2(1-\gamma)\right]. \tag{6.4}$$

Entropy in Eq. (6.4) lies in the range 0 < Ent(Ep) < 1. If entropy is zero, the vector field is fully separable but if entropy equals unity, the vector field is fully (maximally) inseparable. At $\gamma = 1/2$, the Poincare beam is maximally inseparable. In this case, $a = b = 1/\sqrt{2}$, which means that the Poincare beam with maximal inseparability is identical to an nth order cylindrical vector beam:

$$\mathbf{E}_P(\varphi; a = b = 1/\sqrt{2}) = \begin{pmatrix} \cos(n\varphi) \\ \sin(n\varphi) \end{pmatrix}. \tag{6.5}$$

At other values of a and b, the Poincare beams in Eq. (6.1) are partially inseparable, and if one of the parameters equals zero (either $a = 0$ or $b = 0$), the Poincare beams become separable circularly polarized vortex beams. Hereinafter, we analyze what happens with beam (6.1) at the tight focus.

6.1.2 Flow Energy at the Tight Focus of Poincare Beam

Using Richards-Wolf formalism [23], projections of the electric and magnetic fields at the tight focus of Poincare beam (6.1) can be derived in the form:

$$\begin{aligned}
E_x &= \frac{i^{n-1}}{\sqrt{2}}\left[\left(be^{in\varphi} + ae^{-in\varphi}\right)I_{0,n} + \left(be^{i(n-2)\varphi} + ae^{-i(n-2)\varphi}\right)I_{2,n-2}\right], \\
E_y &= \frac{i^{n}}{\sqrt{2}}\left[\left(ae^{-in\varphi} - be^{in\varphi}\right)I_{0,n} - \left(ae^{-i(n-2)\varphi} - be^{i(n-2)\varphi}\right)I_{2,n-2}\right], \\
E_z &= \sqrt{2}i^{n}\left(be^{i(n-1)\varphi} + ae^{-i(n-1)\varphi}\right)I_{1,n-1}, \\
H_x &= \frac{i^{n}}{\sqrt{2}}\left[\left(be^{in\varphi} - ae^{-in\varphi}\right)I_{0,n} + \left(be^{i(n-2)\varphi} - ae^{-i(n-2)\varphi}\right)I_{2,n-2}\right], \\
H_y &= \frac{i^{n-1}}{\sqrt{2}}\left[\left(be^{in\varphi} + ae^{-in\varphi}\right)I_{0,n} - \left(be^{i(n-2)\varphi} + ae^{-i(n-2)\varphi}\right)I_{2,n-2}\right],
\end{aligned}$$

$$H_z = \sqrt{2} i^{n+1} \left(b e^{i(n-1)\varphi} - a e^{-i(n-1)\varphi} \right) I_{1,n-1}. \tag{6.6}$$

In Eq. (6.6), the notation $I_{\nu,\mu}$ stands for integrals that depend just on the radial variable r:

$$I_{\nu,\mu} = \left(\frac{4\pi f}{\lambda} \right) \int_0^{\theta_0} \sin^{\nu+1}\left(\frac{\theta}{2} \right) \cos^{3-\nu}\left(\frac{\theta}{2} \right) \cos^{1/2}(\theta) A(\theta) e^{ikz\cos\theta} J_\mu(x) d\theta, \tag{6.7}$$

where f is the focal length of an aplanatic optical system (an ideal spherical lens), λ is the wavelength, $NA = \sin\theta_0$ is the numerical aperture, $J_\mu(x)$ is the Bessel function of the first kind and μth order, $x = kr \sin\theta$ and (r, φ, z) are the cylindrical coordinates. The function $A(\theta)$ can be in the form of Gaussian, Bessel-Gaussian function or a constant (plane wave).

Using Eq. (6.6), we can write a relationship for the intensity distribution across the focal spot:

$$\begin{aligned} I(r,\varphi) &= |E_x|^2 + |E_y|^2 + |E_z|^2 \\ &= I_{0,n}^2 + I_{2,n-2}^2 + 2I_{1,n-1}^2 + 2\sin\theta\cos(2(n-1)\varphi + \psi)\left(I_{0,n} I_{2,n-2} + 2I_{1,n-1}^2 \right). \end{aligned} \tag{6.8}$$

From Eq. (6.8), it is seen that at $\sin\theta = 0$, the intensity pattern at the focus is radially symmetric and independent of the polar angle φ. Specifically, at $n = 0, 1, 2$, the focal spot forms a circle, but at $n > 2$ or $n < 0$, the focal spot forms a ring. At $\sin\theta \neq 0$, the focal spot is devoid of radial symmetry, depending on the polar angle φ and index n in Eq. (6.8).

By using relationships for the field projections in Eq. (6.8), we obtain Poynting vector projections (energy flow density) in the form [23]:

$$\mathbf{P} = \frac{c}{2\pi} \mathrm{Re}\left(\mathbf{E}^* \times \mathbf{H} \right), \tag{6.9}$$

where **E** and **H** are the electric and magnetic field vectors, * denotes complex conjugation, $\times$ stands for the vector product, and c is the speed of light in vacuum. Hereinafter, the constant $c/(2\pi)$ is neglected. Substituting the relationships for electromagnetic field projections at the focus in Eq. (6.6) into (6.9) yields:

$$\begin{aligned} P_x &= 2\left(|a|^2 - |b|^2 \right) \sin(\varphi) I_{1,n-1} \left(I_{0,n} + I_{2,n-2} \right), \\ P_y &= -2\left(|a|^2 - |b|^2 \right) \cos(\varphi) I_{1,n-1} \left(I_{0,n} + I_{2,n-2} \right), \\ P_z &= I_{0,n}^2 - I_{2,n-2}^2. \end{aligned} \tag{6.10}$$

In view of the relationships for the parameters a and b from Eq. (6.1), the Poynting vector projections of Eq. (6.10) at the focus can be rewritten as follows:

$$P_x = 2\cos(\theta)\sin(\varphi)I_{1,n-1}\left(I_{0,n} + I_{2,n-2}\right),$$
$$P_y = -2\cos(\theta)\cos(\varphi)I_{1,n-1}\left(I_{0,n} + I_{2,n-2}\right),$$
$$P_z = I_{0,n}^2 - I_{2,n-2}^2. \tag{6.11}$$

In Eq. (6.11), the angle θ is a parameter that defines a point on the Poincare sphere of polarization states. The transverse Poynting vector projections in Eq. (6.11) can be converted from the Cartesian to the polar coordinates, with Eq. (6.11) transformed into:

$$P_r = 0,$$
$$P_\varphi = -\cos(\theta)I_{1,n-1}\left(I_{0,n} + I_{2,n-2}\right),$$
$$P_z = I_{0,n}^2 - I_{2,n-2}^2. \tag{6.12}$$

An analysis of Eq. (6.12) allows the following conclusions. With the on-axis Poynting vector component depending only on the radial coordinate, the distribution of the energy flow propagating along the optical axis turns out to be circularly symmetric for any parameters (a, b) or (θ, ψ). Interestingly, at $n = 2$, it follows from Eq. (6.12) that near the optical axis in the focal plane a reverse energy flow occurs:

$$P_{z,2}(r) = -I_{2,0}^2 + I_{0,2}^2,$$
$$P_{z,2}(r = 0) = -I_{2,0}^2 < 0. \tag{6.13}$$

The second equality in Eq. (6.13) stems from the fact that among the integrals entering Eq. (6.7), only those with zero second index ($\mu = 0$) remain nonzero on the optical axis ($r = 0$). This is because only the zero-order Bessel function is nonzero at the origin: $J_0(0) = 1$. Equation (6.13) suggests that an on-axis reverse flow will occur when tightly focusing a multitude of initial optical fields, including a second-order optical vortex with left-handed circular polarization, $\exp(i2\varphi)(1, -i)^T$, where T denotes vector transposition, a minus second-order optical vortex with right-handed circular polarization $\exp(-i2\varphi)(1, i)^T$, a second-order radially polarized vector field $(\cos(2\varphi), \sin(2\varphi))^T$, a second-order azimuthally polarized optical field, $(-\sin(2\varphi), \cos(2\varphi))^T$ and so on. Previously, the reverse energy flow at the tight focus was discussed in Refs. [6, 7].

From Eq. (6.12), it also follows that at $\cos\theta \neq 0$ the near-axis transverse energy flow at the focus is rotating clockwise at $\cos\theta > 0$ and anticlockwise at $\cos\theta < 0$.

At $\cos\theta = 0$, the energy flow at the focal plane is laminar (the on-axis Poynting vector component is zero) and not rotating. Given $\cos\theta \neq 0$ (meaning that elliptic polarization is found in the source field), the transverse energy flow at the tight focus of Poincare beams will rotate around the optical axis at any topological charge n due to spin-orbital conversion. At $\cos\theta > 0$ and $n = 0$, the energy flow will rotate anticlockwise and at $n > 0$ – clockwise.

6.1.3 Spin Angular Momentum at the Tight Focus of Poincare Beams

In this subsection, we derive the spin angular momentum (SAM) of the Poincare beam (6.1) at the focus. The spin density vector is calculated using a familiar formula [24]:

$$\mathbf{S} = \frac{1}{8\pi\omega}\mathrm{Im}\big(\mathbf{E}^* \times \mathbf{E}\big), \tag{6.14}$$

where ω is the cyclic frequency of light. Hereinafter, the constant $1/(8\pi\omega)$ is neglected. From (6.14), the on-axis SAM component (without regard for the constant) is seen to coincide with the third Stokes vector component, s_3:

$$s_3 = S_z = 2\,\mathrm{Im}\big(E_x^* E_y\big), \tag{6.15}$$

Thus, we infer that it is the on-axis SAM component which indicates that the field at the focus is circularly or elliptically polarized. Substituting the electric field projections (6.6) into (6.15) yields:

$$S_z = \big(|a|^2 - |b|^2\big)\big[I_{0,n}^2 - I_{2,n-2}^2\big]. \tag{6.16}$$

For the source Poincare field (1), the on-axis SAM component is given by the following:

$$S_{z,in} = |a|^2 - |b|^2 = \cos\theta. \tag{6.17}$$

From Eqs. (6.16) and (6.17), the initial homogeneous spin density, defined by $\cos\theta$, is seen at the focus to become inhomogeneous, as well as changing sign and taking a zero value at certain radii. The radius-dependent separation of regions with the opposite sign of spin (6.16) may be interpreted as a radial spin Hall effect for all n. If in the initial plane, the spin density is zero ($\cos\theta = 0$), then at the focus the on-axis spin density component of Eq. (6.16) will also be zero. If the original spin in (6.17) is positive ($\cos\theta > 0$), i.e. the initial polarization vector rotates anticlockwise (Eq. (6.12), then, given $n = 0$, the transverse energy flow at the focus will also rotate anticlockwise thanks to the spin-orbital conversion.

6.1.4 Orbital Angular Momentum at the Tight Focus of Poincare Beams

Now, let us analyze the behavior of the angular momentum at the tight focus of a Poincare beam (6.1). The angular momentum is given by [25]:

$$\mathbf{J} = \frac{1}{2c}\mathrm{Re}\big(\mathbf{r} \times (\mathbf{E}^* \times \mathrm{H})\big) = \frac{2\pi}{c^2}(\mathbf{r} \times \mathbf{P}), \tag{6.18}$$

where $\mathbf{r}$ is the radius vector and c is the speed of light in vacuum. Dropping the constant $2\pi/c^2$ in Eq. (6.18) and considering Eq. (6.12), the on-axis projection of the angular momentum vector (6.18) takes the form:

$$J_z = rP_\varphi = -r\cos(\theta)I_{1,n-1}\big(I_{0,n} + I_{2,n-2}\big). \tag{6.19}$$

From Eq. (6.19), the angular momentum is seen to always be zero on the optical axis at $r = 0$, because the "lever" is zero. The angular momentum is the sum of SAM and orbital angular momentum (OAM) [25]:

$$\mathbf{J} = \mathbf{S} + \mathbf{L} = \frac{1}{8\pi\omega}\mathrm{Im}\big(\mathbf{E}^* \times \mathbf{E}\big) + \frac{1}{8\pi\omega}\sum_{p=x,y,z}\mathrm{Im}\Big(E_p^*(\mathbf{r} \times \nabla)E_p\Big). \tag{6.20}$$

Dropping the constant $1/(8\pi\omega)$ in Eq. (6.20) and taking into account Eq. (6.16) and (6.19), the relation for the on-axis component L_z of the OAM density takes the form:

$$L_z = -\cos(\theta)\big[rI_{1,n-1}\big(I_{0,n} + I_{2,n-2}\big) + I_{0,n}^2 - I_{2,n-2}^2\big]. \tag{6.21}$$

Equation (6.21) suggests that if the source Poincare beam has zero spin ($\cos\theta =$ 0), the OAM is zero at the focus ($L_z = 0$). If, however, the source Poincare beam has nonzero spin ($\cos\theta \neq 0$), both the angular momentum (6.19) and OAM (6.21) occur at the focus.

6.1.5 Numerical Modeling

The numerical simulation in this section is based on calculating the Debye integral using a Richards-Wolf method [23]. The focusing is assumed to be conducted with an aplanatic objective with NA = 0.95 for a wavelength of $\lambda = 0.532$ nm. Figure 6.1 depicts intensity patterns $I = |E_x|^2 + |E_y|^2 + |E_z|^2$ at the tight focus of a Poincare beam at the parameters $\theta = \pi/4$, $\psi = \pi/4$, $n = 0$ (non-vortex elliptically polarized beam, Fig. 6.1a); $\theta = 0$, $\psi = \pi/2$, $n = 0$ (non-vortex right-handed circularly polarized beam, Fig. 6.1b); $\theta = \pi/4$, $\psi = \pi/4$, $n = 1$ (elliptically polarized vortex, Fig. 6.1c); and $\theta = 0$, $\psi = 0$, $n =$ -1 (right-handed circularly polarized vortex, Fig. 6.1d).

From Fig. 6.1, beams with elliptical (a), circular (b), and near-radial (c) polarization are seen to produce an elliptic or circular focal spot, whereas a vortex beam with circular polarization (d) produces an annular focal spot. Such types of intensity patterns agree well with Eq. (6.8).

Figure 6.2 depicts transverse Poynting vector components P_x (*a*, *c*) and P_y (*b*, *d*) at the foci of two out of the four beams shown in Fig. 6.1. The energy flow

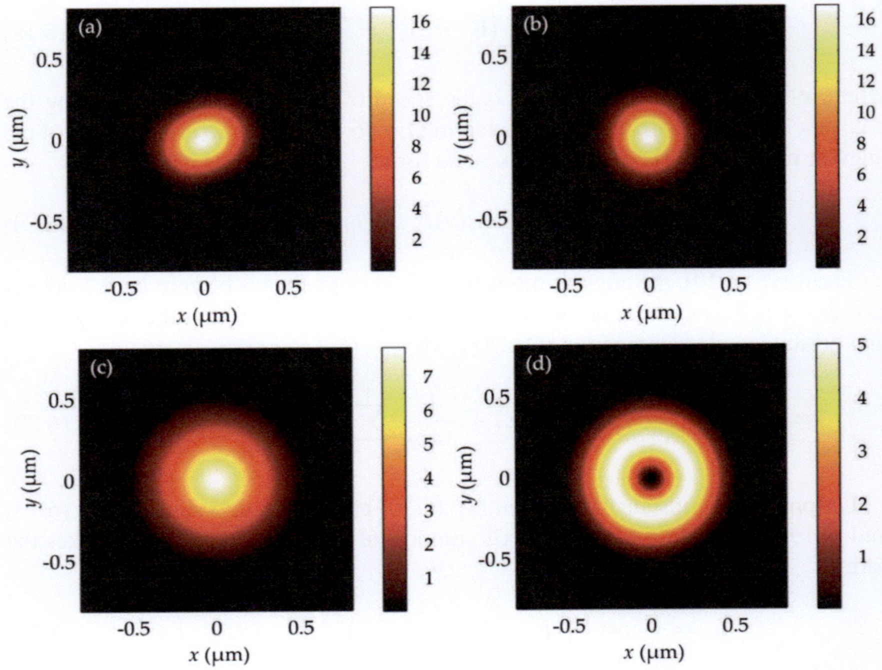

Fig. 6.1 Intensity patterns at the focus for different Poincare beams of Eq. (6.1): $\theta = \pi/4$, $\psi = \pi/4$, $n = 0$ (**a**); $\theta = 0$, $\psi = \pi/2$, $n = 0$ (**b**); $\theta = \pi/4$, $\psi = \pi/4$, $n = 1$ (**c**); $\theta = 0$, $\psi = 0$, $n = -1$ (**d**)

in Fig. 6.1a, b for the beam with the parameters $\theta = \pi/4$, $\psi = \pi/4$, $n = 0$ is identical to those for the beams with $\theta = \pi/4$, $\psi = \pi/4$, $n = 1$ and $\theta = 0$, $\psi = 0$, $n = -1$ (d). From Fig. 6.2a, b, the transverse energy flows at the focus are seen to be rotating anticlockwise for these three beams. Meanwhile, the transverse energy flow produced by the beam with $\pi/4$, $\psi = \pi/4$, $n = 1$ (Fig. 6.2c, d) will experience clockwise rotation. This may be concluded from Eq. (6.12), because at $n = 0, -1$ the angular projection of the near-axis energy flow ($kr < 1$) is positive: $P_\varphi = \cos(\theta) I_{1,1}\left(I_{0,0} + I_{2,2}\right) > 0$, $P_\varphi = \cos(\theta) I_{1,2}\left(I_{0,1} + I_{2,3}\right) > 0$, becoming negative at $n = 1$: $P_\varphi = -\cos(\theta) I_{1,0}\left(I_{0,1} - I_{2,1}\right) < 0$. The only situation when the energy flow does not rotate is at $\cos\theta = 0$.

Figure 6.3 depicts patterns of the on-axis SAM vector component (6.16) at the foci of two beams with the parameters: $\theta = \pi/4$, $\psi = \pi/4$, $n = 0$ (**a, b**); $\theta = \pi/4$, $\psi = \pi/4$, $n = 1$ (**c, d**). We note that the beam with the parameters $\theta = 0$, $\psi = \pi/2$, $n = 0$ produces a SAM pattern similar to that in Fig. 6.3a, with the beam at $\theta = 0$, $\psi = 0$, $n = -1$ producing a pattern of the on-axis SAM projection analogous to that in Fig. 6.3c. This conclusion follows from Eq. (6.16), because at $n = 0$, the on-axis SAM projection does not take zero values: $S_z(r = 0) = \cos\theta\left[I_{0,0}^2 - I_{2,2}^2\right] \neq 0$, meanwhile at $n = 1$, everywhere on the optical axis the SAM projection is zero: $S_z(r = 0) = \cos\theta\left[I_{0,1}^2 - I_{2,1}^2\right] = 0$, as is the case at $n =$ -1: $S_z(r = 0) = \cos\theta\left[I_{0,1}^2 - I_{2,3}^2\right] = 0$.

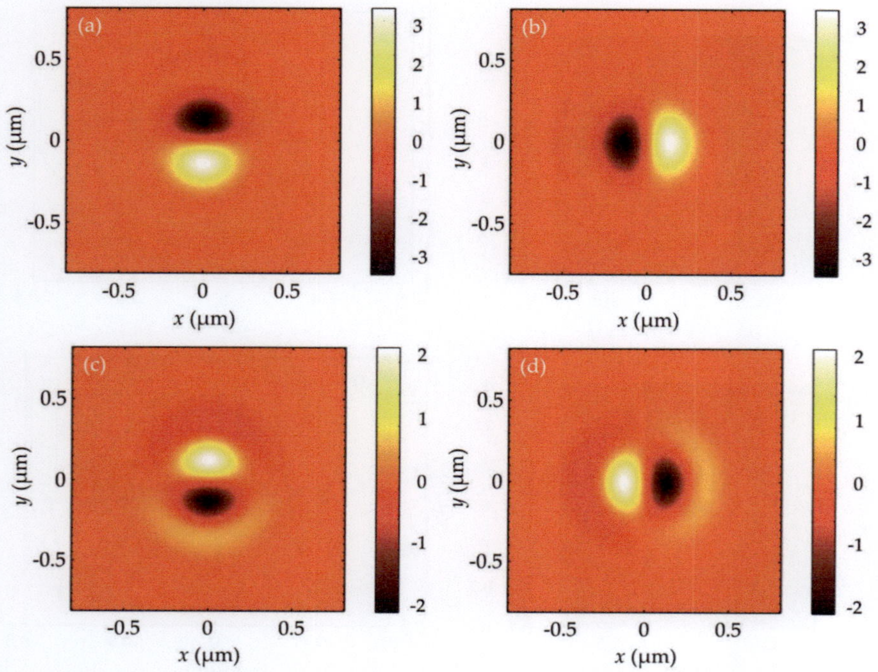

Fig. 6.2 Patterns of the transverse energy flow (Poynting vector projections) at the tight focus: P_x (**a**, **c**), P_y (**b**, **d**) for two different Poincare beams: $\theta = \pi/4$, $\psi = \pi/4$, $n = 0$ (**a**, **b**); $\theta = \pi/4$, $\psi = \pi/4$, $n = 1$ (**c**, **d**)

Figures 6.3b, d, f show sections along the x-axis of the axial projection of the SAM vector. It can be seen from Fig. 6.3b, d that at some distance from the optical axis, the SAM projection becomes negative (see enlarged fragments). In Fig. 6.3f, negative SAM values appear near the optical axis. Negative values of the longitudinal projection of the SAM take place in those places of the focus plane where the axial energy flow is negative, since it follows from (6.12) and (6.16) that

$$S_z = P_z \cos\theta. \tag{6.22}$$

In this section, we obtained the following new results. Using a relationship for von Neumann entropy it has been shown that the family of Poincare beams represents a classical entangled (non-separable) state of light [26]. The entanglement has been shown to become maximal when the Poincare beams are reduced to cylindrical nth order vector beams, Eq. (6.5). Classical entanglement, or non-separability, of a light beam implies that polarization and orbital degrees of freedom cannot be separated. However, at certain parameters, the Poincare beams are fully separable, in which case they are reduced to left- or right-handed circularly polarized vortex beams. The Poincare beams show remarkable properties at the tight focus. Namely, at $n = 2$, a

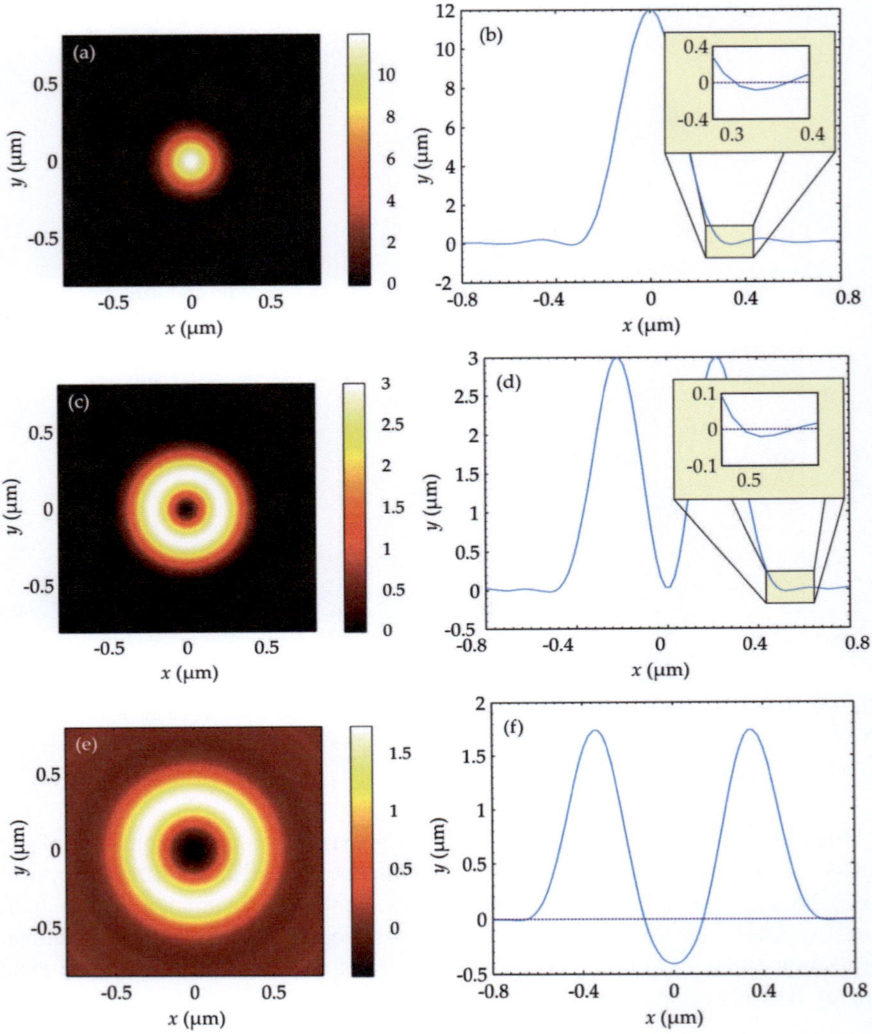

Fig. 6.3 Distributions of the axial projection of the SAM (Sz) and their cross sections along the x-axis for three different Poincaré beams: $\theta = \pi/4$, $\psi = \pi/4$, $n = 0$ (a, b); $\theta = \pi/4$, $\psi = \pi/4$, $n = 1$ (**c**, **d**); $\theta = \pi/4$, $\psi = \pi/4$, $n = 2$ (**e**, **f**)

near-axis reverse energy flow has been found to occur when the on-axis Poynting vector component is negative, Eq. (6.13). Given certain parameters of the Poincare beam ($\cos\theta \neq 0$), the transverse energy flow has been found to rotate around the optical axis at any topological charge n thanks to spin-orbital conversion, Eq. (6.12). More specifically, the energy flow rotates anticlockwise at $\cos\theta > 0$ and $n = 0$, rotating clockwise at $n > 0$. The condition $\cos\theta \neq 0$ means that in the source plane, the on-axis SAM component of Poincare beams is nonzero, Eq. (6.17). In this case, at the focus, a nonzero on-axis SAM component with alternating sign is also observed, Eq. (6.16) that can take zero values on certain radii. Radius-dependent separation of focal regions with alternating sign of spin density (6.16) can be interpreted as a radial spin Hall effect for all n. Finally, it has been revealed that thanks to spin-orbital conversion, a nonzero on-axis component of the OAM vector occurs at the focus of Poincare beams, Eq. (6.21).

6.2 Generalized Poincaré Beams in the Tight Focus

Poincaré beams, whose polarization state is related with the polarization Poincaré sphere [4, 27, 28], are actively studied in optics, starting with works [3, 28–30]. In a general case, a Poincaré beam is a superposition of two optical vortices with different topological charges (TC) p and q and with orthogonal polarizations. As the optical vortices, the conventional Laguerre-Gaussian beams of different indices [31–33] can be chosen, or diffraction-free Bessel beams, or the Bessel-Gaussian beams generated by axicons [34–36]. The Poincaré beams can be generated as all the other vector beams, by using liquid–crystal light modulators, half-wave and quarter-wave plates [34, 37, 38], or by using the lasers and q-plates [39], and metasurfaces [36]. In [19], polarization singularity index (Poincaré-Hopf index) of the Poincaré beams was studied. In [26], the optical Hall effect was theoretically discovered in the tight focus of the Poincaré beams. The optical (or photonic) Hall effect is divided into the spin [9, 18] and orbital [13, 14]. Usually, the Hall effect in optics is observed when a light field is reflected from an interface between the media [13, 14], or when it passes through multilayered media [40], crystals [16, 41], or through a metasurface [42]. There are known works investigating the Hall effect in the tight focus of a laser radiance [24, 43] or in vicinity of the focal plane [44]. We note that the above mentioned works do not contain theoretically obtained, with using the Richards-Wolf formalism [23], key characteristics of the generalized Poincaré beams in the tight focus: amplitudes of the electric and magnetic vectors, intensity distribution, distributions of components of the Poynting vector and of the spin angular momentum (SAM) vector.

In this section, adopting the Richards-Wolf approach, we obtain analytical expressions, describing key characteristics of the generalized Poincaré beams in a case when the topological charges (TC) of the two optical vortices with left and right circular polarization are equal, respectively, to $p = m + 1$ and $q = -m$. We demonstrate that in the focus of such beams, radial spin and orbital Hall effects take place. We note that in [45] we demonstrated the spin Hall effect for fractional-order cylindrical vector beams in the focus plane. In the current work, at $p = m + 1$ and $q = -m$, there is also a cylindrical vector beam with a fractional order $m + 1/2$. Therefore, we can expect that there is also the spin Hall effect in the focus of such generalized Poincaré beam. The work [45] does not contain analytical expressions for electric field components in the focus of fractional-order cylindrical vector beams. In the current work, we derive such analytical expressions.

In our previous work [46], we have shown that the orbital Hall effect occurs before and after the focus of the conventional vectorial cylindrical beams, which are a special case of the Poincaré beams when the optical vortices have the TCs *m* and –*m*, and that local areas in the beam cross section, where the transverse energy flow is rotating clockwise or counterclockwise, reside in pairs on a certain-radius circle with the center on the optical axis. In this work, energy flows, rotating clockwise or counterclockwise, reside on circles with different radii. Therefore, this orbital Hall effect is called radial.

We note that the Richards-Wolf formalism [23] adequately describes the light field only near the focus. The work [47] investigates behavior of light in the focus by using an exact solution of the Helmholtz equation in the spherical coordinates, which is correct in the whole space. However, generating such a light field in the focus requires generating in the initial plane all three components of the electric vector. This is a challenging problem. In our case, only the transverse components of the electric field should be generated in the initial plane, which is easy to implement in practice.

6.2.1 Vector Field in the Initial Plane

We consider here the following Jones vector of the initial light field:

$$E_1 = a\frac{\exp(ip\varphi)}{\sqrt{2}}\begin{pmatrix}1\\ i\end{pmatrix} + b\frac{\exp(iq\varphi)}{\sqrt{2}}\begin{pmatrix}1\\ -i\end{pmatrix},$$
$$a^2 + b^2 = 1, \tag{6.23}$$

with (r, φ) being the polar coordinates in the initial plane, and a, b being complex constants. If $p = -n$ and $q = n$, the beam from Eq. (6.23) reduce to a conventional Poincaré beam [3, 29, 30]. If $a = b = 1/\sqrt{2}$ then the field (6.23) becomes maximally inseparable [26]:

$$E_2 = \exp\left(\frac{i(q+p)\varphi}{2}\right)\begin{pmatrix}\cos\left(\frac{q-p}{2}\right)\varphi\\ \sin\left(\frac{q-p}{2}\right)\varphi\end{pmatrix}. \tag{6.24}$$

When $p = q$, the field (6.23) reduces to a linearly polarized optical vortex with the topological charge (TC) q. When $p = -q$, the field (6.23) is a cylindrical vector beam of the order q [5]. When $p = -m$ and $q = m + 1$, the field (6.24) is given by

$$E_2 = \exp\left(\frac{i\varphi}{2}\right)\begin{pmatrix}\cos\left(m+\frac{1}{2}\right)\varphi \\ \sin\left(m+\frac{1}{2}\right)\varphi\end{pmatrix} = \frac{\exp(i(m+1)\varphi)}{2}\begin{pmatrix}1 \\ -i\end{pmatrix} + \frac{\exp(-im\varphi)}{2}\begin{pmatrix}1 \\ i\end{pmatrix}. \tag{6.25}$$

The field (6.25) is interesting for it is a cylindrical vector beam of a half-integer order. In [48], the beam (6.25) is called a vector vortex beam with a fractional topological charge not quite correctly. As was already shown in [45], in the tight focus of the fractional-order cylindrical vector beams, subwavelength areas are generated with elliptic polarization of different handedness, that is the polarization vector in these areas is rotating clockwise or counterclockwise. We note that the initial light field (6.25) is linearly polarized in each point of its cross section. Therefore, similarly to [45], it should be expected that the focused field (6.25) should also contain the areas with elliptic polarization of different handedness. It is also seen from Eq. (6.25) that the initial field is a coaxial superposition of two optical vortices with left and right circular polarization and with different topological charges of $m + 1$ and $-m$. Since these topological charges do not compensate each other, it is worth to expect circular energy flow in the focus. This means that in the focus, nonzero distribution of the axial component of the angular momentum vector should be present. Below we show that this is indeed so.

6.2.2 Components of the Strength Vector of the Electric Field in the Focus

The Richards-Wolf formalism [23] allows obtaining all the components of the strength vector of the electric field in the tight focus of the initial field (6.25):

$$\begin{aligned}
E_{2x} &= \frac{i^m}{2}\left[e^{i(m+1)\varphi}I_{0,m+1} + e^{i(m-1)\varphi}I_{2,m-1} - ie^{-im\varphi}I_{0,m} - ie^{-i(m-2)\varphi}I_{2,m-2}\right], \\
E_{2y} &= \frac{i^m}{2}\left[-ie^{i(m+1)\varphi}I_{0,m+1} + ie^{i(m-1)\varphi}I_{2,m-1} + e^{-im\varphi}I_{0,m} - e^{-i(m-2)\varphi}I_{2,m-2}\right], \\
E_{2z} &= i^m\left[ie^{im\varphi}I_{1,m} + e^{-i(m-1)\varphi}I_{1,m-1}\right],
\end{aligned} \tag{6.26}$$

where

$$I_{\nu,\mu} = \left(\frac{4\pi f}{\lambda}\right)\int_0^{\theta_0} \sin^{\nu+1}\left(\frac{\theta}{2}\right)\cos^{3-\nu}\left(\frac{\theta}{2}\right)\cos^{1/2}(\theta)A(\theta)e^{ikz\cos\theta}J_\mu(\xi)d\theta, \tag{6.27}$$

where $k = 2\pi/\lambda$ is the wavenumber of light with the wavelength λ, f is the focal length of an aplanatic system (ideal spherical lens), $\nu = 0, 1, 2$, $J_\mu(\xi)$ is the μth-order

Bessel function of the first kind, $\xi = kr\sin\theta$, θ is the polar angle that defines the tilt to the optical axis of rays converging into the focus, θ_0 is the maximal angle that defines the numerical aperture of the aplanatic system ($NA = \sin\theta_0$), (r, φ, z) is the cylindrical coordinate system with the origin in the focus ($z = 0$ is the focus plane), $A(\theta)$ is the amplitude of the initial circularly symmetric field (real-valued function).

6.2.3 Intensity Distribution of the Electric Field in the Focus

From the components of the electric vector (6.26), we can derive the intensity distribution of light field in the focus plane ($z = 0$):

$$\begin{aligned} I &= I_\perp + I_z = I_x + I_y + I_z \\ &= \frac{1}{2}\big[\big(I_{0,m+1}^2 + I_{2,m-2}^2 - 2\sin((2m-1)\varphi)I_{0,m+1}I_{2,m-2}\big) \\ &+ \big(I_{0,m}^2 + I_{2,m-1}^2 - 2\sin((2m-1)\varphi)I_{0,m}I_{2,m-1}\big) \\ &+ 2\big(I_{1,m}^2 + I_{1,m-1}^2 - 2\sin((2m-1)\varphi)I_{1,m}I_{1,m-1}\big)\big], \end{aligned} \tag{6.28}$$

where the first two terms in the round brackets describe the transverse intensity $I_\perp = I_x + I_y$, whereas the third term in the round brackets describes the longitudinal intensity I_z. As seen from Eq. (6.28), the intensity is a non-negative function ($I \geq 0$), since each term in the round brackets in Eq. (6.28) is non-negative, for the sum of two squared numbers is equal or greater than their doubled product. From Eq. (6.28) also follows that the intensity distribution contains $2m - 1$ local maxima and $2m - 1$ local minima (or intensity nulls), that reside on a certain-radius circle with the center on the optical axis. Thus, the number of these intensity maxima and nulls is always odd ($2m - 1$).

6.2.4 Longitudinal Component of the Spin Angular Momentum Vector in the Focus

Using the components of the electric field vector in the focus (6.26), longitudinal component can be derived of the spin angular momentum (SAM) vector of the field (6.25), since the longitudinal SAM component S_z is equal to the third Stokes parameters S_3, whose magnitude indicates the presence of elliptic or circular polarization in the beam cross section. The SAM vector is defined by the following expression [24]:

$$\mathbf{S} = \frac{1}{16\pi\omega}\mathrm{Im}\big(\mathbf{E}^* \times \mathbf{E}\big), \tag{6.29}$$

with ω being the angular frequency of light. Below we omit the constant factor $1/(16\pi\omega)$ for brevity. It is seen from Eq. (6.29) that the longitudinal SAM component (without the constant) coincides with the non-normalized third component of the Stokes vector:

$$S_3 = S_z = 2\,\mathrm{Im}\left(E_x^* E_y\right). \tag{6.30}$$

Substitution of expressions (6.26) for the electric field components into Eq. (6.30) yields

$$\begin{aligned} S_z = \frac{1}{2}\Big[&\left(I_{0,m}^2 + I_{2,m-1}^2 - 2\sin((2m-1)\varphi)I_{0,m}I_{2,m-1}\right) \\ &- \left(I_{0,m+1}^2 + I_{2,m-2}^2 - 2\sin((2m-1)\varphi)I_{0,m+1}\,I_{2,m-2}\right)\Big]. \end{aligned} \tag{6.31}$$

Comparison of Eqs. (6.28) and (6.31) reveals that if the transverse intensity is a sum of two positive terms A and B:

$$\begin{aligned} I_\perp = A + B = \frac{1}{2}\Big[&\left(I_{0,m+1}^2 + I_{2,m-2}^2 - 2\sin((2m-1)\varphi)I_{0,m+1}\,I_{2,m-2}\right) \\ &+ \left(I_{0,m}^2 + I_{2,m-1}^2 - 2\sin((2m-1)\varphi)I_{0,m}\,I_{2,m-1}\right)\Big], \end{aligned} \tag{6.32}$$

then the longitudinal SAM component is a difference of these terms:

$$S_z = B - A. \tag{6.33}$$

According to Eq. (6.31), similarly to the intensity distribution in Eq. (6.28), the SAM distribution also has $2m - 1$ local maxima and $2m - 1$ local minima. As seen from Eq. (6.33), if $B > A$ then $S_z > 0$ (polarization vector is rotating counterclockwise). And vice versa, if $B < A$ then $S_z < 0$ (polarization vector is rotating clockwise). In the areas with $B = A$ ($S_z = 0$), polarization is linear. The points in the beam cross section in the focus, where $S_z = 0$, are called in [49] topological spin defects. Thus, it follows from Eqs. (6.31) and (6.33) that there are areas with different spin in the focus: positive ($S_z > 0$) and negative ($S_z < 0$). Spatial separation of areas with left circular and right circular polarization is called the spin Hall effect [24, 43, 44]. In the Simulation section below, these conclusions are confirmed by concrete examples.

6.2.5 Energy Flow Density in the Focus

Here, we derive the Poynting vector (energy flow density) in the focus of the field (6.25). To do this, we should obtain at first the components of the strength vector of the magnetic field in the focus. The same way as we obtained the components of the electric vector (6.26) by using the Richards-Wolf theory [23], we can obtain also the magnetic vector:

$$\begin{aligned} H_{2x} &= \frac{i^m}{2}\left[ie^{i(m+1)\varphi}I_{0,m+1} + ie^{i(m-1)\varphi}I_{2,m-1} - e^{-im\varphi}I_{0,m} - e^{-i(m-2)\varphi}I_{2,m-2}\right], \\ H_{2y} &= \frac{i^m}{2}\left[e^{i(m+1)\varphi}I_{0,m+1} - e^{i(m-1)\varphi}I_{2,m-1} - ie^{-im\varphi}I_{0,m} + ie^{-i(m-2)\varphi}I_{2,m-2}\right], \\ H_{2z} &= -i^m\left[e^{im\varphi}I_{1,m} + ie^{-i(m-1)\varphi}I_{1,m-1}\right]. \end{aligned} \tag{6.34}$$

The Poynting vector is defined by the well-known formula [23]:

$$\mathbf{P} = [c/(8\pi)]\mathrm{Re}\left[\mathbf{E} \times \mathbf{H}^*\right], \tag{6.35}$$

where c is the vacuum speed of light, Re is the real part of a complex number, $\mathbf{E} \times \mathbf{H}$ is the cross-product, * is the complex conjugation. Below we omit the constant $c/(8\pi)$ for brevity. Substituting the components (6.26) and (6.34) into Eq. (6.35), we get the components of the Poynting vector in the focus of the field (6.25):

$$\begin{aligned} P_x &= Q(r)\sin\varphi, \\ P_y &= -Q(r)\cos\varphi, \\ P_z &= \frac{1}{2}\left(I_{0,m}^2 + I_{0,m+1}^2 - I_{2,m-1}^2 - I_{2,m-2}^2\right), \\ Q(r) &= I_{1,m-1}\left(I_{0,m} + I_{2,m-2}\right) - I_{1,m}\left(I_{0,m+1} + I_{2,m-1}\right). \end{aligned} \tag{6.36}$$

Passing to the polar components P_r and P_φ of the transverse Poynting vector, we get:

$$\begin{aligned} P_r &= 0, \\ P_\varphi &= -Q(r). \end{aligned} \tag{6.37}$$

As seen from Eq. (6.36), the longitudinal component of the Poynting vector in the focus has a circularly symmetric distribution and does not depend on the azimuthal angle φ. It is also seen from Eq. (6.36) that if $m = 1$ or $m = 2$ then there is reverse energy flow on the optical axis, since for $m = 1$ or $m = 2$, we get on the optical axis

$$P_z(r = 0) = -\frac{1}{2}I_{2,0}^2 < 0.$$

Equation (6.37) indicates that the transverse energy flow in the focus is rotating along a closed trajectory with the center on the optical axis, clockwise if $Q(r) > 0$, and counterclockwise if $Q(r) < 0$. Since the function $Q(r)$ is of different signs on different radii r, it can be stated that the radial orbital Hall effect occurs in the focus of the light field (6.25). This also follows from the expression for the longitudinal component of the angular momentum vector $\mathbf{J}$ of the field (6.25), when it is written by definition via the azimuthal component of the energy flow [24]:

$$\mathbf{J} = \mathbf{r} \times \mathbf{P},$$
$$J_z = rP_\varphi = -rQ(r). \tag{6.38}$$

The energy flow in the focus is rotating along a spiral around the optical axis, since the topological charges of the two optical vortices, present in the superposition in the initial field (6.25), do not compensate each other for they have different magnitudes: $m + 1$ and $-m$.

6.2.6 Simulation

Using the Richards-Wolf formalism [23], we computed the distributions of intensity and of the longitudinal component of the SAM vector (spin density) in the tight focus of the light field with the initial distribution given by Eq. (6.25). We supposed that the field amplitude in the initial plane is constant, i.e., $A(\theta) = 1$, wavelength $\lambda =$ 532 nm, focal length $f = 10$ μm, numerical aperture $NA = 0.95$. Shown in Fig. 6.4 are distributions of the longitudinal component of the spin angular momentum S_z (Fig. 6.4a–d) (red and blue colors denote positive and negative values), of intensity I (Fig. 6.4e–h) (black and yellow colors denote zero and maximal values) and of the angular component of the Poynting vector P_φ (Fig. 6.4i–l) (red and blue colors denote positive and negative values) of a light beam with polarization (6.25) of different order m in the tight focus. The beam orders in Fig. 6.1 are $m = 1$ (Fig. 6.4a, e, i), $m = 2$ (Fig. 6.4b, f, j), $m = 3$ (Fig. 6.4c, g, k), $m = 5$ (Fig. 6.4d, h, l). The arrows in Fig. 6.4i–l illustrate the directions of the angular energy flow. The scale mark in each figure denotes 1 μm.

As seen in Fig. 6.4 (2nd row), the number of local maxima in the intensity distribution in the focus is consistent with the theory [Eq. (6.28)] and is equal to $2m - 1$: 1 (Fig. 6.4e), 3 (Fig. 6.4f), 5 (Fig. 6.4g) and 9 (Fig. 6.4h). It is also seen in Fig. 6.4 (1st row), that, according to Eq. (6.31), the SAM distribution also contains $2m - 1$ local maxima (red color in Fig. 6.4a–d), where $S_z > 0$, and which reside on a certain circle with the center on the optical axis. On a circle with a larger radius, (blue color in Fig. 6.4a–d), $S_z < 0$. Black color in Fig. 6.4a–d denotes the areas with zero spin, i.e. where polarization is linear. Since the brightness of the blue color in Fig. 6.4a–d is 2–3 times lower than that of the red color, elliptic polarization in the areas of positive spin is closer to circular polarization, whereas the polarization ellipses in the areas of negative spin are more elongated and close to linear polarization. Nevertheless, space separation of the areas with positive and with negative spin in the focus demonstrates the spin Hall effect.

Figures 6.4i–l (3rd row) confirms theoretical predictions [Eq. (6.37)] and demonstrates that the transverse energy flow in the focus plane rotates. On a circle closer to the optical axis (blue color in Figs. 6.4i–l), $P_\varphi < 0$, i.e. the transverse energy flow is rotating clockwise. On a larger radius circle (red ring in Figs. 6.4i–l), $P_\varphi > 0$, and the energy flow is rotating counterclockwise. Such spatial separation of the orbital

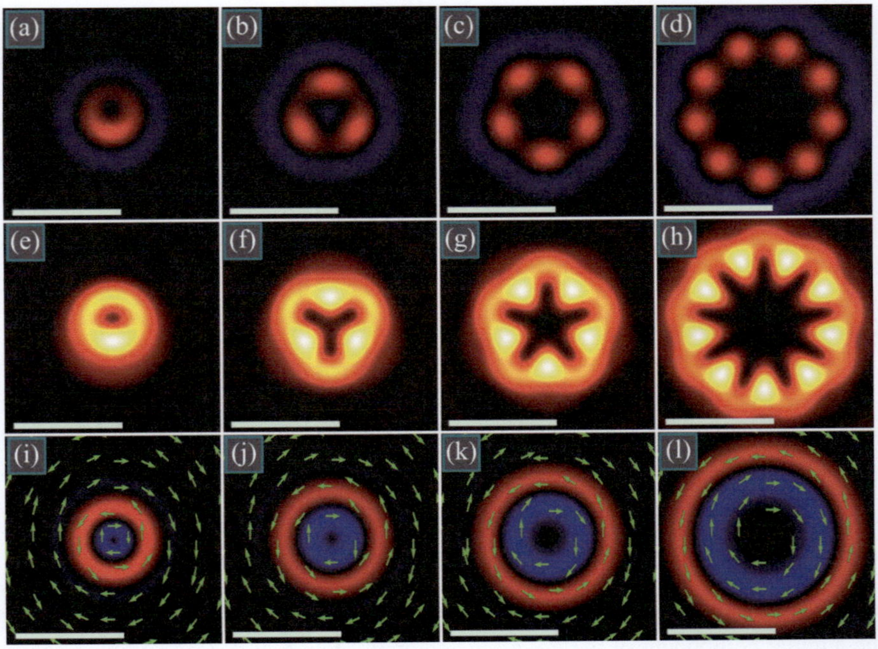

Fig. 6.4 Distributions of the longitudinal component of the spin angular momentum (**a–d**) (red and blue color denotes, respectively, positive and negative values), intensity (**e–h**) (black means zero and yellow means maximum), and angular component of the Poynting vector (**i–l**) (red means positive and blue means negative values) of a light beam with polarization (6.25) and with different order in the tight focus. Arrows (**i–l**) denote the directions of the angular energy flow. Scale mark in each figure denotes 1 μm

energy flux in opposite directions is a manifestation of the radial orbital Hall effect in the focus.

6.2.7 *Discussion of the Results*

Here, we compare the transverse components of the electric field in the initial plane (6.25) and in the focus plane (6.26). Although in the initial plane the components E_x and E_y of the field (6.25) have the same phase, and thus the field has inhomogeneous linear polarization, then, in the focus, the transverse components of the field (6.26) acquire relative phase delay by $\pi/2$ or by $3\pi/2$. This leads to the areas with elliptic polarization in the focus. On the other hand, the longitudinal SAM component (6.29) in the initial plane is equal to zero $S_z = 0$, while the energy flow (6.35) has only one longitudinal component, equal to unit: $P_z = 1$. In the focus plane, the SAM density is given by Eq. (6.31), but if the functions S_z is integrated over the whole focus plane, then it is equal to zero. Thus, the full longitudinal SAM component is conserved and equal to zero. The field (6.25) in the initial plane has a nonzero density of the longitudinal component of the orbital angular momentum (OAM) vector [50]:

$$L_z = \mathrm{Im}\left(E_x^* \frac{\partial}{\partial\varphi} E_x + E_y^* \frac{\partial}{\partial\varphi} E_y\right) = \frac{1}{2}. \tag{6.39}$$

If the OAM density (6.39) is integrated over the angle φ from 0 to 2π, then the full OAM in the initial plane is nonzero and equal to the half of the initial beam power $W/2$.

In the focus plane, longitudinal OAM component can also be obtained:

$$\begin{aligned} L_z &= \mathrm{Im}\left(E_x^* \frac{\partial}{\partial\varphi} E_x + E_y^* \frac{\partial}{\partial\varphi} E_y + E_z^* \frac{\partial}{\partial\varphi} E_z\right) \\ &= \frac{1}{2}\{(m+1)I_{0,m+1}^2 - mI_{0,m}^2 + 2mI_{1,m}^2 + (m-1)I_{2,m-1}^2 - 2(m-1)I_{1,m-1}^2 \\ &\quad -(m-2)I_{2,m-2}^2 + \sin(2m-1)\varphi\left[I_{0,m}\, I_{2,m-1} - 3I_{0,m+1}I_{2,m-2} - 2I_{1,m}\, I_{1,m-1}\right]\} \end{aligned} \tag{6.40}$$

According to Eq. (6.40), the OAM density depends on the angle as $\sin(2m - 1)\varphi$. This means that on a certain-radius circle with the center on the optical axis, the OAM has $(2m - 1)$ local maxima and minima, similarly to the SAM distribution (6.31). It can be shown that if the OAM density is integrated over the whole focus plane, then this also yields the half of the initial beam power $W/2$. Thus, in this case, the full SAM and OAM are conserved separately. Therefore, it can be concluded that the spin Hall effect in the focus of the beam (6.25) arises due to the conservation of the full longitudinal SAM of the beam. Since the longitudinal SAM of the whole beam is zero, the areas with the spin of different sign should arise in pairs. The same way, the radial orbital Hall effect in the focus occurs due to the conservation of the full longitudinal OAM of the beam.

Summing the SAM (6.31) and OAM (6.40) densities, we get:

$$\begin{aligned} L_z + S_z &= \frac{1}{2}\{m\left(I_{0,m+1}^2 + 2I_{1,m}^2 + I_{2,m-1}^2\right) - (m-1)\left(I_{0,m}^2 + 2I_{1,m-1}^2 + I_{2,m-2}^2\right) \\ &\quad - \sin(2m-1)\varphi\left[I_{0,m} I_{2,m-1} + I_{0,m+1}I_{2,m-2} + 2I_{1,m}\, I_{1,m-1}\right]\} \end{aligned} \tag{6.41}$$

Comparison of Eq. (6.38) for the density of the longitudinal component of the angular momentum (AM) vector and of Eq. (6.41) for the sum of the longitudinal SAM and OAM components reveals that they are not equal to each other: $J_z \neq L_z + S_z$. The reason of this inequality we considered earlier in [50].

In conclusion of this section we consider the difference between the Hall effect near the tight focus [26, 45, 46, 50] and the Hall effect occurring when light is reflected off the interface between two media [11, 14]. As was shown in [14], when an optical vortex is reflected from a plane glass surface, the annular intensity distribution becomes inhomogeneous. For the optical vortices with the topological charges m and $-m$, the intensity maxima on the ring appear in different places, i.e., shifted relative to each other (orbital Hall effect). In [11], it was shown that when a p-polarized Gaussian beam (polarization vector is in the incidence plane) is reflected from the glass surface under an angle close to the Brewster angle, the spin Hall effect occurs, when the reflected light is split into two beams with opposite spins in the direction orthogonal to the incidence plane. In the tight focus [26, 45, 46, 50], spin and orbital Hall effects occur due to the conservation of the full angular momentum. The light with opposite spins or (and) with opposite energy rotation in the focus plane is concentrated in different places. A different manifestation of the Hall effect in the focus, investigated in different works [26, 45, 46, 50], is explainable since the different types of the initial vector fields were considered. In [26], focusing of the conventional Poincaré beams was considered, while [45] dealt with focusing of fractional-order cylindrical vector beams. In [46], the Hall effect of the cylindrical vector beams of an integer order arises before and beyond the focus, whereas it is absent in the focus itself. In [50], the Hall effect was studied in the focus of a circularly polarized optical vortex. In contrast to these works, we studied here tight focusing of a generalized Poincaré beam, whose topological charge is equal to 1/2 and whose order of inhomogeneous linear polarization is equal to $m + 1/2$. It is impossible to predict in advance, based on the initial light field, whether or not the Hall effect will arise in the tight focus. Thus, each new type of the initial vector beams should be considered separately.

Based on the Debye integrals [44], we have investigated both theoretically and numerically generalized (hybrid) Poincaré beams in the tight focus [51]. A generalized Poincaré beam is a coaxial superposition of two optical vortices with left and right circular polarization and with the TC of p and q. For certainty, we studied the case when $p = m + 1$ and $q = -m$ [Eq. (6.25)]. Simple analytical expressions have been obtained for the components of the electric and magnetic strength vectors in the focus [Eqs. (6.26) and (6.34)], for the intensity distribution [Eq. (6.28)], for the longitudinal component of the spin angular momentum [Eq. (6.31)], and for the components of the Poynting vector [Eq. (6.36)]. It has been shown that the intensity in the focus has $2m - 1$ local maxima, residing evenly on a certain-radius circle with the center on the optical axis. In addition, radial spin and orbital Hall effects has been demonstrated. This means that the longitudinal SAM component has different signs on circles with different radii in the focal plane, and the azimuthal component of the transverse Poynting vector also has different signs (Fig. 6.4). Such beams can be used for simultaneous trapping into local intensity maxima of several micro- or nanoparticles (Fig. 6.4h) that should simultaneously rotate around their centers of mass (Fig. 6.4d) and move along the ring (Fig. 6.4l). In addition, when moving along the ring, the particles will need to overcome the 'breaks' in the intensity distribution (Fig. 6.4h).

6.3 Controlling the Spin Hall Effect in the Sharp Focus of an Axial Superposition of Two Optical Vortices with Left- and Right-Handed Circular Polarization

The control of a spin–orbit conversion (SOC) in spintronics by managing the spin of electrons is well known (see recent review [52]). In optics, works on the control of SOC parameters have been appeared recently. In [53], SOC was controlled utilizing excitation of a superposition of vortex laser modes with different spins in thin annular fibers with an empty core, although the first works on SOC in multimode fibers was made much earlier [54]. In [55], SOC is the cause of the emission of an optical vortex perpendicular to the plane of a ring cavity with defects, along which two whispering gallery modes with different spins (different polarization states) propagate. It was shown in [56] that SOC can be controlled using q-plates, which transform the light with left-handed circular polarization into an optical vortex with right-handed circular polarization. The theoretical aspects of the spin–orbit interaction of light were considered in [24, 57]. Spin–orbit conversion was observed during light scattering [58], in a sharp focus [59], in an anisotropic medium [60], and upon reflection from an interface of two media [61]. Spin–orbit interaction was also observed in plasmonics [62], in nonlinear optics [63], and in electron optics [64]. One of the clear manifestations of the spin–orbit light conversion is the optical Hall effect, which appears when the light is reflected from multilayer structures [65] and when it passes through metasurfaces [66]. The Hall effect in optics appears as the spatial separation of light beams with different directions of circular polarization (left- and right-handed)—this is the spin Hall effect in optics [9, 18]. Or it leads to the spatial separation of light beams with an orbital angular momentum of a different sign – this is the orbital Hall effect [13, 14]. For further consideration, it is useful to mention works on the optical Hall effect in a sharp focus [50, 51, 67]. The Hall effect at the focus appears not only due to the SOC [68], but also due to the preservation of the total longitudinal SAM equals to zero during beam propagation [69]. As was shown in [43], when a linearly polarized light is sharply focused, the spin is zero in the initial plane and in the focus, but before and after the focus, four local areas appear with nonzero spin of different sign. In [70], SOC was studied in the light reflected from two crossed metasurfaces. In [71], the geometric spin Hall effect was considered. The SOC in a freely propagating vector cylindrical laser beam with a vortex phase was considered in [72]. In [18], the spin-controlled Hall effect was implemented based on a spin-dependent beam splitting into two beams. Sometimes, the Hall effect is understood as a transverse shift of the beam center in the focus because of symmetry violation of a vortex beam or due to the displacement [73].

In this section, we consider a spin–orbit conversion of a superposition of two identical optical vortices with circular polarization of different signs and with different amplitudes at a sharp focus. The ratio of the amplitudes of the beams included in this superposition is a parameter that can be varied in order to manage the reduction in spin angular momentum (SAM) and the increase in the orbital angular momentum (OAM) during focusing. The change in the total spin and orbital angular momentum occurs immediately after the light passes through the spherical lens. Further, during propagation in the free space from the lens to the focus, the total spin and orbital angular momentums remain unchanged.

Controlling the spin–orbit conversion in the focus of a spherical lens can be used for changing the rotation velocity of dielectric microparticles. Continuous variation of the parameter (ratio between the beam amplitudes in the considered superposition) allows, changing the SAM in the focus, changing the rotation velocity of a non-spherical microparticle around its center of mass from zero to maximal. In addition, changing the same parameter allows tuning the OAM in the focus and changing the rotation velocity of a microparticle along a circular path.

Microparticles rotation either around the own axis or along a circular path is needed in micromachines for contactless driving the details by light [74–76].

6.3.1 Projections of the Electric and Magnetic Field Strength Vectors at the Focus

Let us consider an initial light field with a Jones vector satisfies:

$$\mathbf{E} = \frac{\exp(in\varphi)A(\theta)}{\sqrt{2(1+\alpha^2)}}\begin{pmatrix} 1+\alpha \\ i(1-\alpha) \end{pmatrix} = \frac{\exp(in\varphi)A(\theta)}{\sqrt{(1+\alpha^2)}}\left[\frac{1}{\sqrt{2}}\begin{pmatrix} 1 \\ i \end{pmatrix} + \frac{\alpha}{\sqrt{2}}\begin{pmatrix} 1 \\ -i \end{pmatrix}\right]. \tag{6.42}$$

Here (r, φ) are the polar coordinates in the beam section ($r = f \sin\theta$, f is the focal length of a spherical lens), n is an integer topological charge of the optical vortex, $A(\theta)$ is the real function, the amplitude of the radially symmetric initial light field (6.42). The radial coordinate r on the spherical front of the converging wave is related to the angle θ between the optical axis and the segment connecting the center of the focus with a point on the spherical front, $0 < \alpha < 1$ is a real number that regulates the contribution to the superposition of light with left- and right-handed circular polarization. It can be seen from (6.42) that this light field is an axial superposition of two optical vortices with identical topological charges and left- and right-handed circular polarizations. At $\alpha = 0$, the beam (6.42) has right-handed circular polarization, and at $\alpha = 1$ the polarization is linear.

The beam (6.42) can be considered as a generalized Poincare beam [51]:

$$\begin{aligned} \mathbf{E}_P &= a\frac{\exp(ip\varphi)}{\sqrt{2}}\begin{pmatrix} 1 \\ i \end{pmatrix} + b\frac{\exp(iq\varphi)}{\sqrt{2}}\begin{pmatrix} 1 \\ -i \end{pmatrix}, \\ a^2 + b^2 &= 1, \end{aligned} \tag{6.43}$$

at $p = q = n$ and $a = \left(1+\alpha^2\right)^{-1/2}$, $b = \alpha\left(1+\alpha^2\right)^{-1/2}$. In practice, two light modulators are required to form a field (6.43) with different topological charges p

and q. Therefore, we restrict our consideration to the field (6.42), when $p = q$, this beam (6.42) can be formed by only one light modulator. Using the theory [23], we can write components of the electric and magnetic fields vectors in the sharp focus of the beam (6.42) as follows:

$$\begin{aligned}
E_x &= \frac{i^{n-1}\exp(in\varphi)}{\sqrt{2(1+\alpha^2)}}\left((1+\alpha)I_{0,n} + \alpha e^{-2i\varphi}I_{2,n-2} + e^{2i\varphi}I_{2,n+2}\right),\\
E_y &= \frac{i^{n}\exp(in\varphi)}{\sqrt{2(1+\alpha^2)}}\left((1-\alpha)I_{0,n} + \alpha e^{-2i\varphi}I_{2,n-2} - e^{2i\varphi}I_{2,n+2}\right),\\
E_z &= \frac{\sqrt{2}i^{n}\exp(in\varphi)}{\sqrt{(1+\alpha^2)}}\left(\alpha e^{-i\varphi}I_{1,n-1} - e^{i\varphi}I_{1,n+1}\right),\\
H_x &= -\frac{i^{n}\exp(in\varphi)}{\sqrt{2(1+\alpha^2)}}\left((1-\alpha)I_{0,n} - \alpha e^{-2i\varphi}I_{2,n-2} + e^{2i\varphi}I_{2,n+2}\right),\\
H_y &= \frac{i^{n-1}\exp(in\varphi)}{\sqrt{2(1+\alpha^2)}}\left((1+\alpha)I_{0,n} - \alpha e^{-2i\varphi}I_{2,n-2} - e^{2i\varphi}I_{2,n+2}\right),\\
H_z &= \frac{\sqrt{2}i^{n+1}\exp(in\varphi)}{\sqrt{(1+\alpha^2)}}\left(\alpha e^{-i\varphi}I_{1,n-1} + e^{i\varphi}I_{1,n+1}\right).
\end{aligned} \tag{6.44}$$

Functions $I_{\nu,\mu}$ in (6.44) are defined by the following:

$$I_{\nu,\mu} = 2kf\int\limits_0^{\theta_0}\sin^{\nu+1}\left(\frac{\theta}{2}\right)\cos^{3-\nu}\left(\frac{\theta}{2}\right)\cos^{1/2}(\theta)A(\theta)e^{ikz\cos\theta}J_\mu(kr\sin\theta)d\theta, \tag{6.45}$$

where $k = 2\pi/\lambda$ is the wavenumber of the monochromatic light with the wavelength λ, f is a focal distance of the lens, θ_0 is the maximum tilt angle of the rays to the optical axis, which determines the numerical aperture of the aplanatic lens NA = sin θ_0, $J_\mu(x)$ is the Bessel function of the first kind and μ-th order, z is a longitudinal coordinate, the focal plane is situated at $z = 0$. Functions (6.45) depends on the radial and longitudinal coordinates $I_{\nu,\mu}(r, z)$. Numbers of the function (6.45) can be: $\nu =$ 0,1,2; $\mu = n - 2, n - 1, n, n + 1, n + 2$.

Equations (6.44) indicate that the spherical lens generates additional optical vortices in the converging beam. Besides the initial vortex $\exp(in\varphi)$, the field contains four other vortices: $\exp(i(n+2)\varphi)$, $\exp(i(n\text{-}2)\varphi)$, $\exp(i(n+1)\varphi)$, and $\exp(i(n-1)\varphi)$. Therefore, the power of the initial optical vortex as partially transferred to these four additional vortices [77]. Since all the vortices have different amplitudes, left and right circular polarizations can no longer sum up, generating only linear polarization. Elliptic polarization of different sign appears, i.e. the spin Hall effect arises. And vice versa, if initial polarization is right-handed circular then, due to the additional vortices in Eq. (6.44), the SAM of the beam decreases since some portion of the beam power goes for generating transverse energy rotation, i.e., for generating the longitudinal component of the OAM vector.

It can be supposed that if a light field is focused by spherical lenses with aberrations, the OAM spectrum of the transmitted radiance is even broader and thus even more additional vortices with different topological charges are generated. It is also known that the astigmatic transform, implemented by a cylindrical lens, changes the OAM of an initial vortex-free beam [78].

Since the amplitudes of the additional optical vortices $I_{2,n\pm 2}$, $I_{1,n\pm 1}$ are significantly lower than the amplitude of the main optical vortex $I_{0,n}$ [77], then polarization of the main vortex $I_{0,n}$ (for instance, linear) dominates, whereas polarization of the additional vortices $I_{2,n\pm 2}$, $I_{1,n\pm 1}$ (for instance, elliptic with different signs) is weaker for any numerical aperture.

6.3.2 Density of the Longitudinal Component of the Spin Angular Momentum Vector at the Focus

Using the projections of the electric field strength vectors at the focus (6.44), we can find the distribution of the longitudinal projection of the spin angular momentum (SAM) at the focus of the field (6.42). The SAM vector is determined by the relation [57]:

$$\mathbf{S} = \frac{1}{16\pi\omega}\mathrm{Im}\left(\mathbf{E}^* \times \mathbf{E}\right), \tag{6.46}$$

where ω is the angular frequency of the monochromatic light. The constant $1/(16\pi\omega)$ is omitted below. Further, we obtain an expression only for the longitudinal projection of the SAM (6.46), since it coincides with the third component of the Stokes vector and reveals the presence of regions with elliptical polarization in the beam cross section. Taking into account (6.44), the longitudinal projection of the SAM at the focus of the field (6.42) can be written as follows:

$$\begin{aligned} S_{z,\alpha} &= 2\,\mathrm{Im}\left(E_x^* E_y\right) \\ &= \frac{1}{1+\alpha^2}\left[(1-\alpha^2)I_{0,n}^2 + \alpha^2 I_{2,n-2}^2 - I_{2,n+2}^2 + 2\alpha\cos(2\varphi)I_{0,n}\left(I_{2,n-2} - I_{2,n+2}\right)\right]. \end{aligned} \tag{6.47}$$

Equation (6.47) shows that at $\alpha = 1$, field (6.42) becomes linearly polarized and, at the focus, the longitudinal component of the SAM (6.47) is equal to the SAM of an optical vortex with linear polarization obtained in [67]:

$$S_{z,\alpha=1} = \frac{1}{2}\left[I_{2,n-2}^2 - I_{2,n+2}^2 + 2\cos(2\varphi)I_{0,n}\left(I_{2,n-2} - I_{2,n+2}\right)\right]. \tag{6.48}$$

It can be seen from (6.48) that there are four regions at the focus, in two of them the longitudinal SAM is positive at $\varphi = 0$ and $\varphi = \pi$, since near the optical axis $I_{2,n-2}^2 > I_{2,n+2}^2$, and in the other two regions at $\varphi = \pi/2$ and $\varphi = 3\pi/2$ the SAM is negative. On the other circle, where $I_{2,n-2}^2 < I_{2,n+2}^2$, on the contrary, at $\varphi = \pi/2$ and $\varphi = 3\pi/2$ the SAM is positive, and at $\varphi = 0$ and $\varphi = \pi$ the SAM is negative. The presence of regions with the longitudinal projection of the SAM with different signs at the focus demonstrates the spin Hall effect. Where the SAM is positive, the elliptical polarization vector rotates counterclockwise, and in the regions with the negative longitudinal SAM there is a left-handed elliptical polarization and the polarization vector rotates clockwise.

It follows another extreme case from Eq. (6.47), when the parameter $\alpha = 0$. In this case, the longitudinal SAM at the focus will be equal to the SAM of an optical vortex with right-handed circular polarization obtained in [50]:

$$S_{z,\alpha=0} = I_{0,n}^2 - I_{2,n+2}^2. \tag{6.49}$$

It is seen from (6.49) that for an optical vortex with right-handed circular polarization at the focus, the SAM varies only along the radius and can have a different sign at different radii. This effect can be called as the radial spin Hall effect at the focus. Thus, we have shown that, by varying the parameter α of the initial light field, it is possible to control the characteristics of the spin Hall effect at the focus, i.e., obtain different types of spin distribution at the focus, (6.48) and (6.49).

6.3.3 Full Longitudinal SAM at the Focus

Five angular harmonics is seen from (6.44) to involve in the formation of the light field at the focus: $\exp(in\varphi)$, $\exp(i(n+2)\varphi)$, $\exp(i(n-2)\varphi)$, $\exp(i(n+1)\varphi)$, $\exp(i(n-1)\varphi)$. Each such an angular harmonic transfers a certain fraction of the total energy (power) of the beam. It was shown in [77] that the fraction of power transferred by each such harmonic for sharp focusing of light is equal to:

$$\begin{aligned} W_{\nu,\mu} = W_\nu &= \int_0^\infty \int_0^{2\pi} I_{\nu,\mu}^2(r,\varphi)\, r dr d\varphi \\ &= 4\pi f^2 \int_0^\alpha \sin^{2\nu+1}\left(\frac{\theta}{2}\right) \cos^{5-2\nu}\left(\frac{\theta}{2}\right) |A(\theta)|^2 d\theta \end{aligned} \tag{6.50}$$

Using (6.50), we can find the total (averaged over the beam cross section) longitudinal SAM at the beam focus (6.42):

$$\widehat{S}_{z,\alpha} = \int_0^\infty \int_0^{2\pi} S_{z,\alpha} r dr d\varphi = \frac{1}{1+\alpha^2} \int_0^\infty \int_0^{2\pi} r dr d\varphi \big[(1-\alpha^2) I_{0,n}^2 + \alpha^2 I_{2,n-2}^2 - I_{2,n+2}^2 + 2\alpha \cos(2\varphi) I_{0,n} \big(I_{2,n-2} - I_{2,n+2} \big) \big] = \frac{1-\alpha^2}{1+\alpha^2} (W_0 - W_2)$$
$$= \frac{1-\alpha^2}{1+\alpha^2} (W - 2(W_2 + W_1)). \quad (6.51)$$

Let us compare the resulting expression (6.51) with the density of the longitudinal SAM in the initial plane (6.42):

$$\widehat{S}_{z,0,\alpha} = \int_0^\infty \int_0^{2\pi} S_{z,0,\alpha} \, r dr d\varphi = \frac{1-\alpha^2}{1+\alpha^2} \int_0^\infty \int_0^{2\pi} A^2(\theta) \, r dr d\varphi = \frac{1-\alpha^2}{1+\alpha^2} W, \quad (6.52)$$

where W is a full beam power (6.42). To compare (6.51) and (6.52) we derive additional intensity distribution (power density) at the focus of the beam (6.42):

$$I_\alpha = \frac{1}{1+\alpha^2} \big[(1+\alpha^2) I_{0,n}^2 + \alpha^2 I_{2,n-2}^2 + I_{2,n+2}^2 + 2\alpha^2 I_{1,n-1}^2 + 2 I_{1,n+1}^2 + 2\alpha \cos(2\varphi) I_{0,n} \big(I_{2,n-2} + I_{2,n+2} \big) \big], \quad (6.53)$$

from which the total beam power (6.42) can be obtained:

$$W = \int_0^\infty \int_0^{2\pi} I_\alpha(r,\varphi) r dr d\varphi = W_0 + W_2 + 2W_1. \quad (6.54)$$

Comparison of (6.51) and (6.52) shows that when focusing the beam (6.42) the full longitudinal SAM decreases from $(1-\alpha^2)W/(1+\alpha^2)$ to $(1-\alpha^2)(W_0-W_2)/(1+\alpha^2)$, because $W_0 - W_2 = W - 2(W_1 + W_2) < W$. It is also seen from (6.51) and (6.52) that for $\alpha = 1$ the total longitudinal SAM is equal to 0. As α decreases from 1 to 0, the total SAM in the initial plane increases from 0 to W, and in the focal plane also increases from 0 to $W_0 - W_2$. The decrease in the total SAM during focusing is due to the spin–orbit conversion, when part of the spin is converted into an “orbit”. Below we show this in more detail.

6.3.4 *The Density of the Longitudinal Orbital Angular Momentum at the Focus*

To analyze the effect of spin–orbit conversion at the focus, we find the density of the longitudinal component of the orbital angular momentum (OAM) of the field (6.42).

To do this, we use the definition of the OAM vector [79]:

$$\mathbf{L} = \frac{1}{8\pi\omega} \sum_{p=x,y,z} \mathrm{Im}\Big(E_p^*(\mathbf{r} \times \nabla)E_p\Big). \tag{6.55}$$

We consider only the longitudinal projection of the OAM vector in order to compare it with the longitudinal projection of the SAM. Substituting (6.44) into (6.55), we obtain the following:

$$\begin{aligned} L_{z,\alpha} &= \mathrm{Im}\left(E_x^* \frac{\partial}{\partial\varphi} E_x + E_y^* \frac{\partial}{\partial\varphi} E_y + E_z^* \frac{\partial}{\partial\varphi} E_z\right) \\ &= \frac{1}{1+\alpha^2}\big[n(1+\alpha^2)I_{0,n}^2 \\ &+ \alpha^2(n-2)I_{2,n-2}^2 + (n+2)I_{2,n+2}^2 + 2(n+1)I_{1,n+1}^2 \\ &\quad + 2\alpha^2(n-1)I_{1,n-1}^2 + 2\alpha\cos(2\varphi) \\ &\big((n-1)I_{0,n}I_{2,n-2} + (n+1)I_{0,n}I_{2,n+2} - 2nI_{1,n-1}I_{1,n+1}\big)\big]. \end{aligned} \tag{6.56}$$

From (6.56) at $\alpha = 1$, we obtain the longitudinal projection of the OAM at the focus of an optical vortex with linear polarization, previously derived in [50]:

$$\begin{aligned} L_{z,\alpha=1} &= \big[nI_{0,n}^2 + \frac{1}{2}(n-2)I_{2,n-2}^2 + \frac{1}{2}(n+2)I_{2,n+2}^2 + (n+1)I_{1,n+1}^2 \\ &\quad + (n-1)I_{1,n-1}^2 + \cos(2\varphi)((n-1)I_{0,n}I_{2,n-2} \\ &\quad + (n+1)I_{0,n}I_{2,n+2} - 2nI_{1,n-1}I_{1,n+1})\big]. \end{aligned} \tag{6.57}$$

And if we suppose the parameter $\alpha = 0$ in (6.56), then we obtain the longitudinal projection of the OAM at the focus of the optical vortex with right-handed circular polarization, which was previously derived in [50]:

$$L_{z,\alpha=0} = nI_{0,n}^2 + (n+2)I_{2,n+2}^2 + 2(n+1)I_{1,n+1}^2. \tag{6.58}$$

Equation (6.56) by varying the parameter α enables to continuously change the OAM of the beam at the focus from (6.58) to (6.57). In the initial field plane (6.42), the density of the longitudinal component of the OAM can be found from (6.42):

$$L_{z,0,\alpha} = nA^2(\theta). \tag{6.59}$$

6.3.5 Total Longitudinal OAM at the Focus

Utilizing (6.50) and (6.56), we obtain the total (averaged over the beam cross section) longitudinal OAM at the field focus (6.42):

$$\begin{aligned}\hat{L}_{z,\alpha} &= \int_0^{\infty}\int_0^{2\pi} L_{z,\alpha}\, rdrd\varphi \\ &= \frac{1}{1+\alpha^2}\int_0^{\infty}\int_0^{2\pi} rdrd\varphi\big[n(1+\alpha^2)I_{0,n}^2 + \alpha^2(n-2)I_{2,n-2}^2 + (n+2)I_{2,n+2}^2 \\ &\quad + 2(n+1)I_{1,n+1}^2 + 2\alpha^2(n-1)I_{1,n-1}^2 \\ &\quad + 2\alpha\cos(2\varphi)\big((n-1)I_{0,n}I_{2,n-2} + (n+1)I_{0,n}I_{2,n+2} - 2nI_{1,n-1}I_{1,n+1}\big)\big] \\ &= nW_0 + \frac{\alpha^2(n-2)}{1+\alpha^2}W_2 + \frac{(n+2)}{1+\alpha^2}W_2 + \frac{2(n+1)}{1+\alpha^2}W_1 + \frac{2\alpha^2(n-1)}{1+\alpha^2}W_1 \\ &= nW_0 + \left(n + 2\frac{1-\alpha^2}{1+\alpha^2}\right)W_2 + 2\left(n + \frac{1-\alpha^2}{1+\alpha^2}\right)W_1 = nW \\ &\quad + 2\frac{1-\alpha^2}{1+\alpha^2}(W_2 + W_1).\end{aligned} \tag{6.60}$$

From (6.51) and (6.60) follows the expression for the sum of the total longitudinal SAM and OAM at the field focus (6.42):

$$\hat{S}_{z,\alpha} + \hat{L}_{z,\alpha} = \left(n + \frac{1-\alpha^2}{1+\alpha^2}\right)W. \tag{6.61}$$

The total longitudinal OAM in the initial plane is follows from (6.59) to be equal to the expression:

$$\hat{L}_{z,0,\alpha} = \int_0^{\infty}\int_0^{2\pi} L_{z,0,\alpha}\, rdrd\varphi = nW. \tag{6.62}$$

In the initial plane, the sum of the SAM and OAM is follows from (6.52) and (6.62) to be equal to the expression:

$$\hat{S}_{z,0,\alpha} + \hat{L}_{z,0,\alpha} = \left(n + \frac{1-\alpha^2}{1+\alpha^2}\right)W. \tag{6.63}$$

The total longitudinal SAM and OAM for the field (6.42) is seen from comparison of (6.61) and (6.63) to remain unchanged during focusing. By varying the parameter α from 1 to 0, the angular momentum of the field (1) at the focus can be changed in the range:

$$nW \le \hat{S}_{z,\alpha} + \hat{L}_{z,\alpha} \le (n+1)W. \tag{6.64}$$

6.3.6 Simulation

The simulation is carried out utilizing a Richards-Wolf transformation (RW) [23]. The numerical aperture of the spherical lens is assumed to be NA = 0.95, that is, the field is limited by an aperture with a radius of 4 μm, a focal length is $f = 1.31$ μm. The wavelength λ is 532 nm, the incident wave is a Gaussian beam with the radius σ = 1.33 μm, the size of the initial and output fields is 8 × 8 μm, 400 × 400 points, α = 0. The simulation is obtained at a distance from –1 to 1 μm from the focal plane, an apodization is chosen for the case of a spherical lens: $T(\theta) = (\cos\theta)^{1/2}$, and a zone plate (ZP), $T(\theta) = (\cos\theta)^{-3/2}$. In order to avoid dependence on intensity, and also for a more convenient comparison of the results, the OAM and SAM values obtained at a distance z were divided by the total field energy W. The OAM and SAM normalized to the field energy

$$W = \int_0^{\infty}\int_0^{2\pi} |\mathbf{E}(r,\varphi,z)|^2 r dr d\varphi$$

are shown in Fig. 6.5

Shown in Fig. 6.5 is the change in the normalized total longitudinal SAM and OAM for field (6.42) due to a spin–orbit conversion when the field is focused using

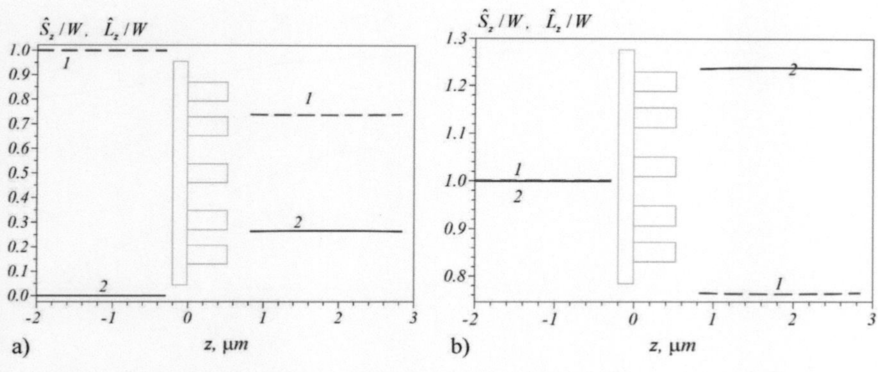

Fig. 6.5 Dependences of $\hat{S}_z/W$ (curve 1) and $\hat{L}_z/W$ (curve 2) on the distance z for a non-vortex field (6.42) with $n = 0$ (**a**) and for an optical vortex with a topological charge $n = 1$ (**b**) and at $\alpha = 0$

a ZP. The numerical aperture of the zone plate is NA = 0.95. The SAM (curve 1) and OAM (curve 2) are calculated for two cases of the topological charge n of the initial beam (6.42): 0 (a) and 1 (b). Shown in Fig. 6.5, the SAM and OAM values are calculated before and after the ZP, but are not calculated inside it. From Fig. 6.5a, the SAM before the lens is seen to be equal to 1, while the OAM is equal to 0, since $n = 0$. After the lens, the SAM decreases and became equal to about 0.76, and the OAM increases up to 0.24. The sum of the SAM and OAM remains unchanged and is equal to 1. Such behavior of the SAM and OAM occurs in accordance with Eqs. (6.51), (6.52) and (6.60), (6.62) at $\alpha = 0$. If the topological charge of the initial field (6.42) is 1, then both the SAM and OAM before the lens are equal to 1 (Fig. 6.5b). Just after the diffractive lens, due to the spin–orbit conversion, the SAM decreases down to 0.76, while the OAM increases and becomes equal to 1.24. The sum of them before and after the lens remains unchanged and is equal to 2. This result confirms the correctness of the Eq. (6.61).

Figure 6.6 shows the change in the normalized total longitudinal SAM and OAM after the lens depending on the α for the initial light field (6.42) at $n = 0$ (a) and $n = 1$ (b). From Fig. 6.6a, at $\alpha = 0$ (circular polarization), the SAM and OAM after the lens is seen to be the same as in Fig. 6.5a. As α increases, the initial circular polarization becomes elliptical and, therefore, the initial SAM decreases (curve 1) and the part of the SAM that converts into the OAM decreases (curve 2). And when $\alpha = 1$ (linear polarization) both the SAM and OAM after the lens (and before the lens) are equal to zero. The behavior of the SAM and OAM after the lens for an optical vortex (6.42) with $n = 1$ is shown in Fig. 6.6b. From it, the maximum SAM for the initial field (6.42) with circular polarization ($\alpha = 0$) is seen to be equal to 0.76. And, the minimal value of the SAM is zero, for the field with the linear polarization ($\alpha = 1$). In this case, the maximal OAM is equal to 1.24 when $\alpha = 0$, and the minimal OAM is equal to 1 at $\alpha = 1$. Such behavior of the SAM and OAM after the lens is described by Eqs. (6.51), (6.60), and (6.61).

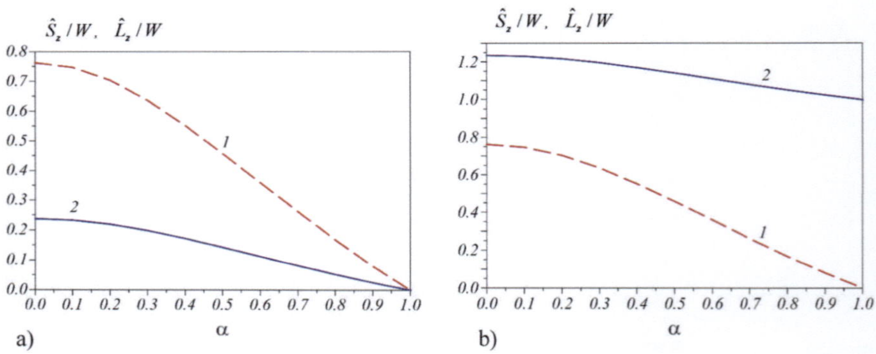

Fig. 6.6 The dependence of $\widehat{S}_z/W$ (curve 1) and $\widehat{L}_z/W$ (curve 2) on α after the spherical lens (or zone plate) for the initial field (6.42) at $n = 0$ (**a**) and $n = 1$ (**b**)

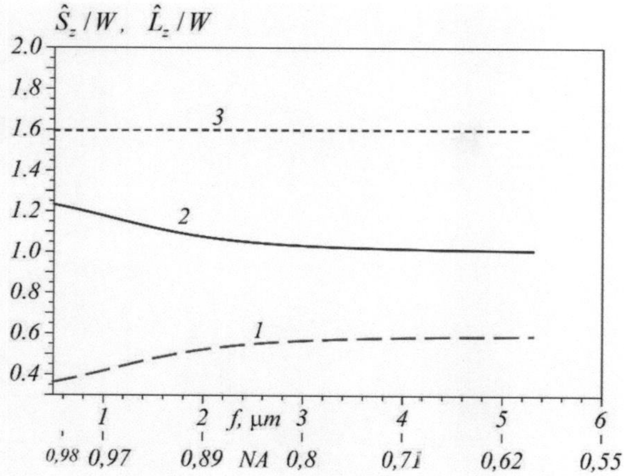

Fig. 6.7 Dependences of $\widehat{S}_z/W$ (curve 1), $\widehat{L}_z/W$ (curve 2) and their sum (curve 3) on the focal length f of a spherical lens or on the numerical aperture NA ($n = 1, \alpha = 0.5$)

Shown in Fig. 6.7 are the normalized total longitudinal SAM and OAM as functions of a numerical aperture and focal distance for the initial field (6.42) at $n = 1$, $\alpha = 0.5$. From Fig. 6.7, the sum of the SAM and OAM is seen to remain unchanged and be equal to 1.6. This follows from Eq. (6.20), since $n + (1-\alpha^2)/(1+\alpha^2) = 1.6$. For the numerical aperture of 0.95 and $\alpha = 0.5$, from Fig. 6.6a, the SAM after the lens is seen to be approximately 0.5, whereas the initial SAM is equal to 0.6. That is, because of the spin–orbit conversion, only a small part of the SAM (about 0.1) transferred into the OAM. From Fig. 6.7, the minimum value of the SAM after the lens is seen to be approximately 0.35 for a numerical aperture close to 1. Therefore, the maximum value of 0.25 is converted from the SAM into the OAM. As the numerical aperture decreases, curve 1 in Fig. 6.7 tends to a value of 0.6, i.e., to the initial SAM value before the lens. That is, it follows from Fig. 6.7 that the effect of the spin–orbit conversion in a spherical lens with a numerical aperture less than 0.7 can be neglected.

Shown in Fig. 6.8 are the intensity I (first and third rows) and the longitudinal SAM S_z (second and fourth rows) for different values of the parameter α: 0 (first column), 0.5 (second column), 0.75 (third column), 0.9 (fourth column), and 1 (fifth column). The topological charges are as follows: $n = 3$ (first and second rows), $n = 5$ (third and fourth rows).

It is seen from Fig. 6.8 that, according to the theory (6.47)–(6.49), the distribution of the longitudinal SAM is radially symmetric (red ring in the first and second columns) in the focus of the field (6.42) for a small parameter α. As the parameter α increases, the red ring in Fig. 6.8 becomes an ellipse (second and third column), and when α is close to 1, one part of the ring remains red, and another part becomes blue (fourth and fifth columns). That is, areas with a negative SAM appear in the focus and the spin Hall effect appears (separation of left- and right-handed elliptical polarizations). When the parameter α increases from 0 to1, the intensity (green rings in Fig. 6.8) gradually transforms from a round ring (at initial circular polarization) into an elliptical one (at initial linear polarization), which is elongated along the

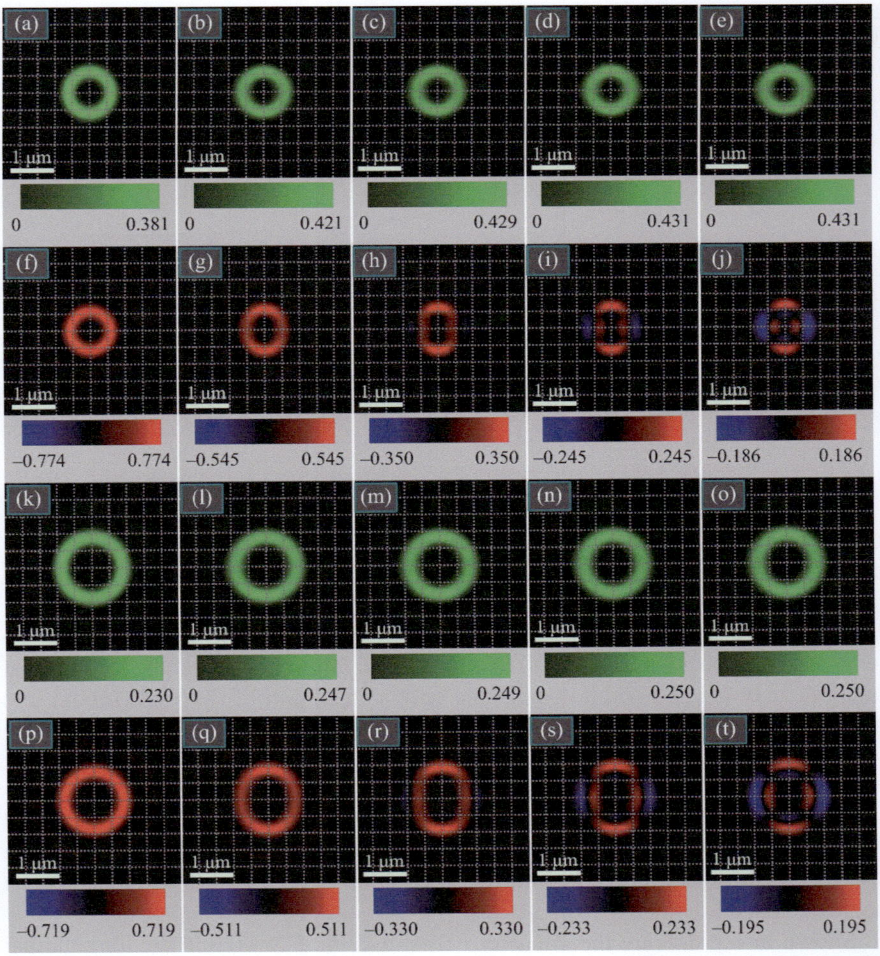

Fig. 6.8 Intensity distributions I (first and third rows) and longitudinal SAM S_z (second and fourth rows) calculated at the beam focus (1) for different α: 0 (first column), 0.5 (second column), 0.75 (third column), 0.9 (fourth column), and 1 (fifth column). The topological charges are as follows: $n = 3$ (first and second rows), $n = 5$ (third and fourth rows)

horizontal axis x. However, since the term with $I_{0,n}^2$ exceeds the term with $\cos(2\varphi)$ in the formula (6.53) for the intensity distribution, the intensity ellipticity in Fig. 6.8 is almost invisible at $\alpha = 1$. With an increase in the topological charge of the optical vortex (1) from $n = 3$ to $n = 5$, the diameter of the light ring shown in Fig. 6.4 increases. This follows from the intensity Eq. (6.53).

In this section, we study a spin–orbit conversion in a spherical lens with a high numerical aperture. Utilizing the Richards-Wolf theory [23], analytical formulas are obtained for the density of the longitudinal SAM and OAM projections at the focus of the lens for the initial light field (6.42). The amplitude of the field depends on the real parameter α. The initial light field (6.42) is an axial superposition of two optical vortices with the same topological charge n and right- and left-handed circular polarizations. Amplitudes of both beams in the superposition are different. The parameter α is the ratio of the amplitudes of the beams. In this section, analytical formulas are obtained for the total longitudinal projections of the SAM and OAM at the focus of the field (6.42). It follows from the formulas (6.51), (6.52) and (6.60), (6.62) that when passing through a spherical lens (or a diffractive lens), the total SAM decreases due to a spin–orbit conversion, while the OAM increases by the same amount. In this case, the sum of the SAM and OAM remains unchanged during focusing (Eq. (6.61)). The simulations are also carried out using the Debye and Richards-Wolf integrals [23]. When the light field (6.42) propagates before a spherical lens, the total SAM and OAM is shown to remain unchanged separately. After passing through the lens, the total SAM decreases, and the total OAM increases by the same value. When propagating after the lens, the total SAM and OAM remain unchanged separately too. The sum of the SAM and OAM before the lens and after it does not change (Fig. 6.5). When the parameter α increases from 0 to 1, the spin–orbit conversion in a spherical lens is shown numerically to decrease from its maximum value to zero. With a numerical aperture of 0.95 and with $\alpha = 0$, the initial normalized total longitudinal SAM is equal to 1 before the spherical lens, and equal to 0.76 after it. That is, only a quarter of the SAM transferred to the OAM (Fig. 6.6). The maximum spin–orbit effect in a spherical lens is shown numerically to occur at a numerical aperture close to 1 (Fig. 6.7). For smaller aperture values, the effect decreases. It is shown that, at a numerical aperture less than 0.7, the spin–orbit conversion can be neglected, since the total SAM remains almost unchanged when light passes through a spherical lens. Theoretically (Eq. (6.48)) and numerically (Fig. 6.8) it is shown that there is a spin Hall effect at the focus of the light field (6.42). This effect leads to appearance of regions with elliptical polarizations with different signs in the sharp focus. That is, at the focus, light with left- and right-handed elliptical polarizations is separated in space. In this case, in the initial plane only linearly polarized light present (Fig. 6.8). By changing the parameter α the spin Hall effect at the focus can be controlled. When $\alpha = 0$, the initial field is circularly polarized and the Hall effect is minimal at the focus. When $\alpha = 1$, there is a linearly polarized initial field, and the Hall effect achieve its maximum at the focus.

Measuring the magnitude of the spin-orbital conversion effect in the sharp focus of a microobjective with a high numerical aperture requires measuring the third component of the Stokes vector averaged over the beam transverse section. The value of this quantity gives the full longitudinal SAM component. Besides, the third component of the Stokes vector can be measured immediately in the microobjective's exit pupil with a diameter of several millimeters, rather than in the focus itself whose size can be less than a micron.

The effects of spin–orbit conversion in a spherical lens and the spin Hall effect at the focus considered in this work can be used to control the rotation of microparticles [80, 81]. Studies on the control of the Hall effect in focus have been done in [82].

6.4 Optical Helicity of Light in the Tight Focus

Optical helicity of circularly polarized light is characterized by a parameter $\sigma = \pm 1$, similar to the spin magnitude. A light field with left circular polarization has the helicity of $h = -1$, while a field with right circular polarization has the helicity of $h = 1$. For linearly polarized light, the helicity is zero, whereas for elliptically polarized light, the helicity modulus is less than unity. In the general form, the optical helicity was defined in several works (e.g., see [83–85]), meanwhile the magnetic helicity was considered back in 1958 [86]. The full helicity, defined based on the electromagnetic democracy principle, is given by

$$\hat{h} = \frac{1}{2} \int\limits_{3D} (\mathbf{AB} - \mathbf{CE})\, dV, \quad \mathbf{E} = -\nabla \times \mathbf{C}, \quad \mathbf{B} = \nabla \times \mathbf{A}, \tag{6.65}$$

where **E**, **B** are the strength vectors of the electric and magnetic fields, **A**, **C** are the electric and magnetic vector potentials. The helicity density is the integrand:

$$h = \frac{1}{2}(\mathbf{AB} - \mathbf{CE}). \tag{6.66}$$

The gauge transformation for the helicity is given by $\mathbf{A} \to \mathbf{A} + \nabla f, \mathbf{C} \to \mathbf{C} + \nabla g$ with f and g being arbitrary scalar functions. This transformation conserves the helicity since the curl of a gradient is zero.

The work [87] has shown a relation between the helicity and chirality, with the latter density defined as follows:

$$\chi = \frac{\varepsilon_0}{2}\left[\mathbf{E}(\nabla \times \mathbf{E}) + c^2\mathbf{B}(\nabla \times \mathbf{B})\right], \tag{6.67}$$

where ε_0 is the vacuum electric permittivity, c is the speed of light in vacuum. As demonstrated in a general form in [88], full helicity and chirality are propagation invariants. The work [89] has shown the difference between the helicity and the chirality. In a circularly polarized light field, each photon has the helicity of $\pm\hbar$ and the chirality of $\pm\hbar\omega^2/c$, where $\hbar$ is the Planck constant, ω is the angular frequency, c is the speed of light in vacuum. In [90], a general relation was considered between the helicity, chirality, and spin density. The helicity and spin densities are written as follows:

$$\begin{aligned} h &= \frac{1}{2}\left[\sqrt{\frac{\varepsilon_0}{\mu_0}}(\mathbf{E}(\nabla \times \mathbf{E})) + \sqrt{\frac{\mu_0}{\varepsilon_0}}(\mathbf{B}(\nabla \times \mathbf{B}))\right], \\ S &= \frac{1}{2}[\varepsilon_0(\mathbf{E} \times \mathbf{A}) + (\mathbf{B} \times \mathbf{C})], \end{aligned} \tag{6.68}$$

with μ_0 being the vacuum permeability. In free space, for monochromatic light with angular frequency ω, the magnetic helicity is equal to the electric helicity, and thus the helicity in free space can be written in the following form [91]:

$$h = -\frac{\sqrt{\varepsilon_0 \mu_0}}{2\omega} \mathrm{Im}\big(\mathbf{E}^* \mathbf{H}\big), \tag{6.69}$$

with **H** being the magnetic field strength vector in free space. Hereinafter, we use the above definition of the helicity density. Spin-to-orbit conversion discussed in [25] is also a matter of interest in this work thanks to the helicity property. In [91], the authors computed densities of some quantities (spin, energy flow, and helicity) for optical vortices propagating in various waveguides (planar waveguide, circular step-index waveguide). It was possible to carry out such an analysis as the components of the electric and magnetic field vectors of modes of those waveguides are well known. The said work has prompted us to investigate the helicity at the tight focus of laser beams. Obtaining explicit expressions for space distributions of light field characteristics, such as the energy density (intensity), energy flow (Poynting vector), spin angular momentum (SAM), orbital angular momentum (OAM), helicity, and others, is an interesting and important problem, allowing these quantities to be analyzed without numerical simulation. For many vector light fields, the Richards-Wolf theory [23] allows deriving exact analytical expressions to describe all field characteristics near the focus.

In this section, we analyze four different initial vector fields and derive relationships for helicity densities in the initial plane and at the focus, as well as for full helicities. The full helicities are obtained by integration only over the beam section rather than over the whole space. We consider the following initial fields: a linearly polarized optical vortex, a circularly polarized optical vortex, superposition of a cylindrical vector beam and a linearly polarized beam, and a hybrid light field with circular-azimuthal polarization. Previously, based on the Richards-Wolf theory [23], expressions for all components of the electric and magnetic field vectors at the tight focus of all fields of interest have been derived [46, 67, 69, 77, 92]. Using these expressions for the light field components we can derive analytical expressions for the helicity of these fields at the focus. In this work, we show that the helicity density at the focus is related to the longitudinal component of the SAM density. Note, however, that the full longitudinal SAM does not conserve when a light field passes through a spherical lens since as thanks to the spin-orbital conversion, the spin is partly converted to the OAM. On the contrary, the full helicity (helicity density averaged over the beam section) is the invariant and conserves upon free-space propagation and focusing of light.

6.4.1 Helicity at the Focus of a Linearly Polarized Optical Vortex

If the initial vector field is a linearly polarized optical vortex, its Jones vectors for the electric and magnetic field are given by the following:

$$\mathbf{E}(\theta,\varphi)=A(\theta)\exp(in\varphi)\begin{pmatrix}1\\0\end{pmatrix},\ \mathbf{H}(\theta,\varphi)=A(\theta)\exp(in\varphi)\begin{pmatrix}0\\1\end{pmatrix}, \tag{6.70}$$

where φ is the azimuthal angle in the beam cross section, θ is the angle between the optical axis and a line drawn from the focus center to a spherical wavefront point in the initial plane, n is integer topological charge of the optical vortex, $A(\theta)$ is a circularly symmetric real-valued function defining the beam amplitude in the initial plane. Using a Richards-Wolf method [23], expressions were obtained in [67] for the intensity distribution and the spin angular momentum (SAM) at the focus of field (6.70). For comparison purposes, below we give those expressions, analyzing them in relation to the helicity distribution at the focus of field (6.70), derived herein. The distributions of the intensity $I=|E_x|^2+\left|E_y\right|^2+|E_z|^2$ and the longitudinal SAM component $S_z=2\,\mathrm{Im}\left(E_x^*E_y\right)$ are given by the following:

$$\begin{aligned}I=\frac{1}{2}\big[&2I_{0,n}^2+I_{2,n+2}^2+I_{2,n-2}^2+2I_{1,n+1}^2+2I_{1,n-1}^2\\&+2\cos(2\varphi)\left(I_{0,n}I_{2,n+2}+I_{0,n}I_{2,n-2}-2I_{1,n+1}I_{1,n-1}\right)\big],\end{aligned} \tag{6.71}$$

$$S_z=\frac{1}{2}\left(I_{2,n+2}-I_{2,n-2}\right)\left(I_{2,n+2}+I_{2,n-2}+2\cos(2\varphi)I_{0,n}\right). \tag{6.72}$$

Equations (6.71) and (6.72) contain functions $I_{\nu,\mu}$ that depend only on the radial and longitudinal variables, r and z:

$$I_{\nu,\mu}=2kf\int\limits_0^{\alpha}\sin^{\nu+1}\left(\frac{\theta}{2}\right)\cos^{3-\nu}\left(\frac{\theta}{2}\right)\cos^{1/2}(\theta)A(\theta)e^{ikz\cos\theta}J_{\mu}(kr\sin\theta)d\theta, \tag{6.73}$$

where $k=2\pi/\lambda$ is the wavenumber of monochromatic light of wavelength λ; f is the lens focal length; α is the maximal tilt angle of light rays to the optical axis, which defines the numerical aperture of an aplanatic lens, $NA=\sin\alpha$; $J_{\mu}(kr\sin\theta)$ is the μth-order Bessel function of the first kind. In Eq. (6.72) and throughout the paper, the indices ν and μ can take the following values: $\nu=0,1,2$; $\mu=n-2,n-1,n,n+1,n+2$.

Further, we obtain the helicity distribution at the focus of field (6.70), using the helicity definition from [91]:

$$h=-\frac{\sqrt{\varepsilon_0\mu_0}}{2\omega}\mathrm{Im}\left(\mathbf{E}^*\mathbf{H}\right), \tag{6.74}$$

where ε_0 and μ_0 are the vacuum electric permittivity and the vacuum magnetic permeability, ω is the angular frequency of light, **E** and **H** are the electric and magnetic field vectors. Below, we omit the constant $\sqrt{\varepsilon_0\mu_0}/(2\omega)$ for brevity. We note that the helicity (6.74) has the same dimension as the SAM (6.72). Using expressions for the components of the electric and magnetic vectors at the focus of field (6.70), derived in [67], Eq. (6.74) is rearranged to:

$$h(r) = \frac{1}{2}\left(I_{2,n+2}^2 - I_{2,n-2}^2 + 2I_{1,n+1}^2 - 2I_{1,n-1}^2\right). \tag{6.75}$$

As seen from Eq. (6.75), the helicity distribution is circularly symmetric. We note that despite zero helicity of field (6.70) in the initial plane ($h = 0$), in the focal plane the helicity has a nonzero density (helicity at each point of the beam cross section). Comparison of Eqs. (6.72) and (6.75) indicates that if the numerical apertures are small, i.e., when the longitudinal field component can be neglected ($2\left|I_{1,n+1}^2 - I_{1,n-1}^2\right| \ll 1$), the helicity coincides with the longitudinal SAM (6.72) at the polar angles $\varphi = \pi/4 + \pi p/2$ ($p = 0,1,2,3$), at which $\cos 2\varphi = 0$. Then, the helicity and the SAM coincide: $h = S_z$. From Eq. (6.72), it follows that if $n = -2$, then, the light is right-handed elliptically polarized near the optical axis and the positive helicity is equal to the SAM, coinciding with the intensity: $h(r = 0) = S_z(r = 0) = I(r = 0) = I_{2,0}^2/2 > 0$. And, vice versa, at $n = 2$, the helicity and the SAM are negative at the focus center: $h(r = 0) = S_z(r = 0) = -I_{2,0}^2/2 < 0$. Thus, the helicity sign coincides with the SAM sign, or, in other words, the helicity is positive in focal regions with right-handed elliptic polarization, being negative in focal regions with left-handed elliptic polarization. The difference between the helicity and the SAM is that at $n = 1$ or $n = -1$, the on-axis helicity (6.75) is nonzero, whereas the on-axis SAM (6.72) is zero. Thus, the helicity (6.75) is not only related to spin and indicates focal areas with left- or right-handed elliptic polarization, but also indicates the transverse helicity of the focal field, related to the longitudinal component of the electric field vector. It can be shown that the full helicity, averaged over the beam cross section, equals zero and conserves upon focusing. Indeed, as shown in [77], the following expression holds:

$$W_{\nu,\mu} = 2\pi \int\limits_0^{\infty} \left|I_{\nu,\mu}(r)\right|^2 r dr = 4\pi f^2 \int\limits_0^{\alpha} \sin^{2\nu+1}\left(\frac{\theta}{2}\right)\cos^{5-2\nu}\left(\frac{\theta}{2}\right)|A(\theta)|^2 d\theta = W_\nu. \tag{6.76}$$

From Eq. (6.76), we get the averaged helicity:

$$\begin{aligned}\hat{h} &= \int\limits_0^{\infty}\int\limits_0^{2\pi} h(r) r dr d\varphi = \frac{1}{2}\int\limits_0^{\infty}\int\limits_0^{2\pi}\left(I_{2,n+2}^2 - I_{2,n-2}^2 + 2I_{1,n+1}^2 - 2I_{1,n-1}^2\right) r dr d\varphi \\ &= \frac{1}{2}(W_2 - W_2 + 2W_1 - 2W_1) = 0.\end{aligned} \tag{6.77}$$

The full longitudinal SAM, averaged over the focal plane based on Eq. (6.72), is also zero:

$$\hat{S}_z = \frac{1}{2}\int_0^\infty\int_0^{2\pi} rdrd\varphi\left(I_{2,n+2} - I_{2,n-2}\right)\left(I_{2,n+2} + I_{2,n-2} + 2\cos(2\varphi)I_{0,n}\right) = W_2 - W_2 = 0. \quad (6.78)$$

6.4.2 Helicity at the Focus of a Circularly Polarized Optical Vortex

In this section, we investigate the helicity at the focus of a circularly polarized optical vortex in a similar way. The Jones vectors for the initial electric and magnetic fields read as

$$\mathbf{E}(\theta,\varphi) = \frac{A(\theta)}{\sqrt{2}}\exp(in\varphi)\begin{pmatrix}1\\ i\end{pmatrix},\ \mathbf{H}(\theta,\varphi) = \frac{A(\theta)}{\sqrt{2}}\exp(in\varphi)\begin{pmatrix}-i\\ 1\end{pmatrix}. \quad (6.79)$$

The amplitudes in the initial field (6.79) are given for right circular polarization. In the initial plane, the longitudinal SAM component of field (6.79) equals $S_z = A^2(\theta)$, whereas the full spin in the initial plane is equal to the full beam energy:

$$\hat{S}_z = 2\pi\int_0^\infty\int_0^{2\pi} A^2(\theta)\, rdr = W. \quad (6.80)$$

In [92], adopting the Richards-Wolf method [23], expressions were obtained for the intensity and the longitudinal SAM component at the focus. We write these expressions here for a comparison with a helicity expression that is derived below. Distributions of the intensity and the longitudinal SAM component at the focus of field (6.79) are given by the following:

$$I(r,\varphi) = |E_x|^2 + \left|E_y\right|^2 + |E_z|^2 = I_{0,n}^2 + I_{2,n+2}^2 + 2I_{1,n+1}^2, \quad (6.81)$$

$$S_z = 2\,\mathrm{Im}\left(E_x^* E_y\right) = I_{0,n}^2 - I_{2,n+2}^2. \quad (6.82)$$

Due to circular polarization of the initial field, both the intensity distribution (6.81) and the spin density distribution (6.82) have circular symmetry at the focus. The helicity distribution (6.74) at the focus of field (6.79) is equal to the intensity:

$$h_R(r) = I(r) = I_{0,n}^2 + I_{2,n+2}^2 + 2I_{1,n+1}^2. \quad (6.83)$$

Similar to Eq. (6.77), the full helicity of beam (6.79) at the focus is given by the following:

$$\begin{aligned}\hat{h}_R &= \int_0^\infty\int_0^{2\pi} h_R(r)rdrd\varphi = W = \int_0^\infty\int_0^{2\pi} \left(I_{0,n}^2 + I_{2,n+2}^2 + 2I_{1,n+1}^2\right)rdrd\varphi \\ &= W_0 + W_2 + 2W_1,\end{aligned} \tag{6.84}$$

with W being the total beam power/energy. The full longitudinal SAM component at the focus of field (6.79) is given by the following:

$$\hat{S}_z = \int_0^\infty\int_0^{2\pi} rdrd\varphi\left(I_{0,n}^2 - I_{2,n+2}^2\right) = W_0 - W_2. \tag{6.85}$$

In the initial plane, the helicity density and the full helicity of field (6.79) read as follows:

$$h_{0R}(r) = |A(\theta)|^2,\ \hat{h}_{0R} = W. \tag{6.86}$$

A comparison of Eqs. (6.84) and (6.86) reveals that the full helicity conserves upon focusing, which is in contrast with the full spin. As seen from comparison of Eqs. (6.80) and (6.85), the full spin is not conserved. Instead, it decreases and, due to the spin–orbit conversion, the spin is partly transferred to the orbital angular momentum (OAM).

It can also be seen that for circular polarization, the helicity achieves its maximal value and equals the beam power. For left circular polarization, the helicity in the initial plane and at the focus changes sign:

$$\begin{aligned}&h_L(r) = -I(r) = -\left(I_{0,n}^2 + I_{2,n-2}^2 + 2I_{1,n-1}^2\right),\\ &h_{0L}(r) = -|A(\theta)|^2,\quad \hat{h}_{0L} = \hat{h}_L = -W,\ r = f\sin\theta,\end{aligned} \tag{6.87}$$

We note that the on-axis helicity magnitude depends on the topological charge and, with increasing modulus of n, decays from the maximum (at $n = 0$) to zero (at $n > 2$ or $n < -2$):

$$\begin{aligned}&h_R(r=0) = I_{0,0}^2,\quad n = 0,\\ &h_R(r=0) = I_{2,0}^2,\quad n = -2,\\ &h_R(r=0) = 2I_{1,0}^2,\quad n = -1,\\ &h_R(r=0) = 0,\quad |n| > 2.\end{aligned} \tag{6.88}$$

The full helicity for the circularly polarized optical vortex is independent of the topological charge n and equal to the beam power, W, for right-handed circular polarization, taking the opposite sign, $-W$, for left-handed circular polarization.

6.4.3 Helicity at the Focus of a Cylindrical Vector Beam

In the initial plane, a high-order cylindrical vector beam has the following Jones vectors:

$$\mathbf{E}(\theta,\varphi)=A(\theta)\begin{pmatrix}\cos m\varphi\\ \sin m\varphi\end{pmatrix},\quad \mathbf{H}(\theta,\varphi)=A(\theta)\begin{pmatrix}-\sin m\varphi\\ \cos m\varphi\end{pmatrix}. \tag{6.89}$$

As shown in [46], the longitudinal component S_z of the SAM vector at the focus of field (6.89) is zero. It can be shown that the helicity density and the full helicity in the initial plane (6.89) and at the focus is also zero ($h(r)=0$). However, if the cylindrical vector field (6.89) is coherently superimposed with a linearly polarized field, we obtain an initial light field

$$\mathbf{E}(\theta,\varphi)=A(\theta)\begin{pmatrix}\cos m\varphi-a\\ \sin m\varphi\end{pmatrix},\quad \mathbf{H}(\theta,\varphi)=A(\theta)\begin{pmatrix}-\sin m\varphi\\ \cos m\varphi-a\end{pmatrix}, \tag{6.90}$$

with a nonzero helicity density at the focus. Indeed, in [69], analytical expressions were obtained for the electric and magnetic field components at the focus of the initial field (6.90). Using these expressions and the definition in (6.74), we can derive a formula for the helicity density at the focus of field (6.90):

$$h(r,\varphi)=\begin{cases}-2a(-1)^p\big(I_{0,0}I_{0,m}+I_{2,2}I_{2,m-2}-2I_{1,1}I_{1,m-1}\big)\sin(m\varphi),\\ m=2p+1,\\ 0,\ m=2p.\end{cases} \tag{6.91}$$

According to Eq. (6.91), the helicity of a cylindrical vector beam at the focus is nonzero only for an odd order ($m=2p+1$). We note that at the focus of beam (6.90), the spin density is also nonzero only for an odd order $m=2p+1$ [69]:

$$S_z(r,\varphi)=\begin{cases}2a(-1)^p\big[\sin((m-2)\varphi)\big(I_{0,0}I_{2,m-2}-I_{2,2}I_{0,m}\big)\\ -\sin(m\varphi)\big(I_{0,0}I_{0,m}-I_{2,2}I_{2,m-2}\big)\big],\quad m=2p+1,\\ 0,\quad m=2p,\quad p=0,1,2,\ldots\end{cases} \tag{6.92}$$

A comparison of Eqs. (6.91) and (6.92) indicates that the helicity changes sign in approximately the same areas of the focal spot at which the spin density changes sign, since both quantities depend on $\sin(m\varphi)$. Besides, with the terms with $I_{0,0}I_{0,m}$ contribute the most [77], the magnitudes h and S_z have the same sign. The difference is, in particular, that the expression for the helicity in (6.91) includes the longitudinal component of the electric field, whereas the spin density in (6.92) includes only the transverse components of the electric field. The full helicity $\hat{h}$, i.e., the helicity density (6.91) averaged over the focal plane, as well as the full spin $\widehat{S}_z$, i.e., the averaged spin density (6.92), are both equal to zero ($\hat{h} = \widehat{S}_z = 0$), since the integrals over the polar angle in Eqs. (6.91) and (6.92) yield zeros.

6.4.4 Helicity at the Focus of a Field with Hybrid Circular-Azimuthal Polarization

Here, we consider the tight focusing of light with initial hybrid circular-azimuthal polarization. At different polar angles, polarization of this field changes from linear to elliptic and to circular polarization. The Jones vectors of the initial electric and magnetic fields of such a hybrid field read as

$$\mathbf{E}(\theta, \varphi) = A(\theta)\begin{pmatrix} -i\sin(m\varphi) \\ \cos(m\varphi) \end{pmatrix}, \ \mathbf{H}(\theta, \varphi) = A(\theta)\begin{pmatrix} -\cos(m\varphi) \\ -i\sin(m\varphi) \end{pmatrix}. \tag{6.93}$$

The initial field (6.93) has the spin density equal to $S_z = A^2(\theta)\sin(2m\varphi)$, whereas the full spin, averaged over the entire beam section, is equal to zero:

$$\widehat{S}_z = \int_0^\infty \int_0^{2\pi} A^2(\theta)\sin(2m\varphi) r dr d\varphi = 0. \tag{6.94}$$

As shown in [93], at the focus of field (6.93), the density of the longitudinal component of the SAM vector is nonzero and reads as follows:

$$S_z = \frac{1}{4}\left[I_{0,m}\left(I_{2,m+2} - I_{2,m-2}\right)\sin 2\varphi + \left(I_{0,m}^2 - I_{2,m-2}I_{2,m+2}\right)\sin(2m\varphi)\right]. \tag{6.95}$$

According to Eq. (6.76), the full longitudinal component of the SAM vector is zero:

$$\widehat{S}_z = \frac{1}{4}\int_0^\infty \int_0^{2\pi} r dr d\varphi \left[I_{0,m}\left(I_{2,m+2} - I_{2,m-2}\right)\sin 2\varphi + \left(I_{0,m}^2 - I_{2,m-2}I_{2,m+2}\right)\sin(2m\varphi)\right] = 0. \tag{6.96}$$

The integrals over the angle are equal to zero in Eqs. (6.94) and (6.96), since the integration of the periodic function is done over an integer number of periods. From Eqs. (6.94) and (6.96), it follows that the full spin is zero and conserves upon focusing. Using the components of the strength vectors of field (6.93) at the focus, computed in [93], an expression can be obtained for the helicity density:

$$h = \left(I_{0,m}^2 + I_{2,m+2}I_{2,m-2} - 2I_{1,m+1}I_{1,m-1}\right)\sin(2m\varphi). \tag{6.97}$$

As seen from Eq. (6.97), the helicity, as the SAM (6.95), changes sign $4m$ times along a certain-radius circle around the optical axis. Since the terms with $I_{0,m}^2$ contribute the most [77], approximate expressions for the SAM (6.95) and for the helicity (6.97) are almost identical and equal to $S_z \approx h \approx I_{0,m}^2 \sin(2m\varphi)$. From Eq. (6.97), it also follows that the full helicity at the focus is zero, like the full spin in (6.96). In the initial plane of field (6.93), the helicity density is also coincident with the longitudinal SAM density: $h = S_z = A^2(\theta)\sin(2m\varphi)$, whereas the full helicity and the full SAM in the initial plane are equal to zero: $\hat{h} = \widehat{S}_z = 0$. Thus, the full helicity and the full longitudinal SAM component are zero and conserve upon focusing.

6.4.5 Numerical Simulation

Shown in Fig. 6.9 are the intensity, helicity, and SAM density distributions for linearly polarized beam (6.70) at the tight focus for different values of the topological charge. The helicity and SAM density distributions in Fig. 6.9 and in all the figures below are obtained by formulae $S_z = 2\,\mathrm{Im}\left(E_x^* E_y\right)$ and $h = -\mathrm{Im}(\mathbf{E}^*\mathbf{H})$ with the components of the strength vectors **E** and **H** obtained using the Richards-Wolf formulae, as well as Eqs. (6.75) and (6.72). The computed images are visually indistinguishable.

As seen from Fig. 6.9, indeed, in contrast to the SAM density, the helicity distribution is circularly symmetric. Besides, Fig. 6.9 confirms the negative helicity in the focus center at $n = 2$, as well as the nonzero helicity on the optical axis at $n = \pm 1$.

Figure 6.10 also illustrates the intensity, helicity, and SAM density distributions at the tight focus for different values of the topological charge, but for a beam with right circular polarization (6.79).

Figure 6.10 confirms that if the initial field is circularly polarized, then the intensity, helicity, and the SAM density distributions at the focus are circularly symmetric. In addition, Fig. 6.10 demonstrates that the helicity distribution at the focus coincides with the intensity distribution. The maximal helicity value is equal to the maximal intensity value, whereas the maximal SAM density value is smaller due to the spin–orbit conversion, when the SAM is partly converted into the OAM. Nevertheless, the maximal SAM density value is much higher than when the initial field is linearly polarized (Fig. 6.9).

Shown in Fig. 6.11 are the intensity, helicity, and SAM density distributions at the tight focus for different values of the polarization singularity index for the superposition of two beams with cylindrical and linear polarization states (6.90).

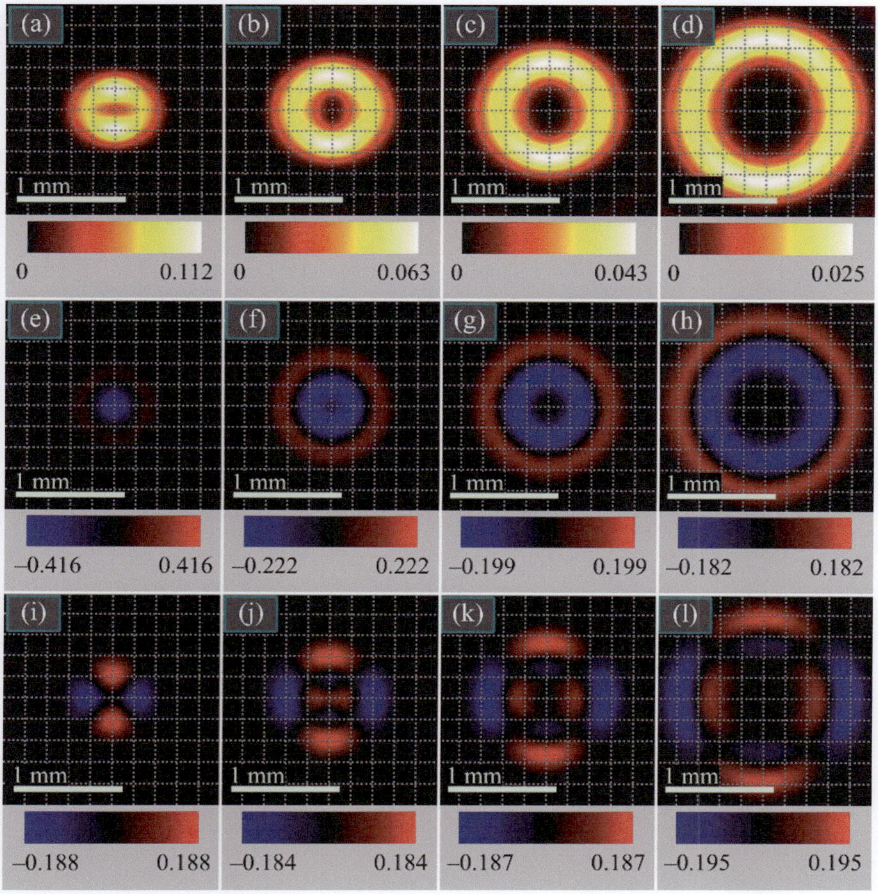

Fig. 6.9 Intensity (**a–d**), helicity (**e–h**), and SAM density (**i–l**) distributions of linearly polarized optical vortex (6.70) at the tight focus for the following parameters: wavelength $\lambda = 532$ nm, focal length of the lens $f = 10$ μm, numerical aperture of the lens NA = 0.95, topological charges $n =$ 1 (**a**, **e**, **i**), 2 (**b**, **f**, **j**), 3 (**c**, **g**, **k**), 5 (**d**, **h**, **l**), radial amplitude function $A(\theta) \equiv 1$ (plane beam). The numbers near the colorbars under each figure denote the maximal and minimal values. Maximal intensity values are given in relative units and proportional to the maximal intensities for different topological charges. Maximal values of the helicity and the SAM density are normalized to the maximal intensity values

According to Fig. 6.11, the helicity and the SAM density at the focus are indeed nonzero only for an odd-order cylindrical vector beam, $m = 2p + 1$. It is also seen that the helicity changes sign at near-same polar angles in the focal plane, as does the SAM density (the outer rings in Fig. 6.11g, k, h, l).

Finally, Fig. 6.12 demonstrates the intensity, helicity, and SAM density distributions at the tight focus for different values of the polarization singularity index, but for a beam with hybrid circular-azimuthal polarization (6.93).

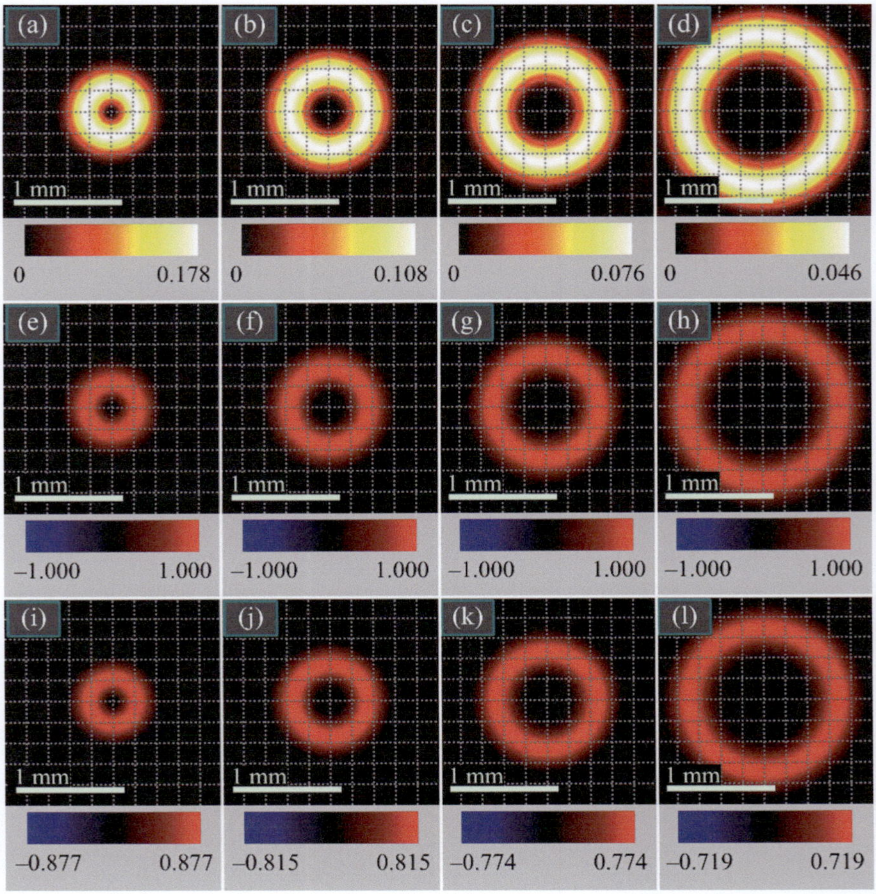

Fig. 6.10 Intensity (**a–d**), helicity (**e–h**), and SAM density (**i–l**) distributions of a optical vortex with right circular initial polarization (6.79) at the tight focus for the following parameters: wavelength $\lambda = 532$ nm, focal length of the lens $f = 10$ μm, numerical aperture of the lens NA = 0.95, topological charges $n = 1$ (**a**, **e**, **i**), 2 (**b**, **f**, **j**), 3 (**c**, **g**, **k**), 5 (**d**, **h**, **l**), radial amplitude function $A(\theta) \equiv 1$ (plane beam). The numbers near the colorbars under each figure denote the maximal and minimal values. Maximal intensity values are given in relative units and proportional to the maximal intensities for different topological charges. Maximal values of the helicity and of the SAM density are normalized to the maximal intensity values

As Fig. 6.12 reveals, in contrast to the superposition of beams with cylindrical and linear polarization, the helicity and the SAM at the focus change sign $4m$, rather than $2m$, times along a certain-radius circle around the optical axis. Besides, Fig. 6.12 also confirms that if the numerical aperture is high enough, then the approximate expressions for the SAM and for the helicity are very similar.

Thus, the numerical simulation confirms the theoretical expressions and the properties of the helicity and the SAM density. As seen in Figs. 6.9, 6.10, 6.11 and 6.12,

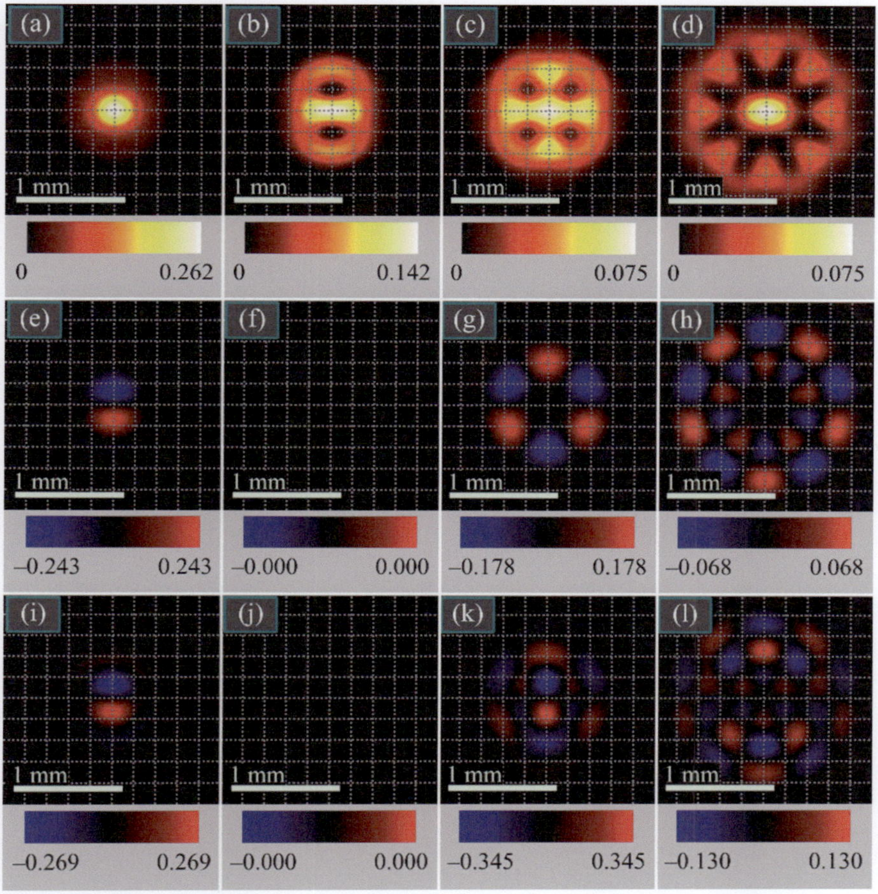

Fig. 6.11 Intensity (**a–d**), helicity (**e–h**), and SAM density (**i–l**) distributions of superposition of beams with cylindrical and linear polarization (6.90) at the tight focus for the following parameters: wavelength $\lambda = 532$ nm, focal length of the lens $f = 10$ μm, lens numerical aperture, NA = 0.95, polarization singularity index $m = 1$ (**a**, **e**, **i**), 2 (**b**, **f**, **j**), 3 (**c**, **g**, **k**), 5 (**d**, **h**, **l**), radial amplitude function $A(\theta) \equiv 1$ (plane beam), amplitude of the linearly polarized term $a = 0.5$. The numbers near the colorbars under each figure denote the maximal and minimal values. Maximal intensity values are given in relative units and proportional to the maximal intensities for different index. Maximal values of the helicity and of the SAM density are normalized to the maximal intensity values

much greater values of the helicity and the SAM density are achieved for fields that are nonlinearly polarized in the initial plane, as is the case for the circularly polarized beams (Fig. 6.10) and beams with hybrid circular-azimuthal polarization (Fig. 6.12).

For four different vector fields, we have obtained exact expressions for the helicity density at the tight focus. The expressions derived are similar to the expressions for the longitudinal SAM component at the focus, with the difference being that the helicity expressions include terms consisting of angular harmonics that are present

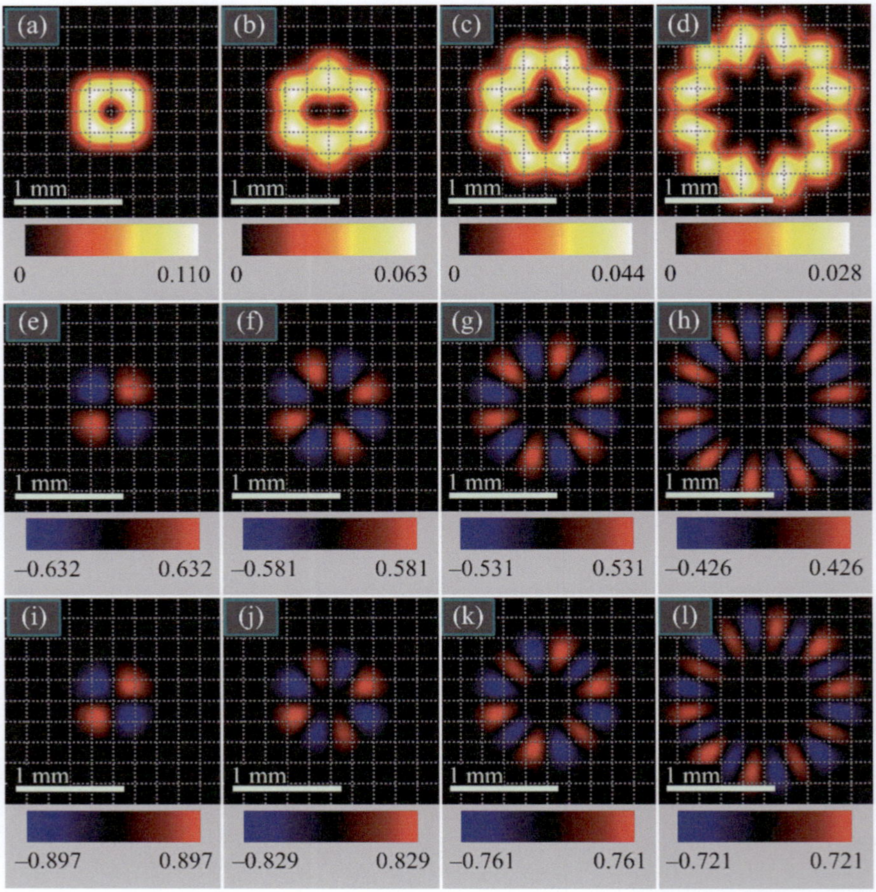

Fig. 6.12 Intensity (**a–d**), helicity (**e–h**), and SAM density (**i–l**) distributions of a beam with hybrid circular-azimuthal initial polarization (6.93) at the tight focus for the following parameters: wavelength $\lambda = 532$ nm, focal length of the lens $f = 10$ μm, numerical aperture of the lens NA = 0.95, polarization singularity index $m = 1$ (**a, e, i**), 2 (**b, f, j**), 3 (**c, g, k**), 5 (**d, h, l**), radial amplitude function $A(\theta) \equiv 1$ (plane beam). The numbers near the colorbars under each figure denote the maximal and minimal values. Maximal intensity values are given in relative units and proportional to the maximal intensities for different topological charges. The maximal values of the helicity and SAM density are normalized to the maximal intensity values

only in the longitudinal components of the field strength vectors. If the full SAM in the initial plane is zero, then it conserves upon propagation and equals zero at the focus. In this case, the full helicity in the initial plane and in the focus is also zero and conserves upon focusing. If, however, the full SAM in the initial plane is nonzero, then does not conserve upon focusing, but is partially converted to the longitudinal OAM component due to the spin–orbit conversion. In this case, the full helicity is also nonzero, but conserves upon focusing. This means that the helicity, like the beam power, is propagation invariant. For a circularly polarized vortex, the full helicity achieves a maximal value and is equal in magnitude to the full beam energy /power.

Similar to the Poynting vector components (energy flow), measuring the helicity density in practice is also challenging. The longitudinal component of the spin density S_z can be measured by using a well-known technique of determining the Stokes vector components [94, 95], since the longitudinal component of the spin density is equal to the third component of the Stokes vector S_3. However, determining the helicity density h [96, 97], as well as the Poynting vector components, requires knowing the amplitude and phase of the electric and magnetic field components [98, 99]. Transverse and longitudinal components of the spin density vector were investigated in an interesting work [100] for plasmons propagating on a cylindrical or conical surface. The helicity density in this work was considered as a spin density. Studies on helicity in focus were carried out in [101].

References

1. D. Bouwmeester, J.W. Pan, K. Mattle, M. Eibl, H. Weinfurter, A. Zeilinger, Experimental quantum teleportation. Phil. Trans. R. Soc. A **356**, 1733–1737 (1998)
2. R. Fickler, G. Campbell, B. Buchler, P.K. Lam, A. Zeilinger, Quantum entanglement of angular momentum states with quantum numbers up to 10,010. Proc. Natl. Acad. Sci. **113**, 13642–13647 (2016)
3. A.M. Beckley, T.G. Brown, M.A. Alonso, Full Poincaré beams. Opt. Express **18**, 10777 (2010)
4. S. Chen, X. Zhou, Y. Liu, X. Ling, H. Luo, S. Wen, Generation of arbitrary cylindrical vector beams on the higher order Poincare sphere. Opt. Lett. **39**, 5274–5276 (2014)
5. Q. Zhan, Cylindrical vector beams: from mathematical concepts to applications. Adv. Opt. Photon. **1**, 1–57 (2009)
6. V.V. Kotlyar, A.A. Kovalev, A.G. Nalimov, Energy density and energy flux in the focus of an optical vortex: reverse flux of light energy. Opt. Lett. **43**, 2921–2924 (2018)
7. V.V. Kotlyar, S.S. Stafeev, A.A. Kovalev, Reverse and toroidal flux of light fields with both phase and polarization higher-order singularities in the sharp focus area. Opt. Express **27**, 16689–16702 (2019)
8. C. Leyder, M. Romanelli, J.P. Karr, E. Giacobino, T.C.H. Liew, M.M. Glazov, A.V. Kavokin, G. Malpuech, A. Bramati, Observation of the optical spin Hall effect. Nat. Phys. **3**(9), 628–631 (2007)
9. X. Yin, Z. Ye, J. Rho, Y. Wang, X. Zhang, Photonic spin Hall effect at metasurfaces. Science **339**, 1405 (2013)
10. M. Nath Baitha, K. Kim, All angle polarization-independent photonic spin Hall effect. Opt. Laser Technol., **156**, 108458 (2022)
11. S.-M. Li, J. Chen, Spin Hall effect of reflected light from an air-glass interface around the Brewster's angle. Appl. Phys. Lett. **100**, 071109 (2012)
12. B. Roy, N. Ghosh, A. Banerjee, S.D. Gupta, S. Roy, Manifestations of geometric phase and enhanced spin Hall shifts in an optical trap. New J. Phys. **16**, 083037 (2014)
13. R.N. Kumar, Yatish, S.D. Gupta, N. Ghosh, A. Banerjee, Probing the rotational spin-Hall effect in a structured Gaussian beam. Phys. Rev. A, **105**, 023503 (2022)
14. J. Zhang, X.-X. Zhou, X.-H. Ling, S.-Z. Chen, H.-L. Luo, S.-C. Wen, Orbit-orbit interaction and photonic orbital Hall effect in reflection of a light beam. Chinese Phys. B **23**, 064215 (2014)

15. Y. He, Z. Xie, B. Yang, X. Chen, J. Liu, H. Ye, X. Zhou, Y. Li, S. Chen, D. Fan, Controllable photonic spin Hall effect with phase function construction. Phot. Res. **8**, 963–971 (2020)
16. S. Fu, C. Guo, G. Liu, Y. Li, H. Yin, Z. Li, Z. Chen, Spin-orbit optical Hall effect. Phys. Rev. Lett. **123**, 243904 (2019)
17. X. Ling, X. Zhou, K. Huang, Y. Liu, C. Qiu, H. Luo, S. Wen, Recent advances in the spin Hall effect of light. Rep. Prog. Phys.Prog. Phys. **80**, 066401 (2017)
18. X. Ling, X. Yi, X. Zhou, Y. Liu, W. Shu, H. Luo, S. Wen, Realization of tunable spin-dependent splitting in intrinsic photonic spin Hall effect. Appl. Phys. Lett. **105**, 151101 (2014)
19. V.V. Kotlyar, A.A. Kovalev, S.S. Stafeev, V.D. Zaitsev, Index of the Polarization Singularity of Poincare Beams. Bull. Russ. Acad. Sci. Phys. **86**, 1158–1163 (2022)
20. C.V.S. Borges, M. Hor-Meyll, J.A.O. Huguenin, A.Z. Khoury, Bell-like inequality for the spin-orbit separability of a laser beam. Phys. Rev. A **82**, 033833 (2010)
21. E. Otte, C. Rosales-Guzmán, B. Ndagano, C. Denz, A. Forbes, Entanglement beating in free space through spin–orbit coupling. Light Sci. Appl. **7**, 18009 (2018)
22. M. McLaren, T. Konrad, A. Forbes, Measuring the nonseparability of vector vortex beams. Phys. Rev. A **92**, 023833 (2015)
23. B. Richards, E. Wolf, Electromagnetic Diffraction in Optical Systems. II. Structure of the Image Field in an Aplanatic System. Proc. R. Soc. Lond. ALond. A **253**, 358–379 (1959)
24. K.Y. Bliokh, E.A. Ostrovskaya, M.A. Alonso, O.G. Rodriguez-Herrera, D. Lara, C. Dainty, Spin-to-orbital angular momentum conversion in focusing, scattering, and imaging systems. Opt. Express **19**, 26132–26149 (2011)
25. K. Bliokh, A. Bekshaev, F. Nori, Extraordinary momentum and spin in evanescent waves. Nat. Commun.Commun. **5**, 3300 (2014)
26. V.V. Kotlyar, S.S. Stafeev, V.D. Zaitsev, A.M. Telegin, Poincaré Beams at the Tight Focus: Inseparability, Radial Spin Hall Effect, and Reverse Energy Flow. Photonics **9**(12), 969 (2022)
27. A. Hollezek, A. Aiello, C. Gabriel, C. Morquargt, G. Leuchs, Classical and quantum properties of cylindrically polarized states of light. Opt. Express **19**, 9714–9736 (2011)
28. X. Yi, Y. Liu, X. Ling, X. Zhou, Y. Ke, H. Luo, S. Wen, D. Fan, Hybrid-order Poincaré sphere. Phys. Rev. A **91**, 023801 (2015)
29. G. Milione, H.I. Sztul, D.A. Nolan, R.R. Alfano, Higher-order Poincaré sphere, Stokes parameters, and the angular momentum of light. Phys. Rev. Lett. **107**, 053601 (2011)
30. G. Milione, S. Evans, D.A. Nolan, R.R. Alfano, Higher Order Pancharatnam-Berry Phase and the Angular Momentum of Light. Phys. Rev. Lett. **108**, 190401 (2012)
31. E.J. Galvez, S. Khadka, Poincare modes of light. Proc. SPIE **8274**, 82740Y (2012)
32. E.J. Galvez, S. Khadka, W.H. Schubert, S. Nomoto, Poincare-beam patterns produced by nonseparable superpositions of Laguerre-Gauss and polarization modes of light. Appl. Opt. **51**, 2925 (2012)
33. D. Lopez-Mago, On the overall polarisation properties of Poincaré beams. J. Opt. **21**, 115605 (2019)
34. Z. Gu, D. Yin, F. Gu, Y. Zhang, S. Nie, S. Feng, J. Ma, C. Yuan, Generation of concentric perfect Poincaré beams. Sci. Rep. **9**, 15301 (2019)
35. D. Li, S. Feng, S. Nie, C. Chang, J. Ma, C. Yuan, Generation of arbitrary perfect Poincaré beams. J. Appl. Phys. **125**, 073105 (2019)
36. M. Liu, P. Huo, W. Zhu, C. Zhang, S. Zhang, M. Song, S. Zhang, Q. Zhou, L. Chen, H.J. Lezec, A. Agrawal, Y. Lu, T. Xu, Broadband generation of perfect Poincaré beams via dielectric spin-multiplexed metasurface. Nat. Commun.Commun. **12**, 2230 (2021)
37. C. Alpmann, C. Schlickriede, E. Otte, C. Denz, Dynamic modulation of Poincaré beams. Sci. Rep. **7**, 8076 (2017)
38. S. Fu, C. Gao, T. Wang, Y. Zhai, C. Yin, Anisotropic polarization modulation for the production of arbitrary Poincaré beams. J. Opt. Soc. Am. B **35**, 1–7 (2018)
39. D. Naidoo, F.S. Roux, A. Dudley, I. Litvin, B. Piccirillo, L. Marrucci, Controlled generation of higher-order Poincaré sphere beams from a laser. Nat. Photonics **10**, 327–333 (2016)
40. A. Kavokin, G. Malpuech, M. Glazov, Optical Spin Hall Effect. Phys. Rev. Lett. **95**(13), 136601 (2005)

41. T.A. Fadeyeva, A.F. Rubass, A.V. Volyar, Transverse shift of a high-order paraxial vortex-beam induced by a homogeneous anisotropic medium. Phys. Rev. A **79**, 053815 (2009)
42. F. Zhang, Y. Guo, M. Pu, X. Li, X. Ma, X. Luo, Metasurfaces enabled by asymmetric photonic spin-orbit interactions. Opto-Electr. Eng. **47**, 200366 (2020)
43. W. Shu, C. Lin, J. Wu, S. Chen, X. Ling, X. Zhou, H. Luo, S. Wen, Three-dimensional spin Hall effect of light in tight focusing. Phys. Rev. A **101**, 023819 (2020)
44. Z. Man, Z. Xi, X. Yuang, R.E. Burge, P.H. Urbach, Dual coaxial longitudinal polarization vortex structures. Phys. Rev. Lett. **124**, 103901 (2020)
45. S.S. Stafeev, A.G. Nalimov, V.D. Zaitsev, V.V. Kotlyar, Tight focusing cylindrical vector beams with fractional order. J. Opt. Soc. Am. B **38**, 1090–1096 (2021)
46. V.V. Kotlyar, S.S. Stafeev, A.A. Kovalev, V.D. Zaitsev, Spin hall effect before and after the focus of a high-order cylindrical vector beam. Appl. Sci. **12**(23), 12218 (2022)
47. A.V. Volyar, V.G. Shvedov, T.A. Fadeeva, Structure of a nonparaxial Gaussian beam near the focus. III. Stability, eigenmodes and vortices. Opt. Spectr. **91**, 235–245 (2001)
48. X. Weng, Y. Miao, G. Wang, Q. Zhan, X. Dong, J. Qu, X. Gao, S. Zhuang, Light beam carrying natural non-integer orbital angular momentum in free space. ArXiv preprint, arXiv: 2105.11251 (2021)
49. H. Wang, C.C. Wojcik, S. Fan, Topological spin defects of light. Optica **9**, 1417–1423 (2022)
50. V.V. Kotlyar, A.A. Kovalev, A.M. Telegin, Angular and orbital angular momenta in the tight focus of a circularly polarized optical vortex. Photonics **10**(2), 160 (2023)
51. V.V. Kotlyar, A.A. Kovalev, A.M. Telegin, Generalized poincaré beams in tight focus. Photonics **10**(2), 218 (2023)
52. K. Premasiri, X.P. Gao, Tuning spin–orbit coupling in 2D materials for spintronics: a topical review. J. Phys.: Condens. Matter, **31**, 193001 (2019)
53. P. Gregg, P. Kristensen, A. Rubano, S. Golowich, L. Marrucci, S. Ramachandran, Enhanced spin orbit interaction of light in highly confining optical fibers for mode division multiplexing. Nat. Commun.Commun. **10**, 4707 (2019)
54. V.S. Liberman, B.Y. Zel'dovich, Spin-orbit polarization effects in isotropic multimode fibres. Pure Appl. Opt. **2**, 367 (1993)
55. Z. Shao, J. Zhu, Y. Chen, Y. Zhang, S. Yu, Spin-orbit interaction of light induced by transverse spin angular momentum engineering. Nat. Commun.Commun. **9**, 926 (2018)
56. L. Marrucci, E. Karimi, S. Slussarenko, B. Piccirillo, E. Santamato, E. Nagali, F. Sciarrino, Spin-to-orbital conversion of the angular momentum of light and its classical and quantum applications. J. Opt. **13**, 064001 (2011)
57. K. Bliokh, F. Rodríguez-Fortuño, F. Nori, A.V. Zayats, Spin–orbit interactions of light. Nat. Photon. **9**, 796–808 (2015)
58. A. Dogariu, C. Schwartz, Conservation of angular momentum of light in single scattering. Opt. Express **14**, 8425–8433 (2006)
59. Y. Zhao, J.S. Edgar, G.D.M. Jeffries, D. McGloin, D.T. Chiu, Spin-to-orbital angular momentum conversion in a strongly focused optical beam. Phys. Rev. Lett. **99**, 073901 (2007)
60. A. Ciattoni, G. Cincotti, C. Palma, Angular momentum dynamics of a paraxial beam in a uniaxial crystal. Phys. Rev. E **67**, 036618 (2003)
61. O. Hosten, P. Kwiat, Observation of the spin Hall effect of light via weak measurements. Science **319**, 787–790 (2008)
62. D. Pan, H. Wei, L. Gao, H. Xu, Strong spin-orbit interaction of light in plasmonic nanostructures and nanocircuits. Phys. Rev. Lett. **117**, 166803 (2016)
63. G. Li, S. Chen, N. Pholchai, B. Reineke, P.W.H. Wong, E.Y.B. Pun, K.W. Cheah, T. Zentgraf, S. Zhang, Spin-orbit interaction of light in nonlinear optics, in *Conference on Lasers and Electro-Optics 2016, OSA Technical Digest (online)* (Optica Publishing Group, 2016), paper FW4A.1.
64. N. Tancogne-Dejean, F.G. Eich, A. Rubio, Effect of spin-orbit coupling on the high harmonics from the topological Dirac semimetal Na3Bi. Npj Comput. Mater. **8**, 145 (2022)
65. M. Onoda, S. Marakami, N. Nagaosa, Hall effect of light. Phys. Rev. Lett. **93**, 083901 (2004)

66. K.Y. Bliokh, A. Niv, V. Kleiner, E. Hasman, Geometrodynamics of spinning light. Nat. Photon. **2**, 748–753 (2008)
67. V.V. Kotlyar, A.A. Kovalev, E.S. Kozlova, A.M. Telegin, Hall effect at the focus of an optical vortex with linear polarization. Micromachines **14**, 788 (2023)
68. J.-X. Guo, W.-Y. Wang, T.-Y. Cheng, J.-Q. Lü, Interaction of spin-orbit angular momentum in the tight focusing of structured light. Front. Phys. **10**, 1079265 (2022)
69. V. Kotlyar, S. Stafeev, V. Zaitsev, E. Kozlova, Spin-orbital conversion with the tight focus of an axial superposition of a high-order cylindrical vector beam and a beam with linear polarization. Micromachines **13**, 1112 (2022)
70. W. Zhang, Y. Wang, D. Xu, H. Luo, Spin-orbit interaction of light: when twisted light meets twisted metasurfaces. Phys. Rev. A **107**, 043502 (2023)
71. A. Aiello, N. Lindlein, C. Marquardt, G. Leuchs, Transverse angular momentum and geometric spin hall effect of light. Phys. Rev. Lett. **103**, 100401 (2009)
72. Y. Zhang, P. Li, S. Liu, J. Zhao, Unveiling the photonic spin Hall effect of freely propagating fan-shaped cylindrical vector vortex beams. Opt. Lett. **40**, 4444–4447 (2015)
73. S.N. Khonina, I. Golub, Vectorial spin hall effect of light upon tight focusing. Opt. Lett. **47**, 2166–2169 (2022)
74. P.-K. Andrew, M.A.K. Williams, E. Avci, Optical micromachines for biological studies. Micromachines **11**, 192 (2020)
75. I.A. Favre-Bulle, S. Zhang, A.V. Kashchuk, I.C.D. Lenton, L.J. Gibson, A.B. Stilgoe, T.A. Nieminen, H. Rubinsztein-Dunlop, Optical tweezers bring micromachines to biology. Opt. Photonics News **29**, 40–47 (2018)
76. Y.-J. Liu, Y.-H. Lee, Y.-S. Lin, C. Tsou, P.L. Baldeck, C.-L. Lin, Optically driven mobile integrated micro-tools for a lab-on-a-chip. Actuators **2**, 19–26 (2013)
77. A.A. Kovalev, V.V. Kotlyar, Spin hall effect of double-index cylindrical vector beams in a tight focus. Micromachines **14**, 494 (2023)
78. V.V. Kotlyar, A.A. Kovalev, A.P. Porfirev, Astigmatic laser beams with a large orbital angular momentum. Opt. Express **26**, 141–156 (2018)
79. J. Humblet, Sur le moment d'impulsion d'une onde électromagnétique. Physica **10**, 585–603 (1943)
80. K. Cheng, J. Guo, Y. Fu, J. Guo, Active microparticle manipulation: recent advances. Sens. Actuators A: Phys. **322**, 112616 (2021)
81. M.F. Jamil, M. Pokharel, K. Park, Optical manipulation of microparticles in fluids using modular optical tweezers. In *2022 International Symposium on Medical Robotics (ISMR)* (GA, USA, 2022), pp. 1–7.
82. V.V. Kotlyar, A.G. Nalimov, A.A. Kovalev, Controlling the spin hall effect in the sharp focus of an axial superposition of two optical vortices with left- and right-handed circular polarization. Appl. Sci. **13**, 8466 (2023)
83. G.N. Afanasiev, Y.P. Stepanovsky, The helicity of the free electromagnetic field and its physical meaning. Nuov. Cim. A **109**, 271–279 (1996)
84. J.L. Trueba, A.F. Rañada, The electromagnetic helicity. Eur. J. Phys. **17**, 141–144 (1996)
85. R.P. Cameron, S.M. Barnett, A.M. Yao, Optical helicity, optical spin and related quantities in electromagnetic theory. New J. Phys. **14**, 053050 (2012)
86. L. Woltjer, A theorem on force-free magnetic fields. Proc. Natl. Acad. Sci. U.S.A. **44**, 489–491 (1958)
87. F. Crimin, N. Mackinnon, J.B. Götte, S.M. Barnett, Optical helicity and chirality: conservation and sources. Appl. Sci. **9**, 828 (2019)
88. D.M. Lipkin, Existence of a new conservation law in electromagnetic theory. J. Math. Phys. **5**, 696–700 (1964)
89. N. Mackinnon, On the differences between helicity and chirality. J. Opt. **21**, 125402 (2019)
90. A. Aiello, Helicity, chirality, and spin of optical fields without vector potentials. Phys. Rev. A **106**, 043519 (2022)
91. D.R. Abujetas, J.A. Sánchez-Gil, Spin angular momentum of guided light induced by transverse confinement and intrinsic helicity. ACS Photonics **7**, 534–545 (2020)

92. V.V. Koltyar, A.G. Nalimov, S.S. Stafeev, Exploiting the circular polarization of light to obtain a spiral energy flow at the subwavelength focus. J. Opt. Soc. Am. B **36**, 2850–2855 (2019)
93. V.V. Kotlyar, S.S. Stafeev, V.D. Zaitsev, A.A. Kovalev, Multiple optical spin-orbit Hall effect at the tight focus. Phys. Lett. A, **458**, 128596 (2023)
94. H.G. Berry, G. Gabrielse, A.E. Livingston, Measurement of the Stokes parameters of light. Appl. Opt. **16**, 3200–3205 (1977)
95. N.V. D'yachkov, A.P. Bogatov, Measuring the Stokes parameters of radiation of semiconductor lasers. Quantum Electron, **41**, 869–874 (2011)
96. K.A. Forbes, G.A. Jones, Measures of helicity and chirality of optical vortex beams. J. Opt. **23**, 115401 (2021)
97. K. Koksal, M. Babiker, V.E. Lembessis, J. Yuan, Chirality and helicity of linearly-polarised Laguerre-Gaussian beams of small beam waists. Opt. Commun.Commun. **490**, 126907 (2021)
98. I.A. Litvin, The behavior of the instantaneous Poynting vector of symmetrical laser beams. J. Opt. Soc. Am. A **29**, 901–907 (2012)
99. J. Leach, S. Keen, M.J. Padgett, C. Saunter, G.D. Love, Direct measurement of the skew angle of the Poynting vector in a helically phased beam. Opt. Express **14**, 11919–11924 (2006)
100. D. Garoli, P. Zilio, F. De Angelis, Y. Gorodetski, Helicity locking of chiral light emitted from a plasmonic nanotaper. Nanoscale **9**, 6965–6969 (2017)
101. A.A. Kovalev, V.V. Kotlyar, A.M. Telegin, Optical helicity of light in the tight focus. Photonics **10**, 719 (2023)

Chapter 7
Hall Effect in Paraxial Laser Beams

7.1 Spin Hall Effect in the Paraxial Light Beams with Multiple Polarization Singularities

In micromachines, elements can be driven by light and optical vortex beams are widely adopted in optical tweezers. One of the natural generalizations of optical vortex beams are light fields with multiple vortices. Rather general expressions for describing such fields have been obtained by Indebetouw [1] and Abramochkin [2]. Such fields propagate in free space without changing their transverse intensity shape, up to scale and rotation around the optical axis. In addition to the optical vortices, which are phase singularities or points with an uncertain phase, vector light fields can have polarization singularities, where uncertain is polarization. Polarization singularities were investigated in a huge number of papers. Recently, a review about polarization singularities was published [3]. Most widely studied polarization singularities are radial and azimuthal polarization, and their superposition known as cylindrical polarization [4]. Such light fields can be constructed as superpositions of optical vortices with opposite circular polarizations and with opposite topological charges of $\pm$ 1 [5]. Based on this scheme, we investigated in 2018 light fields with multiple polarization singularities residing on a circle with the center on the optical axis [6, 7]. Such fields have locally linear polarization. We found that, in contrast to the phase singularities, which conserve in an arbitrary transverse plane, polarization singularities appear only in a discrete number of transverse planes. We discovered that the singularities can transform, for instance, from radial and azimuthal polarization in the initial plane to, respectively, azimuthal and radial polarization in the far field. However, we did not investigate what happens with polarization between the planes where polarization singularities reconstructed, i.e., between the planes with locally linear polarization. In [8], it was noticed that along the propagation direction, such singularities generically split into a pair of C-points. As it turns out, such splitting is a manifestation of the optical spin Hall effect, which means that areas with nonzero spin angular momentum appear, despite linear polarization in the initial plane. Such an

V. V. Kotlyar et al., *Optical Hall Effect in the Sharp Focus of Laser Light*,
https://doi.org/10.1007/978-3-031-64683-6_7

effect was observed, for instance, in microcavities [9], metamaterials [10], dielectric gratings [11], and in free space in the tight-focusing conditions [12].

In this section, we investigate this phenomenon in the paraxial Gaussian beam with multiple polarization singularities from [6, 7]. Such a beam is a superposition of a cylindrically polarized Laguerre-Gaussian (LG) beam with a linearly polarized Gaussian beam. We found that maximal SAM density magnitudes appear on a circle and the radius of this circle has been derived. It was obtained that the maximal SAM density is generated in a certain transverse plane, the distance to which has been obtain approximately. It was derived that the maximal SAM density can be achieved when the singularities circle has a definite radius, such that the energy of the Gaussian beam is equal to the energy of the LG beam. We revealed the mechanism of the spin Hall effect in such beams. This effect arises due to the different divergence of the LG beam and of the Gaussian beam.

7.1.1 Paraxial Light Fields with Multiple Phase or Polarization Singularities

It has been shown (Eq. (17) in [2]) that any function given by the following:

$$E(r, \varphi, z) = \frac{1}{q} \exp\left(-\frac{r^2}{qw_0^2}\right) f\left(\frac{re^{i\varphi}}{qw_0}\right), \tag{7.1}$$

where (r, φ, z) are the cylindrical coordinates, w_0 is the waist radius of the Gaussian beam, and $q = 1 + iz/z_0, f(re^{i\varphi})$ is an arbitrary entire analytical function, describes a solution of a paraxial Schrödinger-type Helmholtz equation: $2ik(\partial E/\partial z) + (\partial^2 E/\partial x^2) + (\partial^2 E/\partial y^2) = 0$.

The light field described by Eq. (7.1) propagates in free space without changing its shape. The parameter q defines the scaling and rotation of the light field: at a distance z the field becomes $|q| = [1 + (z/z_0)^2]^{1/2}$ times wider and rotates around the optical axis by an angle equal to the Gouy phase $\psi = \arg q = \arctan(z/z_0)$.

Using Eq. (7.1), it is possible to obtain a solution of the paraxial Helmholtz equation with optical vortices located in arbitrary points with their polar coordinates in the initial plane (r_p, φ_p) $(p = 0, 1, \ldots, m - 1)$. In an arbitrary transverse plane, the complex amplitude of such a field takes the form [1]:

$$E(r, \varphi, z) = \frac{1}{qw_0^m} \exp\left(-\frac{r^2}{qw_0^2}\right) \prod_{p=0}^{m-1} \left(\frac{re^{i\varphi}}{q} - r_p e^{i\varphi_p}\right). \tag{7.2}$$

Choosing the vortices on a circle with the radius a_0, i.e., $r_p = a_0$, $\varphi_p = 2\pi p/m$, we get

$$E(r,\varphi,z)=\frac{1}{qw_0^{|m|}}\exp\left(-\frac{r^2}{qw_0^2}\right)\left[\left(\frac{r}{q}\right)^{|m|}e^{im\varphi}-a_0^{|m|}\right]. \tag{7.3}$$

This field is a superposition of a single-ringed mth-order Laguerre-Gaussian (LG) beam with a fundamental Gaussian beam.

It has been known [5] that radially polarized light can be expressed as a superposition of optical vortices of the ± 1st order and of the opposite handedness since the Jones vector can be presented as follows:

$$\begin{pmatrix}\cos\varphi\\ \sin\varphi\end{pmatrix}=\frac{1}{2}e^{i\varphi}\begin{pmatrix}1\\ -i\end{pmatrix}+\frac{1}{2}e^{-i\varphi}\begin{pmatrix}1\\ i\end{pmatrix}. \tag{7.4}$$

If in this expression $e^{i\varphi}$ and $e^{-i\varphi}$ are replaced by the field (7.3) of the orders, respectively, m and $-m$, we get a vector light field with m polarization singularities located on a circle of the radius a_0 [6, 7]:

$$\mathbf{E}(r,\varphi,z)=\frac{1}{q^{m+1}w_0^m\sqrt{W_0}}\exp\left(-\frac{r^2}{qw_0^2}\right)\begin{bmatrix}r^m\cos m\varphi-a_0^mq^m\\ r^m\sin m\varphi\end{bmatrix}, \tag{7.5}$$

where the multiplier

$$W_0=\frac{\pi w_0^2}{2}\left[\frac{m!}{2^m}+\left(\frac{a_0}{w_0}\right)^{2m}\right] \tag{7.6}$$

is introduced for normalizing the beam energy (making it equal to unit and thus equal for all values m and a_0).

Examples of such field with $m = 2, 3, 4$ are shown in Fig. 7.1. Since the vortices reside on a circle with the radius a_0, from now on, we call this parameter a singularities circle radius.

We note that the field (7.5) can be treated as a superposition of two single-ringed Laguerre-Gaussian beams with opposite topological charges and with circular polarizations, and of a linearly polarized Gaussian beam:

$$\begin{aligned}\mathbf{E}(r,\varphi,z)&=\frac{1}{2\sqrt{W_0}}\mathrm{LG}_m(r,\varphi,z)\begin{bmatrix}1\\ -i\end{bmatrix}+\frac{1}{2\sqrt{W_0}}\mathrm{LG}_{-m}(r,\varphi,z)\begin{bmatrix}1\\ i\end{bmatrix}\\ &-\frac{1}{\sqrt{W_0}}\left(\frac{a_0}{w_0}\right)^m\mathrm{LG}_0(r,\varphi,z)\begin{bmatrix}1\\ 0\end{bmatrix},\end{aligned} \tag{7.7}$$

with $\mathrm{LG}_m(r, \varphi, z)$ being the scalar mth-order single-ringed Laguerre-Gaussian beam:

$$\mathrm{LG}_m(r,\varphi,z)=\frac{1}{q}\left(\frac{r}{qw_0}\right)^{|m|}\exp\left(-\frac{r^2}{qw_0^2}+im\varphi\right). \tag{7.8}$$

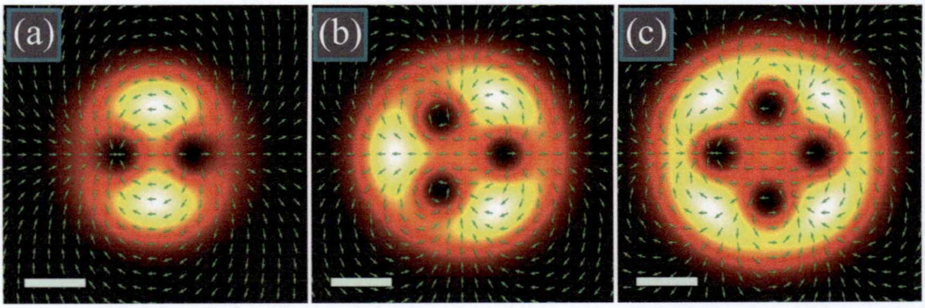

Fig. 7.1 Intensity distributions of the light field (7.5) in the initial plane ($z = 0$) for the following parameters: wavelength $\lambda = 532$ nm, Gaussian beam waist radius $w_0 = 1$ mm, number of polarization singularities $m = 2$ (**a**), $m = 3$ (**b**) and $m = 4$ (**c**), radius of the singularities circle $a_0 = 600$ μm (**a**), $a_0 = 700$ μm (**b**), $a_0 = 800$ μm (**c**). Size of all figures is 5×5 mm^2, scale mark in each figure denotes 1 mm. Arrows show the directions of linear polarization

7.1.2 Intensity Distribution

From Eq. (7.5), the intensity distribution is given by the following:

$$I(r, \varphi, z) = |E_x(r, \varphi, z)|^2 + \left|E_y(r, \varphi, z)\right|^2$$
$$= \frac{1}{|q|^{2m+2} w_0^{2m} W_0} \exp\left(-\frac{2r^2}{|q|^2 w_0^2}\right)\left[r^{2m} + a_0^{2m}|q|^{2m} - 2a_0^m |q|^m r^m \cos(m\psi)\cos(m\varphi)\right], \tag{7.9}$$

with $\psi = \arctan(z/z_0)$ being the Gouy phase. It is seen that the intensity nulls can appear only in a discrete set of transverse planes, where $\cos(m\psi) = \pm 1$, i.e., $\tan(m\psi) = 0$, which is consistent with [6, 7]. In the initial plane, the intensity is

$$I(r, \varphi, 0) = \frac{1}{w_0^{2m} W_0} \exp\left(-\frac{2r^2}{w_0^2}\right)\left[r^{2m} + a_0^{2m} - 2a_0^m r^m \cos(m\varphi)\right]. \tag{7.10}$$

However, the beam from Eq. (7.5) is a superposition of circularly polarized single-ringed LG vortex beams of the orders $\pm m$ and of a linearly polarized Gaussian beam. At a small singularities circle radius a_0, the LG beam overwhelms and the intensity looks like a ring (Fig. 7.2a, d). At large a_0, vice versa, the Gaussian beam is brighter and the intensity looks more like a spot (Fig. 7.2b, e). In some applications, however, it is desirable to confine the intensity nulls between the light walls. For instance, in 2008, Dienerowitz et al. showed that a vortex beam with annular profile can confine metal nanoparticles in the dark region of the beam center [13]. Thus, the intensities from Fig. 7.2a–e are undesirable. Now we try to find the radius a_0 such that the intensity in the beam center and in the edges, beyond the intensity nulls, are nearly the same. Since the first intensity null is at $\varphi = 0$, this condition can be written as follows:

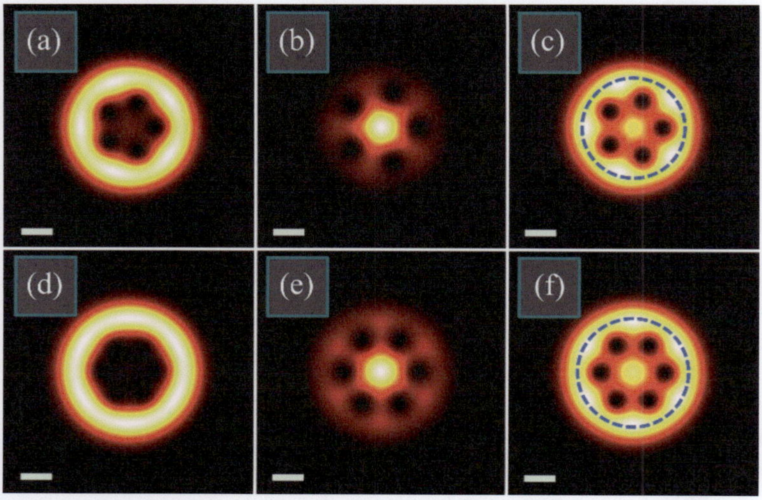

Fig. 7.2 Intensity distributions of the light field (7.5) in the initial plane ($z = 0$) for the following parameters: wavelength $\lambda = 532$ nm, Gaussian beam waist radius $w_0 = 1$ mm, number of polarization singularities $m = 5$ (**a–c**) and $m = 6$ (**d–f**), radius of the singularities circle $a_0 = 800$ μm (**a**, **d**), $a_0 = 1200$ μm (**b**, **e**), $a_0 = 942$ μm (**c**), $a_0 = 1041$ μm (**f**). Size of all figures is 8×8 mm^2, scale mark in each figure denotes 1 mm. Blue-dashed circles (**c**, **f**) denote the radius of the maximal peripheral intensity computed by Eq. (7.20)

$$I(0, 0, 0) = \max_{r>a_0} I(r, 0, 0), \tag{7.11}$$

or, after taking the square root of both parts,

$$a_0^m = \max_{r>a_0}\left\{\exp\left(-\frac{r^2}{w_0^2}\right)\left(r^m - a_0^m\right)\right\}. \tag{7.12}$$

Thus, we need to determine the maximal peripheral intensity, beyond the null. Taking the derivative of the right part of Eq. (7.12) with respect to r yields an equation for the radial coordinate $r_{\max}$ of the maximal intensity:

$$2\left(r_{\max}^m - a_0^m\right) = m w_0^2 r_{\max}^{m-2}. \tag{7.13}$$

This equation can be solved only for small values m. However, we do not need to solve it, since we are interested in a_0 rather than in $r_{\max}$. Expressing a_0 via $r_{\max}$ and substituting it into Eq. (7.12), we get

$$r_{\max}^m - \frac{m w_0^2}{2} r_{\max}^{m-2} = \frac{m w_0^2}{2} r_{\max}^{m-2} \exp\left(-\frac{r_{\max}^2}{w_0^2}\right), \tag{7.14}$$

Division of both parts by $\left(m w_0^2/2\right) r_{\max}^{m-2}$ yields a simple equation:

$$\frac{2}{m}\xi - 1 = \exp(-\xi). \tag{7.15}$$

with $\xi = (r_{\max}/w_0)^2$. Since for large m an approximate solution is $\xi \approx m/2$, we denote $\xi = m/2 + \eta$ and get

$$\frac{2}{m}\eta \approx e^{-m/2}(1 - \eta) \tag{7.16}$$

and, therefore,

$$\eta \approx \frac{e^{-m/2}}{2/m + e^{-m/2}}. \tag{7.17}$$

Returning back to ξ, we obtain the solution

$$\xi = \frac{m}{2}\left(1 + \frac{1}{m/2 + e^{m/2}}\right). \tag{7.18}$$

Thus, we get the singularities circle radius a_0, for which the intensity in the center and at the edge (near the dark spot) is nearly the same:

$$a_0 = \left(r_{\max}^m - \frac{mw_0^2}{2} r_{\max}^{m-2}\right)^{1/m}, \tag{7.19}$$

with

$$r_{\max} = w_0\sqrt{\frac{m}{2}}\sqrt{1 + \frac{1}{m/2 + e^{m/2}}}. \tag{7.20}$$

It is seen that the first two multipliers are equal to the maximal-intensity radius of a single-ringed mth-order LG beam with the waist radius w_0, while the third multiplier tends to unit with growing number of singularities m. Shown in Fig. 7.2c,f are the intensity distributions with the singularities circle radius obtained by Eq. (7.19). These figures confirm that Eq. (7.19) allows making the intensities in the center and in the periphery nearly equal.

In optical tweezers, the intensity distribution affects where the particles are trapped. However, the motion of particles is governed by the spin and orbital angular momenta.

7.1.3 Spin Angular Momentum Density

In paraxial light fields, only the longitudinal component of the SAM vector can be significant. It is equal to

$$S_z = 2\mathrm{Im}\left\{E_x^* E_y\right\}, \tag{7.21}$$

Substituting here Eq. (7.5) for the light field, we get

$$S_z = \frac{2}{|q|^2 W_0}\left(\frac{a_0 r}{|q| w_0^2}\right)^m \exp\left(-\frac{2r^2}{|q|^2 w_0^2}\right)\sin(m\psi)\sin(m\varphi), \tag{7.22}$$

where $\psi = \arctan(z/z_0)$ is the Gouy phase.

It is seen from this expression that there are transverse planes where the SAM is zero, i.e., polarization is linear. In these planes, $\sin(m\psi) = 0$, i.e., they are located at the following distances [6, 7]:

$$z = z_0 \tan\left(\frac{\pi p}{m}\right), \tag{7.23}$$

with $p = 0, 1, \ldots, [m/2]$, where [.] means the integer part of a fractional number. In other planes, the SAM is generally nonzero, but in each plane, it equals to zero at the polar angles $\varphi_p = \pi p/m$ with $p = 0, \ldots, m - 1$. Comparison of the expressions for the SAM and for the intensity reveals that in an arbitrary transverse plane, the light field has C-points, where polarization is circular [14, 15]. Equation $S_z(r, \varphi, z) = \pm I(r, \varphi, z)$ leads to the following C-points coordinates:

$$\begin{cases} r = a_0|q|, \\ \varphi = \pm\psi + \dfrac{2\pi p}{m}, \end{cases} \tag{7.24}$$

where $p = 0, \ldots, m - 1$. Thus, there are m points with right circular polarization (at $\varphi = \psi + 2\pi p/m$) and m points with left circular polarization (at $\varphi = -\psi + 2\pi p/m$). It is seen that on propagation, C-points with right and left circular polarization are rotated around the optical axis in opposite directions. When passing through the planes given by Eq. (7.23), coordinates of these C-points coincide, they annihilate each other and polarization becomes linear. Evolution of the C-points is illustrated in Fig. 7.3.

Now we try to determine where the SAM achieves zero or maximal magnitudes. If $\sin(m\psi) > 0$ in Eq. (7.22), then the maximal and minimal SAM density is achieved, respectively, at the polar angles $\varphi_p = (\pi + 4\pi p)/(2m)$ and $\varphi_p = (-\pi + 4\pi p)/(2m)$ with $p = 0, \ldots, m - 1$, and these angles are independent of the propagation distance and on the singularities circle radius a_0. However, after passing the planes with linear polarization and with the polarization singularities (Eq. 7.23), the angles of the maximal and minimal SAM density are swapped. Differentiating Eq. (7.22) by

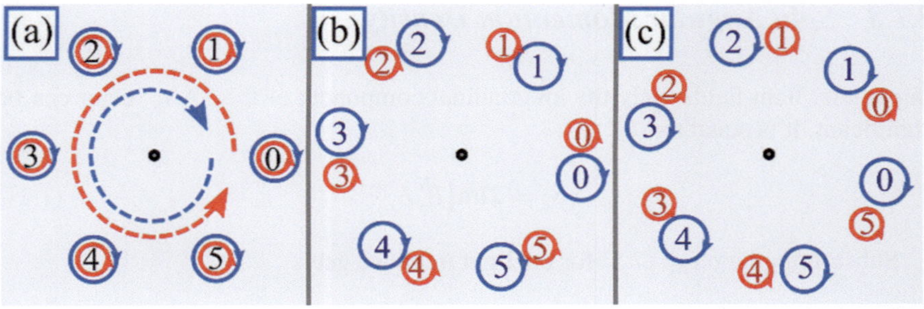

Fig. 7.3 Evolution of the C-points in the light field (7.5). In the initial plane (**a**), polarization is linear, i.e., opposite C-points reside in the same points and compensate each other. When starting to propagate (**b**), C-points with right and left circular polarization rotate in the transverse plane in opposite directions. When approaching the next plane with linear polarization (**c**), given by Eq. (7.23), C-points with right and left circular polarization merge again

r yields that at a fixed propagation distance z and at the angles φ_p, maximal SAM density is achieved on a circle with the radius

$$r = \frac{w_0|q|}{2}\sqrt{m}, \tag{7.25}$$

This radius is $\sqrt{2}$ times smaller than the radius of maximal intensity of a single-ringed mth-order LG beam. Substituting the radius from Eq. (7.25) into Eq. (7.22) for the SAM, we get the maximal SAM density in a transverse plane at a distance z from the initial plane:

$$\max_{r,\varphi} S_z = \frac{2}{|q|^2 W_0}\left(\frac{a_0\sqrt{m}}{2w_0}\right)^m \exp\left(-\frac{m}{2}\right)\sin(m\psi). \tag{7.26}$$

Now, after obtaining the maximal SAM in each transverse plane, we find the planes with where the SAM achieves the maximal value. For this plane, the following condition should be fulfilled:

$$\frac{\partial}{\partial z}\left(\max_{r,\varphi} S_z\right) = 0, \tag{7.27}$$

Substituting here the maximal SAM in the plane [Eq. (7.26)], we get

$$\frac{\partial}{\partial z}\left\{\left(1+\frac{z^2}{z_0^2}\right)^{-1}\sin\left[m\arctan\left(\frac{z}{z_0}\right)\right]\right\} = 0. \tag{7.28}$$

This equation leads us to the following distance $z_{\max}$ to the plane with the maximal SAM:

$$\tan\left(m\arctan\left(\frac{z_{\max}}{z_0}\right)\right)=\frac{mz_0}{2z_{\max}}. \tag{7.29}$$

This equation can also be written in a short form via the Gouy phase $\psi_{\max}$ of this plane:

$$\tan(\psi_{\max})\tan(m\psi_{\max})=m/2, \tag{7.30}$$

Both these equations indicate that the distance to the plane with maximal SAM is independent of the radius of the singularities circle a_0. These equations can be solved analytically only for small values m. For instance, $z = z_0/\sqrt{2}$ at $m = 1$ and $z = z_0/\sqrt{3}$ at $m = 2$. For larger values m, we try to estimate their solutions. Since

$$\sin(m\psi_{max})=\frac{1}{\sqrt{1+1/\tan^2(m\psi_{max})}}=\frac{m}{\sqrt{4\tan^2(\psi_{max})+m^2}}=\frac{m}{\sqrt{4(z_{max}/z_0\,)^2+m^2}}, \tag{7.31}$$

we get the following expression for the SAM in the planes, where it achieves extreme magnitudes:

$$\max_{r,\varphi}S_z(z=z_{\max})=\frac{2}{W_0}\left(\frac{a_0\sqrt{m}}{2w_0}\right)^m\frac{e^{-m/2}}{\left[1+(z_{\max}/Z_0)^2\right]\left[1+(2/m)^2(z_{\max}/Z_0)^2\right]^{1/2}}. \tag{7.32}$$

This expression indicates that in each such plane with locally maximal SAM, the SAM achieves lower and lower magnitude, i.e., the strongest SAM is in the first plane given by Eqs. (7.29) and (7.30). The left and right parts of Eq. (7.29) are shown in Fig. 7.4. The right part is always positive and decays hyperbolically. The left part is a discontinuous function with the zeros in the points $z_{1,p} = z_0 \tan(\pi p/m)$ $(p = 0, 1, \ldots)$ and with the discontinuities in the points $z_{2,p} = z_0 \tan(\pi(2p + 1)/(2m))$ $(p = 0, 1, \ldots)$. The roots of Eq. (7.29) are thus in the intervals $[z_{1,p}, z_{2,p}]$. On the other hand, the left part of Eq. (7.30) grows slower than the function $\tan(m\psi)$ without the multiplier $\tan(\psi)$. Thus, the first root of Eq. (7.30) should be between the values $\psi = m^{-1}\arctan(m/2)$ and the discontinuity $\psi = \pi/(2m)$.

Thus, the first root of Eq. (7.29) is in the interval from $z = z_0\tan[m^{-1}\arctan(m/2)]$ to $z = z_0 \tan(\pi/(2m))$. In our work, we will use the average value

$$z_{\max}\approx\frac{z_0}{2}\tan\left[\frac{\arctan(m/2)}{m}\right]+\frac{z_0}{2}\tan\left[\frac{\pi}{2m}\right]. \tag{7.33}$$

The Simulation section below confirms that this is a good approximation at $m > 3$.

We note that the SAM magnitude is different for different values a_0. In two extreme cases, when all the vortices merge in the center ($a_0 = 0$) and when they

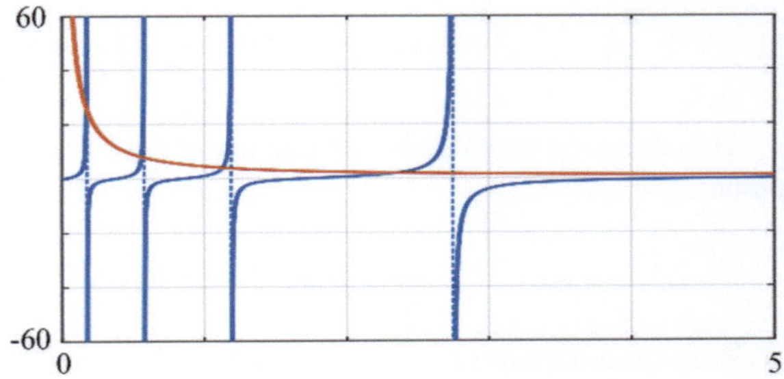

Fig. 7.4 Left and right parts of Eq. (7.29) as functions of z/z_0 at $m = 9$

move to infinity ($a_0 \to \infty$) the SAM should be equal to zero. In the first case, the light field reduces to a LG beam with mth-order cylindrical polarization which is not destroyed on propagation and the field has inhomogeneous linear polarization in an arbitrary transverse plane. In the second case, the superposition (7.5) consists only of the fundamental Gaussian beam while the portion of the LG beam with mth-order polarization vortex tends to zero. The Gaussian beam is linearly polarized and therefore the SAM should tend to zero.

To derive the radius of the polarization singularities that yield the maximal SAM, we should differentiate Eq. (7.26) by a_0. Thus, we get

$$a_{0,\max} = (m!)^{1/(2m)} \frac{w_0}{\sqrt{2}} \tag{7.34}$$

At this value, the maximal SAM in an arbitrary transverse plane is then equal to

$$\max_{r,\varphi} S_z(\alpha_0 = \alpha_{0,\max}) = \frac{2}{\pi w_0^2} \frac{1}{|q|^2 \sqrt{m!}} \left(\frac{m}{2}\right)^{m/2} \exp\left(-\frac{m}{2}\right) \sin(m\psi). \tag{7.35}$$

Applying the Stirling's approximation $m! \sim (2\pi m)^{1/2}(m/e)^m$ [16], we get

$$\max_{r,\varphi} S_z(\alpha_0 = \alpha_{0,max}) \approx \frac{2}{\pi w_0^2} \frac{1}{|q|^2 (2\pi)^{1/4}} \frac{1}{2^{m/2} m^{1/4}} \sin(m\psi) \le \frac{2}{\pi w_0^2 (2\pi)^{1/4}} \frac{1}{2^{m/2} m^{1/4}}. \tag{7.36}$$

This estimation indicates that the maximal achievable SAM decreases with increasing number of the polarization singularities. We note that the linearly polarized Gaussian beam in the whole field has the following initial intensity distribution

$$I_{\mathrm{GB}}(r, \varphi, 0) = \frac{1}{W_0} \left(\frac{a_0}{w_0}\right)^{2m} \exp\left(-\frac{2r^2}{w_0^2}\right) \tag{7.37}$$

and its energy fraction in the whole energy is

$$\begin{aligned} W_{\mathrm{GB}} &= 2\pi \int_0^{\infty} I_{\mathrm{GB}}(r, \varphi, 0) r dr \\ &= \frac{2\pi}{W_0}\left(\frac{a_0}{w_0}\right)^{2m} \int_0^{\infty} \exp\left(-\frac{2r^2}{w_0^2}\right) r dr = \left(\frac{a_0}{w_0}\right)^{2m}\left[\frac{m!}{2^m} + \left(\frac{a_0}{w_0}\right)^{2m}\right]. \end{aligned} \tag{7.38}$$

At the singularities circle radius from Eq. (7.34), this energy reduces to

$$W_{\mathrm{GB}}\left(a_0 = a_{0,\max}\right) = \frac{1}{2}. \tag{7.39}$$

Thus, the maximal SAM density is achieved when the energy of the linearly polarized Gaussian beam in the superposition (7.5) is equal to the energy of the cylindrically polarized LG beam, i.e., equal to the half of the energy of the whole light field.

7.1.4 Orbital Angular Momentum Density

In paraxial light fields, only the longitudinal component of the OAM vector can be significant. It is equal to [17, 18]:

$$J_z = \operatorname{Im}\left\{E_x^* \frac{\partial E_x}{\partial \varphi} + E_y^* \frac{\partial E_y}{\partial \varphi}\right\}. \tag{7.40}$$

Substituting here the light field from Eq. (7.5), we get

$$\begin{aligned} J_z &= \frac{1}{|q|^{2m+2} w_0^{2m} W_0} \exp\left(-\frac{2r^2}{|q|^2 w_0^2}\right) \\ &\times \operatorname{Im}\left\{\left(r^m \cos m\varphi - a_0^m q^m\right)^* \frac{\partial}{\partial \varphi}\left(r^m \cos m\varphi - a_0^m q^m\right) + \left(r^m \sin m\varphi\right)^* \frac{\partial}{\partial \varphi}\left(r^m \sin m\varphi\right)\right\}. \end{aligned} \tag{7.41}$$

The second term is real and its imaginary part is zero. Then, the OAM density reduces to

$$\begin{aligned} J_z &= \frac{1}{|q|^{2m+2} w_0^{2m} W_0} \exp\left(-\frac{2r^2}{|q|^2 w_0^2}\right) \operatorname{Im}\left\{\left(r^m \cos m\varphi - a_0^m q^{*m}\right)\left(-m r^m \sin m\varphi\right)\right\} \\ &= \frac{-m}{|q|^2 W_0} \exp\left(-\frac{2r^2}{|q|^2 w_0^2}\right)\left(\frac{a_0 r}{|q| w_0^2}\right)^m \sin(m\psi) \sin(m\varphi). \end{aligned} \tag{7.42}$$

It is seen that the OAM is equal to the SAM from Eq. (7.22), but multiplied by $-m/2$. It is in contrast with the conventional vortex beams with homogeneous circular polarization, whose OAM exceeds SAM m of $-m$ times.

7.1.5 Analogy with Plane Wave and Revealing the Mechanism

According to Eq. (7.5), the light field includes two opposite-charge circularly polarized LG beams. On propagation in free space, they rotate clockwise and counterclockwise. Thus, an angular analogue of standing wave is generated. This leads to a natural question whether the above described effect can occur with the conventional standing wave if it is composed of two plane waves that have opposite tilt to the optical axis and opposite circular polarization. When superimposed with a plane wave without the tilt, such a field has the following complex amplitude:

$$\mathbf{E}(x,y,z) = \frac{1}{\sqrt{W_1}}\exp(ik_x x + ik_z z)\begin{bmatrix}1\\-i\end{bmatrix} + \frac{1}{\sqrt{W_1}}\exp(-ik_x x + ik_z z)\begin{bmatrix}1\\i\end{bmatrix} + \frac{a_0}{\sqrt{W_1}}\exp(ikz)\begin{bmatrix}1\\0\end{bmatrix}, \tag{7.43}$$

where $k_x^2 + k_z^2 = k^2 = (2\pi/\lambda)^2$ and $k_x = k\cos\alpha$ with α being the tilt angle. The multiplier a_0 defines the relative strength of the plane wave without the tilt. The field (7.43) is of infinite energy, but to make the energy equal for different values a_0, we introduced the multiplier $W_1^{-1/2}$ with $W_1 = 4 + a_0^2$ (four scalar tilted plane waves and one wave without the tilt and with the amplitude proportional to a_0). It turns out that, on propagation in space, such a field also acquires nonzero SAM density:

$$S_z(x,y,z) = 2\mathrm{Im}\{E_x^* E_y\} = \frac{-4a_0}{4 + a_0^2}\sin(k_x x)\sin[(k - k_z)z]. \tag{7.44}$$

Due to infinite energy, plane waves do not spread on propagation. Therefore, the SAM density is repeated periodically and does not decay, in contrast to the realistic finite-energy beam (7.5), whose SAM density decays.

It is seen in Eq. (7.44) that no matter how the beam without the tilt is polarized, the nonzero SAM would not occur without the difference $k - k_z$. Due to circular polarizations, electric vectors of the tilted plane waves rotate, but in opposite directions. These rotations cancel each other and common polarization of tilted waves is linear (Fig. 7.5a). Adding linearly polarized non-tilted beam changes polarization direction, but leaves it linear in the initial plane, where all the waves are superimposed in phase. But on propagation, tilted waves become retarded compared to the non-tilted wave (Fig. 7.5b). Thus, a phase delay appears between the tilted waves

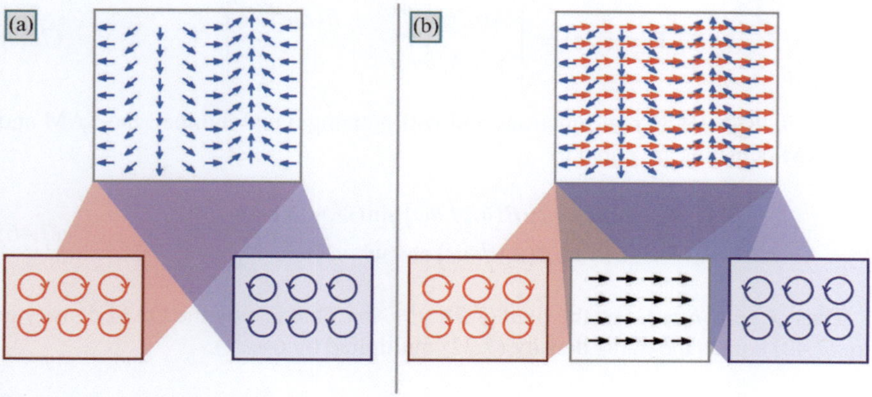

Fig. 7.5 Generating linearly polarized field in a superposition of two tilted plane waves with opposite tilts and with opposite circular polarizations (a), generating nonzero SAM density in a superposition of two tilted circularly polarized plane waves with a linearly polarized wave without the tilt: Tilted waves acquire phase retard and polarization becomes elliptic

and non-tilted one. This delay means elliptic polarization. However, if the electric field of the linearly polarized wave greatly exceeds or, vice versa, much weaker than the electric field of tilted circularly polarized waves, than, despite the phase delay between them, elliptic polarization is close to linear and the SAM density is small. Thus, a question arises about the energies of the tilted waves and of the non-tilted wave that leads to the maximal SAM. From Eq. (7.44) follows that the maximal SAM magnitude is achieved at $a_0 = 2$. This means that the energy of the linearly polarized wave is half of the energy of the whole superposition. Thus, we obtained just the same result as for the field (7.5) with multiple polarization singularities: The energy of the linearly polarized term should amount half the energy of the whole field.

Elliptic polarization is not generated when the linear polarization of the two tilted circularly polarized plane waves is parallel to linear polarization of the non-tilted wave. Thus, according to Eq. (7.44), if $k_x x = \pi p$ (p is an integer), then polarization is linear. On the contrary, when these vectors are orthogonal, polarization is closest to circular. This happen when $\cos(k_x x) = 0$. Thus, the maximal SAM magnitude of the field (7.5) should be achieved when $\mathrm{LG}_m(r, \varphi, z) + \mathrm{LG}_{-m}(r, \varphi, z) = 0$, i.e., when $\cos(m\varphi) = 0$. This occurs exactly at the above obtained polar angles $\varphi_p = (\pm\pi + 4\pi p)/(2m)$ ($p = 0, \ldots, m - 1$).

The above explanation of the nonzero SAM also explains the proportionality between the OAM and SAM densities. It has no special physical meaning, but it is a consequence of the special-type complex amplitude (7.5). Indeed, the SAM density is due to the phase delay between the terms $r^m \sin(m\varphi)$ and $(a_0 q)^m$, whereas the OAM is contributed only by the E_x component and it is due to the phase delay between the terms $r^m \cos(m\varphi)$ and $(a_0 q)^m$. It can be shown that for an arbitrary vector light field given by the following:

$$\mathbf{E}(r,\varphi,z)=\begin{bmatrix} A(r,\varphi)e^{i\Psi(r)}+B(r)e^{iX(r)} \\ \gamma(\partial A/\partial\varphi)e^{i\Psi(r)} \end{bmatrix}, \tag{7.45}$$

with A, B, Ψ, X being real functions and with γ being a real number, the SAM and the OAM densities are equal to

$$\begin{aligned} S_z &= 2\gamma B\left(\partial A/\partial\varphi\right)\sin(\Psi - X), \\ J_z &= B\left(\partial A/\partial\varphi\right)\sin(\Psi - X). \end{aligned} \tag{7.46}$$

Thus, $J_z = S_z/(2\gamma)$. For the field (7.5), $\gamma = -m$. That is why the OAM density in Eq. (7.42) equals the SAM density (7.22) multiplied by ($-m/2$).

7.1.6 Simulation

Figure 7.6 depicts the initial intensity distributions of the light field (7.5) for several orders m as well as the intensity and SAM density distributions on propagation in space to the plane with the maximal SAM. The singularities circle radius a_0 was chosen so as to equalize the intensities in the center and in the periphery, i.e., by Eq. (7.19). The intensity distributions were obtained by Eq. (7.9), but were compared with those obtained by the numerical Fresnel transform implemented as a convolution with using the fast Fourier transform. All the figures looked identical. To make the magnitudes of the order of units, all field components were multiplied by a constant factor $C_0 = 3000$.

The radii a_0 were computed by Eq. (7.19) to equalize the intensities in the center and in the periphery. Circles on the SAM distributions (k–o) show the maximal SAM radii obtained by Eq. (7.25). All quantities (maximal intensity and SAM magnitude) are given in arbitrary units. Scale mark in each figure denotes 1 mm. Figure 7.6 confirms that the singularities circle radius computed by Eq. (7.19) allows equalizing the beam intensity in the center and in the periphery. Figure 7.6 also confirms that in the transverse plane the maximal SAM magnitude is achieved at a circle of the radius given by Eq. (7.25), which is $\sqrt{2}$ times smaller than the radius of maximal intensity of a single-ringed mth-order LG beam. If is also seen in Fig. 7.6 that the transverse plane with the maximal SAM is closer and closer to the initial plane with increasing number of singularities m. This is because this plane should be close than the first transverse plane with linear polarizations, which is also closer and closer, according to Eq. (7.23).

To verify the approximate expression (7.33) for finding the plane with the maximal SAM, Fig. 7.7 illustrates the SAM dependence on the propagation distance for the beams shown in Fig. 7.6. The SAM was computed by Eq. (7.26). Dots in the top of each plot indicate the maximal SAM distance obtained by Eq. (7.33).

Figure 7.8 depicts the initial intensity distributions of the light field (7.5) as well as the intensity and SAM density distributions on propagation in space to the plane

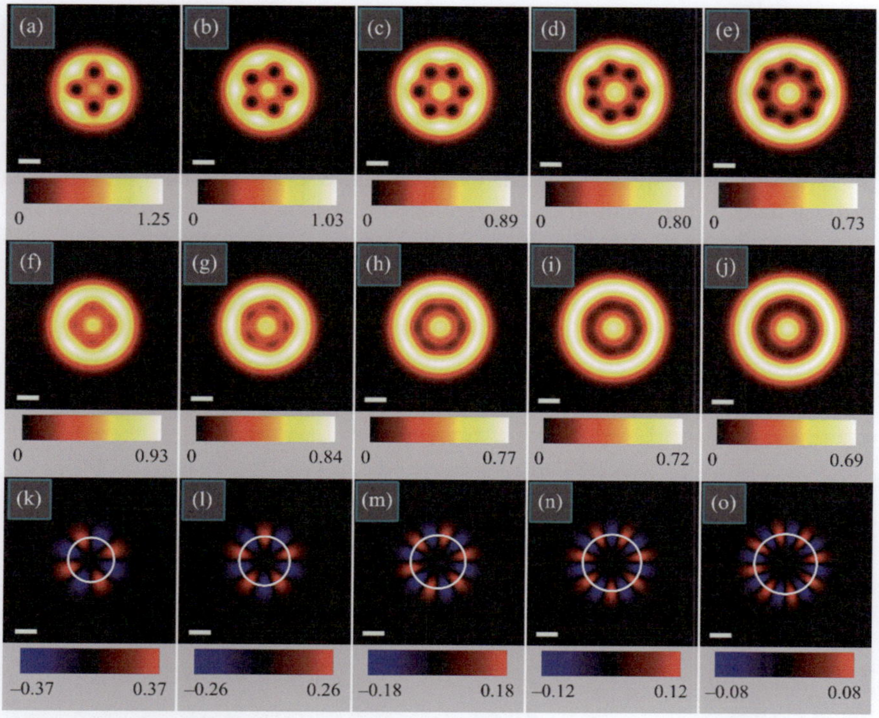

Fig. 7.6 Intensity distributions in the initial plane (**a–e**) and at the maximal SAM distance (7.33) (**f–j**), as well as SAM density distributions at the maximal SAM distance (**k–o**) for the following parameters: wavelength $\lambda = 532$ nm, Gaussian beam waist radius $w_0 = 1$ mm, number of the polarization singularities in the initial plane $m = 4$ (**a, f, k**), $m = 5$ (**b, g, l**), $m = 6$ (**c, h, m**), $m = 7$ (**d, i, n**), $m = 8$ (**e, j, o**), radius of the singularities circle $a_0 = 828$ μm (**a, f, k**), $a_0 = 942$ μm (**b, g, l**), $a_0 = 1041$ μm (**c, h, m**), $a_0 = 1129$ μm (**d, i, n**), $a_0 = 1210$ μm (**e, j, o**), propagation distance is $z = 0.349z_0$ (**a, f, k**), $z = 0.284z_0$ (**b, g, l**), $z = 0.240z_0$ (**c, h, m**), $z = 0.208z_0$ (**d, i, n**), $z = 0.183z_0$ (**e, j, o**)

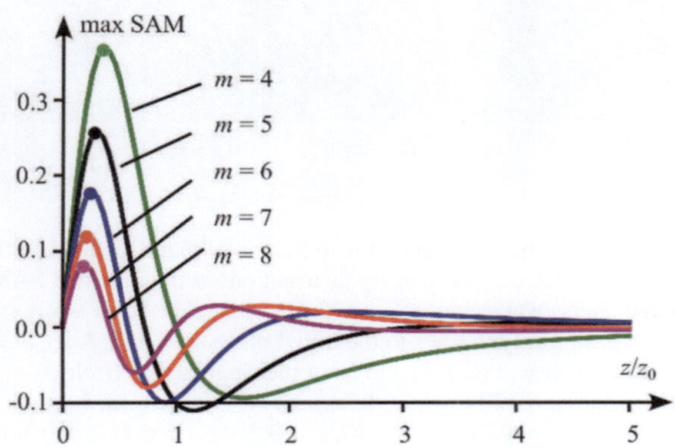

Fig. 7.7 Maximal SAM for several numbers of the polarization singularities m when the singularities circle radius is chosen so as to equalize the intensities in the center and in the periphery. Dots in the top of each plot indicate the maximal SAM distance obtained by the approximate formula (7.33)

with the maximal SAM with all parameters being the same as in Fig. 7.6, but the singularities circle radius is chosen so as to maximize the SAM density.

The radii a_0 were computed by Eq. (7.34) to maximize the SAM over all other radii a_0. Propagation distances were computed by Eq. (7.33) to maximize the SAM density. Circles on the SAM distributions (k–o) show the maximal SAM radii obtained by Eq. (7.25). All quantities (maximal intensity and SAM magnitude) are given in arbitrary units. Scale mark in each figure denotes 1 mm. It is seen in Fig. 7.8 that for each number of singularities m, the SAM density achieves magnitudes nearly 1.5 times higher than those in Fig. 7.6.

In addition, it is seen that the initial fields have the same maximal intensity independently of m. This is because the maximal SAM is achieved, according to the above theory, when the energy of the Gaussian beam is equal to the half of whole

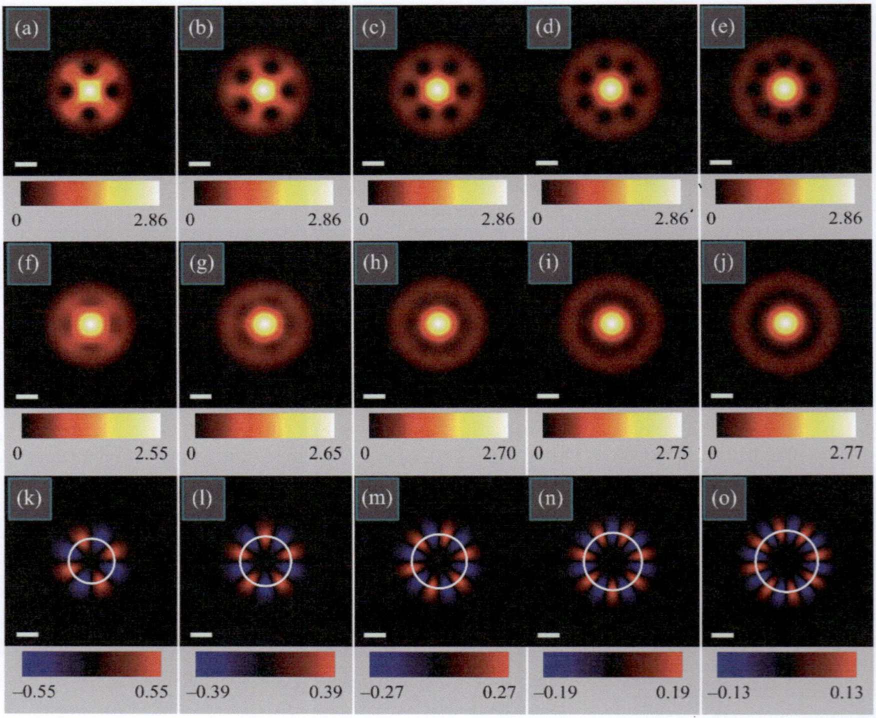

Fig. 7.8 Intensity distributions in the initial plane (**a–e**) and at the maximal SAM distance (7.33) (**f–j**), as well as SAM density distributions at the maximal SAM distance (**k–o**) for the following parameters: wavelength $\lambda = 532$ nm, Gaussian beam waist radius $w_0 = 1$ mm, number of the polarization singularities in the initial plane $m = 4$ (**a, f, k**), $m = 5$ (**b, g, l**), $m = 6$ (**c, h, m**), $m = 7$ (**d, i, n**), $m = 8$ (**e, j, o**), radius of the singularities circle $a_0 = 1052$ μm (**a, f, k**), $a_0 = 1141$ μm (**b, g, l**), $a_0 = 1223$ μm (**c, h, m**), $a_0 = 1300$ μm (**d, i, n**), $a_0 = 1371$ μm (**e, j, o**), propagation distance is $z = 0.349z_0$ (**a, f, k**), $z = 0.284z_0$ (**b, g, l**), $z = 0.240z_0$ (**c, h, m**), $z = 0.208z_0$ (**d, i, n**), $z = 0.183z_0$ (**e, j, o**)

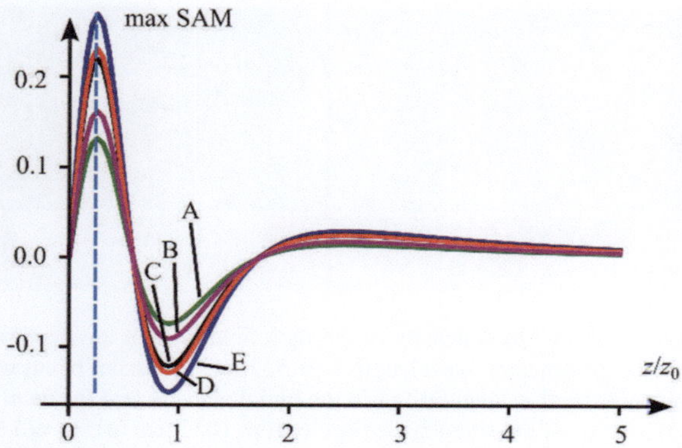

Fig. 7.9 Maximal SAM of the light field (7.5) at $m = 6$ for several values of the singularities circle radius a_0: $a_0 = a_{0,\max}$ given by Eq. (7.34) (curve E), $a_0 = 0.8a_{0,\max}$ (curve A), $a_0 = 1.2a_{0,\max}$ (curve B), $a_0 = 0.9a_{0,\max}$ (curve C), $a_0 = 1.1a_{0,\max}$ (curve D). Dashed line indicates the maximal SAM distance obtained by the approximate formula (7.33), which is independent on a_0

beam energy. The rest energy of the same amount goes into the light ring and the intensity of this ring is weaker than the central intensity of the Gaussian beam. Thus, the central part of the intensity patterns of all the beams in Fig. 7.8 is the Gaussian beam of the same energy, i.e., of the same amplitude. That is why the central intensity is the same.

Now we verify that indeed the singularities ring radius a_0 from Eq. (7.34) yields the maximal SAM density over other radii. Figure 7.9 illustrates the longitudinal SAM distributions of the light field (7.5) at a_0 given by Eq. (7.34) and at some other values a_0.

Figure 7.9 confirms that the maximal SAM is achieved at a_0 given by Eq. (7.34).

Finally, we compute the OAM density of the light field (7.5).

Figure 7.10 depicts the OAM density distributions of the light field shown in Fig. 7.8. The OAM distributions look like inverted SAM distributions, but they were obtained by a quite different way: by Eq. (7.40) where the angular derivative was represented as $\partial/\partial\varphi = x\partial/\partial y - y\partial/\partial x$ and was computed by using finite differences.

The radii a_0 were computed by Eq. (7.34) to maximize the SAM (and thus the OAM) over all other radii a_0. Propagation distances were computed by Eq. (7.33) to maximize the SAM (OAM) density. All quantities are given in arbitrary units. Scale mark in each figure denotes 1 mm. The maximal OAM magnitudes confirm that the OAM exceeds the SAM $-m/2$ times.

We investigated paraxial light beams with multiple polarization singularities residing evenly on a circle (singularities circle) [19]. In the initial plane, polarization of such light beam is linear in all points, and the beam has zero spin and orbital angular momenta (SAM and OAM). When such a beam is propagating in free space, there are several transverse planes where polarization is also linear and polarization singularities occur. However, between such planes, polarization is, in general, elliptic, and there are alternating areas with the positive and negative SAM, i.e., the spin Hall

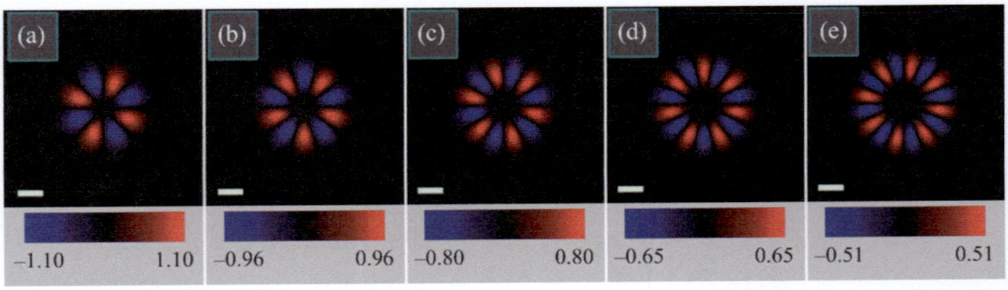

Fig. 7.10 Distributions of the OAM density of the light field (7.5) at the maximal SAM distance (7.33) for the following parameters: wavelength $\lambda = 532$ nm, Gaussian beam waist radius $w_0 =$ 1 mm, number of the polarization singularities in the initial plane $m = 4$ (**a**), $m = 5$ (**b**), $m = 6$ (**c**), $m = 7$ (**d**), $m = 8$ (**e**), radius of the singularities circle $a_0 = 1052$ μm (**a**), $a_0 = 1141$ μm (**b**), $a_0 =$ 1223 μm (**c**), $a_0 = 1300$ μm (**d**), $a_0 = 1371$ μm (**e**), propagation distance is $z = 0.349z_0$ (**a**), $z =$ $0.284z_0$ (**b**), $z = 0.240z_0$ (**c**), $z = 0.208z_0$ (**d**), $z = 0.183z_0$ (**e**)

effect arises. We found the intensity distribution and obtained an approximate expression [Eqs. (7.19) and (7.20)] for the singularities circle radius, when the intensity in the center and on the beam edge is nearly equal. In this case, radius of the maximal intensity in the beam edge tends (with increasing number of singularities m) to the maximal-intensity radius of the single-ringed Laguerre-Gaussian (LG) beam. When the beam propagates, in each transverse plane, the maximal SAM density magnitudes are achieved in $2m$ points (m points of maximal SAM density and m points of minimal SAM density) on a ring with a radius independent on the singularities circle radius. It turns out that this maximal SAM radius is equal to the half of the Gaussian beam radius multiplied by the square root from the number of singularities (i.e., $\sqrt{2}$ times smaller than the radius of maximal intensity of a single-ringed mth-order LG beam). In each transverse plane, there are $2m$ C-points with circular polarization (m points with left polarization and m points with right polarization). These points reside on a circle with a radius proportional to that of the singularities circle. We obtained an approximate expression [Eq. (7.33)] for the propagation distance where the SAM density achieves maximal magnitudes. It turned out that this distance is independent of the singularities circle radius and is shorter than the distance to the first transverse plane with linear polarization and polarization singularities. We derived an exact expression [Eq. (7.34)] for the singularities circle radius that maximizes the SAM density. It turns out that the maximal achievable SAM density decreases with the number of singularities m. In addition, for the beam with the maximal SAM density and for the beam with equal intensities in the center and in the edge, the singularities circle radius is different. The investigated light beam is actually a superposition of a LG beam with cylindrical polarization and of a linearly polarized Gaussian beam. It turns out that the maximal SAM density can be achieved when the energy of the Gaussian beam is equal to the half of the whole beam energy, i.e., to the energy of the LG beam.

We also obtained an expression for the OAM density [Eq. (7.42)]. It is proportional to the SAM density and exceeds it $-m/2$ times, which is in contrast to the conventional vortex beam with circular polarization, whose OAM density exceeds the SAM density $\pm\, m$ times. We considered a simple analogy of the investigated beams – superpositions of three plane waves. Two waves are tilted and circularly polarized, and the third wave is non-tilted and linearly polarized. We found that in such simple field, the spin Hall effect also arises, and this reveals its physical mechanism. This effect arises due to the phase delay between the tilted waves and non-tilted wave on propagation. The same way, for the studied vector light with multiple polarization singularities, the spin Hall effect arises due to different divergence of the cylindrically polarized LG beam and of the linearly polarized Gaussian beam.

7.2 Spin Hall Effect in Paraxial Vectorial Light Beams with an Infinite Number of Polarization Singularities

Properties of light beams, and, in particular, optical vortices, can differ in near and in far field of diffraction. For instance, in [20], fractional-order optical vortices are studied in the near field. Such vortices contain chains of alternating $\pm$ 1st-order vortices, which disappear in the far field. As was demonstrated in [21], an optical vortex generated by a refractive spiral phase plate has an asymmetric shape in the Fresnel diffraction zone. The work [22] investigated near-field diffraction of a Gaussian beam by fork gratings with different topological charges, and it was found that the generated optical vortices reside on spiral lines around the $\pm$ 1st diffraction orders. In the far field of diffraction, transformation was found of the initially fractional topological charge [23, 24]. In addition, in the far field (in the lens focus) of optical vortices, spin Hall effect arises [25, 26], i.e., alternating pairs of areas are generated in the focus with left and right circular polarization, despite linear polarization of the initial beam. This effect was discovered both for homogeneous linear polarization of the initial field [27] and for inhomogeneous one (cylindrical) [28].

A natural generalization of a vortex light beam is a beam with several vortices. A seminal work with multiple vortices [1] investigates Gaussian beams with optical vortices located in the transverse plane arbitrarily. These beams are described by finite products with the number of multipliers equal to the number of vortices. The work [2] describes in a closed form propagation-invariant light fields with an arbitrary envelope analytical function, whose zeros define positions of vortices in the beam. Based on [1], a light field can be constructed that has phase singularities residing on a circle [29]. In our paper [7], we investigated similar field, but with polarization singularities on a circle. Recently, we found that the spin Hall effect also arises in such fields [19], but even in the Fresnel diffraction zone, rather than in the far field.

Besides the finite number of optical vortices, the approach from [2] allows obtaining a light field with an infinite array of optical vortices, residing on a straight line [30]. Such fields have an infinite topological charge [31], can be generated by non-coaxial superposition of two-tilted Gaussian beams [32], and identified by optical vortices density, rather than by the topological charge, which can be measured interferometrically [30].

In this section, we study a vectorial Gaussian beam where, instead of phase singularities, an infinite number of polarization singularities resides on a straight line. The polarization singularities index (Poincaré-Hopf index) [3, 33] of such a beam is shown to be also infinite. We found that the azimuthal angle of the polarization singularities [4] increases in the far field by π/2, i.e., initial radial polarization is converted to azimuthal and vice versa. It is demonstrated that when such a beam propagates in space, two pairs of areas are generated in the Fresnel zone with nonzero density of the longitudinal component of the spin angular momentum (SAM), despite linear polarization in the initial plane, i.e., the optical spin Hall effect arises.

7.2.1 *Paraxial Light Fields with an Infinite Number of Phase or Polarization Singularities*

In [2], the following solution to the paraxial Helmholtz equation has been obtained (Eq. (17) in [2]):

$$E(r,\varphi,z)=\frac{1}{q}\exp\left(-\frac{r^2}{qw_0^2}\right)f\left(\frac{re^{i\varphi}}{qw_0}\right), \tag{7.47}$$

where (r, φ, z) are the cylindrical coordinates, w_0 is the waist radius of the Gaussian beam, $q = 1 + iz/z_0$, and $f(\xi)$ is an arbitrary entire analytical function. The field (1) does not change its intensity structure on propagation in space. It is only widened $|q| = [1 + (z/z_0)]^{1/2}$ times and rotated around the optical axis by an angle arg $q = \arctan(z/z_0)$.

This general expression allows obtaining a solution of the paraxial Helmholtz equation with and infinite or with a finite number of optical vortices. For instance, if $f(\xi) = \cos(w_0\xi/\alpha_0)$, then the vortices reside evenly along a straight line [30]:

$$E(r,\varphi,z)=\frac{1}{q}\exp\left(-\frac{r^2}{qw_0^2}\right)\cos\left(\frac{re^{i\varphi}}{\alpha_0 q}\right). \tag{7.48}$$

This field is an example of light fields with an infinite topological charge [31]. In the initial plane of such a field, optical vortices reside in points with the Cartesian coordinates $x_p = \alpha_0(\pi/2 + \pi p)$, $y = 0$ with p being integer numbers.

It has long been known [5] that cylindrical polarization can be represented as a superposition of ± 1st-order optical vortices with opposite circular polarization. For the Jones vectors, such representation can be written as follows:

$$\begin{bmatrix}\cos(\varphi+\delta)\\ \sin(\varphi+\delta)\end{bmatrix}=\frac{1}{2}\exp(i\varphi+i\delta)\begin{pmatrix}1\\-i\end{pmatrix}+\frac{1}{2}\exp(-i\varphi-i\delta)\begin{pmatrix}1\\i\end{pmatrix}. \tag{7.49}$$

where δ is the azimuthal angle of cylindrical polarization [4] (if δ = 0 or if δ = π/2, polarization is, respectively, radial or azimuthal). Then, if we use the same Jones vectors, but instead of the multipliers $e^{i\varphi}$ and $e^{-i\varphi}$ we substitute the field (7.48) with the cosine argument $re^{i\varphi}$ and $re^{-i\varphi}$, respectively, we construct a vector field with an infinite number of the polarization singularities:

$$\begin{aligned} \mathbf{E}(x, y, z) &= \frac{1}{2q\sqrt{W_0}} \exp\left(-\frac{x^2 + y^2}{qw_0^2}\right) \\ &\times \left[\exp(i\delta)\cos\left(\frac{x + iy}{\alpha_0 q}\right)\begin{pmatrix} 1 \\ -i \end{pmatrix} + \exp(-i\delta)\cos\left(\frac{x - iy}{\alpha_0 q}\right)\begin{pmatrix} 1 \\ i \end{pmatrix}\right], \end{aligned} \tag{7.50}$$

where (x, y) are the Cartesian coordinates in the transverse plane and W_0 is a multiplier introduced for normalizing the beam energy (i.e., in order to make it equal to unit). This multiplier can be obtained from an expression for the energy of the scalar field (7.48) [30]:

$$W_0 = \frac{\pi w_0^2}{2} \cosh\left(\frac{w_0^2}{2\alpha_0^2}\right) \tag{7.51}$$

Superposition (7.50) consists of two beams. Upon propagation in space, one of them is rotated clockwise and another—counterclockwise. For a compact description of such propagation, we introduce two rotated coordinate systems (Fig. 7.11):

$$\begin{cases} x_\pm = (x \cos\psi \pm y \sin\psi) / (\alpha_0 |q|), \\ y_\pm = (y \cos\psi \mp x \sin\psi) / (\alpha_0 |q|), \end{cases} \tag{7.52}$$

with $\psi = \arctan(z/z_0)$ being the Gouy phase.

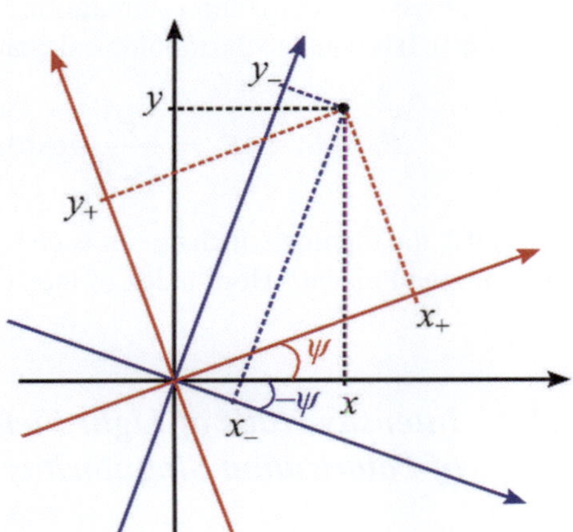

Fig. 7.11 Coordinate systems (x_+, y_+) and (x_-, y_-)

Then the complex amplitude (7.50) can be rewritten as follows:

$$\begin{aligned}\mathbf{E}(x,y,z) &= \frac{1}{2q\sqrt{W_0}}\exp\left(-\frac{x^2+y^2}{qw_0^2}\right)\\ &\times\left[\exp(i\delta)\cos(x_+ + iy_+)\begin{pmatrix}1\\ -i\end{pmatrix} + \exp(-i\delta)\cos(x_- - iy_-)\begin{pmatrix}1\\ i\end{pmatrix}\right],\end{aligned} \tag{7.53}$$

or in a matrix form

$$\mathbf{E}(x,y,z) = \frac{1}{2q\sqrt{W_0}}\exp\left(-\frac{x^2+y^2}{qw_0^2}\right)\begin{bmatrix}\cos\delta & -\sin\delta\\ \sin\delta & \cos\delta\end{bmatrix}\begin{bmatrix}1 & 1\\ -i & i\end{bmatrix}\begin{bmatrix}\cos(x_+ + iy_+)\\ \cos(x_- - iy_-)\end{bmatrix} \tag{7.54}$$

where the matrix

$$S = \begin{bmatrix}1 & 1\\ -i & i\end{bmatrix} \tag{7.55}$$

converts phase singularities (optical vortices) into polarization singularities (radial polarization), while the matrix

$$R = \begin{bmatrix}\cos\delta & -\sin\delta\\ \sin\delta & \cos\delta\end{bmatrix} \tag{7.56}$$

rotates the strength vectors by the azimuthal angle δ.

As seen from the matrix representation, the azimuthal angle δ of cylindrical polarization does not affect the intensity distribution in an arbitrary transverse plane. In the initial plane, polarization is linear in each point. Adopting an approach from [34], we can derive the polarization singularities index (Poincaré-Hopf index) [33] of the field (7.50). It is equal to the topological charge of the following scalar complex field:

$$E_c = E_x + iE_y = \frac{e^{i\delta}}{q\sqrt{W_0}}\exp\left(-\frac{x^2+y^2}{qw_0^2}\right)\cos\left(\frac{x+iy}{\alpha_0 q}\right), \tag{7.57}$$

In [30], the topological charge of such scalar fields was shown to be infinite, and therefore the Poincaré-Hopf index of the vector field (7.50) is also infinite.

7.2.2 *Intensity Nulls of Light Fields with an Infinite Number of Polarization Singularities*

Here we obtain the intensity nulls of the field (7.50). Since the determinants of both matrices are nonzero, for the zero intensity in some point, the following conditions should be fulfilled:

$$\begin{cases} \cos(x_+ + iy_+) = 0, \\ \cos(x_- - iy_-) = 0. \end{cases} \tag{7.58}$$

Both real and imaginary parts should be zero and thus we get

$$\begin{cases} \cos x_+ \cosh y_+ = 0, \\ \sin x_+ \sinh y_+ = 0, \\ \cos x_- \cosh y_- = 0, \\ \sin x_- \sinh y_- = 0. \end{cases} \tag{7.59}$$

The hyperbolic cosine cannot be zero. Thus, from the 1st and 3rd equation in (7.59) we get that $\cos x_+ = \cos x_- = 0$. This means that $\sin y_+ \neq 0$ and $\sin x_- \neq 0$ and, therefore,

$$\begin{cases} \cos x_+ = 0, \\ y_+ = 0, \\ \cos x_- = 0, \\ y_- = 0. \end{cases} \tag{7.60}$$

Since $y_+ = y_- = 0$, we get that $y \cos \psi = x \sin \psi = 0$. In the initial plane, $\psi = 0$, and thus the intensity nulls reside in points with the coordinates

$$\begin{cases} x = \alpha_0(\pi/2 + \pi p), \\ y = 0. \end{cases} \tag{7.61}$$

In the far field, $\psi \to \pi/2$, and the intensity nulls reside on the vertical axis in points with the coordinates

$$\begin{cases} y = \alpha_0|q|(\pi/2 + \pi p), \\ x = 0. \end{cases} \tag{7.62}$$

For finite distances z, $\cos \psi \neq 0$ and $\sin \psi \neq 0$. Therefore, conditions (7.60) cannot be fulfilled and the field (7.50) does not have the intensity nulls.

Now we consider vicinities of the intensity nulls in the far field, i.e., points with the coordinates

$$\begin{cases} x = \rho \cos \theta, \\ y = \alpha_0|q|(\pi/2 + \pi p) + \rho \sin \theta, \end{cases} \tag{7.63}$$

with $\rho << \alpha_0, w_0$. Since $\psi \to \pi/2$ in the far field, then in these points the rotated coordinates (7.52) read as

$$\begin{cases} x_\pm = \pm(\pi/2 + \pi p) \pm \rho \sin\theta / (\alpha_0 |q|), \\ y_\pm = \mp\rho\cos\theta / (\alpha_0|q|). \end{cases} \tag{7.64}$$

Therefore, $x_\pm \pm iy_\pm \approx \pm(\pi/2 + \pi p) - i\rho e^{\pm i\theta}/(\alpha_0|q|)$, $\cos(x_\pm \pm iy_\pm) \approx \pm(-1)^p i\rho e^{\pm i\theta}/(\alpha_0|q|)$, and the field amplitude is equal to:

$$\begin{aligned} &\mathbf{E}(\rho\cos\theta, \alpha_0|q|(\pi/2+\pi p) + \rho\sin\theta, z >> z_0) \\ &\approx (-1)^p \frac{1}{2q\sqrt{W_0}} \frac{i\rho}{\alpha_0|q|} \exp\left(-\frac{\alpha_0^2|q|^2}{qw_0^2}\right)\left[\exp(i\delta + i\theta)\begin{pmatrix}1\\-i\end{pmatrix} - \exp(-i\delta - i\theta)\begin{pmatrix}1\\i\end{pmatrix}\right] \\ &= (-1)^p \frac{1}{q\sqrt{W_0}} \frac{\rho}{\alpha_0|q|} \exp\left(-\frac{\alpha_0^2|q|^2}{qw_0^2}\right)\begin{pmatrix}\cos(\theta+\delta+\pi/2)\\ \sin(\theta+\delta+\pi/2)\end{pmatrix}. \end{aligned} \tag{7.65}$$

This means that the azimuthal angle δ in the far field increases by $\pi/2$, i.e., radial polarization is converted to azimuthal and vice versa.

7.2.3 *Intensity and Spin Angular Momentum Density Distribution of Light Fields with an Infinite Number of Polarization Singularities*

The intensity distribution of the field (7.50) is given by the following:

$$\begin{aligned} I(x,y,z) &= |E_x(x,y,z)|^2 + \left|E_y(x,y,z)\right|^2 = \\ &= \frac{1}{4|q|^2 W_0} \exp\left(-2\frac{x^2+y^2}{|q|^2 w_0^2}\right)(\cos 2x_+ + \cos 2x_- + \cosh 2y_+ + \cosh 2y_-), \end{aligned} \tag{7.66}$$

Derivation of the intensity distribution (7.66). Since the azimuthal angle δ of cylindrical polarization does not affect the intensity distribution, we suppose for simplicity that $\delta = 0$. Then, transverse field components are given by the following:

$$\begin{cases} E_x(x,y,z) = \dfrac{1}{2q\sqrt{W_0}} \exp\left(-\dfrac{x^2+y^2}{qw_0^2}\right)\left[\cos(x_+ + iy_+) + \cos(x_- - iy_-)\right], \\ E_y(x,y,z) = \dfrac{-i}{2q\sqrt{W_0}} \exp\left(-\dfrac{x^2+y^2}{qw_0^2}\right)\left[\cos(x_+ + iy_+) - \cos(x_- - iy_-)\right], \end{cases} \tag{7.67}$$

and the intensity distribution is

$$I(x,y,z) = |E_x(x,y,z)|^2 + \left|E_y(x,y,z)\right|^2 = \\ = \frac{1}{2W_0|q|^2}\exp\left(-2\frac{x^2+y^2}{|q|^2 w_0^2}\right)\left[|\cos(x_+ + iy_+)|^2 + |\cos(x_- - iy_-)|^2\right]. \tag{7.68}$$

Using an identity $\cos(x+iy) = \cos x \cosh y - i \sin x \sinh y$, we get

$$I(x,y,z) = \frac{1}{2W_0|q|^2}\exp\left(-2\frac{x^2+y^2}{|q|^2 w_0^2}\right) \\ \times\left[\cos^2 x_+ \cosh^2 y_+ + \sin^2 x_+ \sinh^2 y_+ + \cos^2 x_- \cosh^2 y_- + \sin^2 x_- \sinh^2 y_-\right]. \tag{7.69}$$

Finally, using the formulae for trigonometric and hyperbolic functions with a double argument, we obtain an expression for the intensity distribution (7.66):

$$I(x,y,z) = \frac{1}{4W_0|q|^2}\exp\left(-2\frac{x^2+y^2}{|q|^2 w_0^2}\right)\times \\ \times\left[\cos 2x_+ + \cos 2x_- + \cosh 2y_+ + \cosh 2y_-\right]. \tag{7.70}$$

The same way we can derive the distribution of the longitudinal component of the SAM density:

$$S_z(x,y,z) = \frac{1}{4|q|^2 W_0}\exp\left(-2\frac{x^2+y^2}{w_0^2|q|^2}\right)(\cos 2x_- + \cosh 2y_- - \cos 2x_+ - \cosh 2y_+). \tag{7.71}$$

Hence, both the SAM density and intensity distributions are independent of the azimuthal angle δ of cylindrical polarization. These expressions allow obtaining the coordinates of C-points of the field (7.50). For instance, right circular polarization appears in points where $S_z = I$. In these points, $\cos 2x_+ + \cosh 2y_+ = 0$ and, therefore, $\cos 2x_+ = -1$ and $\cosh 2y_+ = 1$, i.e., $y_+ = 0$ and $x_+ = (\pi/2)(2p+1)$, with p being an integer number. Then the coordinates of the C-points are as follows:

$$\begin{pmatrix} x_{\mathrm{RCP}} \\ y_{\mathrm{RCP}} \end{pmatrix} = \frac{\pi}{2}\alpha_0|q|(1+2p)\begin{pmatrix} \cos\psi \\ \sin\psi \end{pmatrix}. \tag{7.72}$$

Similarly, left circular polarization appears in points with $S_z = -I$. Coordinates of these points are equal to

$$\begin{pmatrix} x_{\mathrm{LCP}} \\ y_{\mathrm{LCP}} \end{pmatrix} = \frac{\pi}{2}\alpha_0|q|(1+2p)\begin{pmatrix} \cos\psi \\ -\sin\psi \end{pmatrix}. \tag{7.73}$$

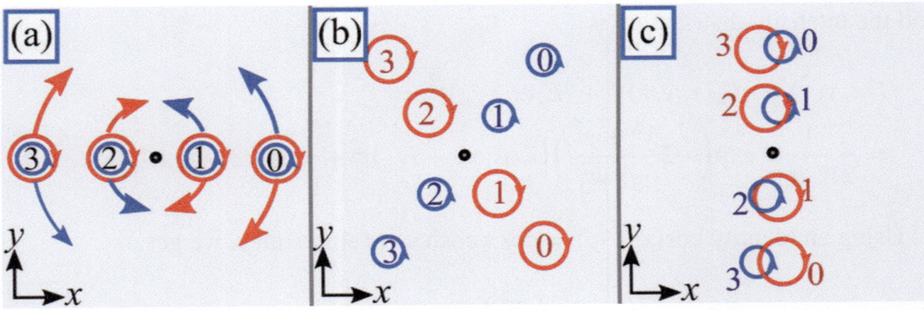

Fig. 7.12 Mechanism of destruction of the polarization singularities after the initial plane and of their reconstruction in the far field. In the initial plane, points with left and right circular polarization coincide (**a**). Then, on propagation, they rotate around the optical axis in different directions (**b**) and in the far field they merge again (**c**)

The dynamics of the C-points explains destroying the polarization singularities after the initial plane and their reconstruction in the far field (Fig. 7.12). Due to the splitting of left and right circular polarization, the beam (7.50) acquires nonzero SAM density upon propagation, and appearing areas with alternating SAM density manifests about the spin Hall effect.

We failed to derive exact expressions for the points of maximal SAM density. However, expressions (7.66) and (7.71) are simplified when $w_0 \gg \alpha_0$. Indeed, using identities for the sums and differences of trigonometric and hyperbolic functions, we get

$$I(x,y,z) = \frac{1}{2|q|^2 W_0}\exp\left(-2\frac{x^2+y^2}{|q|^2 w_0^2}\right) \times\left[\cos\left(\frac{2x\cos\psi}{\alpha_0|q|}\right)\cos\left(\frac{2y\sin\psi}{\alpha_0|q|}\right)+\cosh\left(\frac{2x\sin\psi}{\alpha_0|q|}\right)\cosh\left(\frac{2y\cos\psi}{\alpha_0|q|}\right)\right], \tag{7.74}$$

$$S_z(x,y,z) = \frac{1}{2|q|^2 W_0}\exp\left(-2\frac{x^2+y^2}{w_0^2|q|^2}\right) \times\left[\sin\left(\frac{2x\cos\psi}{\alpha_0|q|}\right)\sin\left(\frac{2y\sin\psi}{\alpha_0|q|}\right)+\sinh\left(\frac{2x\sin\psi}{\alpha_0|q|}\right)\sinh\left(\frac{2y\cos\psi}{\alpha_0|q|}\right)\right]. \tag{7.75}$$

Products of two trigonometric or hyperbolic functions can be represented as a sum of four exponents. Thus, Eqs. (7.66) and (7.71) contain eight exponential terms. The first four terms do not exceed the value $(8|q|^2 W_0)^{-1}$, whereas the other four terms describe off-axis Gaussian beams:

$$GB_{\pm\pm}(x, y, z) = \frac{1}{8|q|^2 W_0} \exp\left(-2\frac{x^2+y^2}{w_0^2|q|^2}\right) \exp\left(\pm\frac{2x\sin\psi}{\alpha_0|q|} \pm \frac{2y\sin\psi}{\alpha_0|q|}\right) =$$
$$= \frac{1}{8|q|^2 W_0} \exp\left\{\frac{-2}{w_0^2|q|^2}\left[(x - x_{c,\pm})^2 + (y - y_{c,\pm})^2\right] + \frac{w_0^2}{2\alpha_0^2}\right\}, \tag{7.76}$$

with

$$\begin{aligned} x_{c,\pm} &= \pm\frac{w_0^2|q|}{2\alpha_0}\sin\psi, \\ y_{c,\pm} &= \pm\frac{w_0^2|q|}{2\alpha_0}\cos\psi. \end{aligned} \tag{7.77}$$

These terms achieve values $(8|q|^2 W_0)^{-1}\exp[(w_0/\alpha_0)^2/2]$. Thus, if $w_0 >> \alpha_0$, the first four terms can be neglected. Then the intensity and the SAM density (7.47) are equal to the intensity of the four Gaussian beams:

$$I(x, y, z) = GB_{++} + GB_{--} + GB_{+-} + GB_{-+}, \tag{7.78}$$

$$S_z(x, y, z) = GB_{++} + GB_{--} - GB_{+-} - GB_{-+}. \tag{7.79}$$

If these four beams are far enough from each other (i.e., $|x_{c+} - x_{c-}| >> w_0|q|$, $|y_{c+} - y_{c-}| >> w_0|q|$), then these terms almost do not affect each other and, obviously, the points with the maximal SAM density coincide with the points of maximal intensity. In this case, the dependence of the maximal SAM density on the propagation distance z is given by

$$\max_{x,y} S_z(x, y, z) \approx \frac{1}{8|q|^2 W_0} \exp\left(\frac{w_0^2}{2\alpha_0^2}\right). \tag{7.80}$$

Hence, the SAM density decreases with the propagation distance from the initial plane in a similar law as does the intensity in the Gaussian beam center. However, polarization is everywhere linear in the initial plane, i.e., the maximal SAM density is zero. This means that it at first increases near the initial plane and then, when the Gaussian beams are split, begins to decrease, i.e., the maximal SAM density is achieved in the near field of diffraction.

7.2.4 *Identification of Light Fields with an Infinite Number of Polarization Singularities*

In our work [30], we studied an analogy between scalar fields with an infinite topological charge and conventional circularly symmetric optical vortices. Similarly, we can consider circular fields with cylindrical polarization and fields with an infinite number of polarization singularities. We suppose that a field with cylindrical polarization is composed of two circularly polarized single-ringed LG beams with opposite topological charges:

$$\begin{aligned}\mathbf{E}(r,\varphi,0) &= \left(\frac{r}{w_0}\right)^m \exp\left(-\frac{r^2}{w_0^2}+im\varphi\right)\begin{pmatrix}1\\-i\end{pmatrix}+\left(\frac{r}{w_0}\right)^m \exp\left(-\frac{r^2}{w_0^2}-im\varphi\right)\begin{pmatrix}1\\i\end{pmatrix}\\ &= 2\left(\frac{r}{w_0}\right)^m \exp\left(-\frac{r^2}{w_0^2}\right)\begin{bmatrix}\cos(m\varphi)\\ \sin(m\varphi)\end{bmatrix}.\end{aligned} \tag{7.81}$$

where (r, φ) are the polar coordinates. Such a field can be easily identified using a polarizer. If it transmits only one polarization, then a multi-petal intensity distribution is obtained, which allows determining the order of cylindrical polarization by counting the petals (Fig. 7.13a, b).

Similarly, registering the intensity of only one transverse component of the field (7.50) allows determining the density of polarization singularities. Indeed, in the initial plane, the field (7.50) can be written as follows:

$$\begin{aligned}E_x(x,y,0) &= \frac{1}{2\sqrt{W_0}}\exp\left(-\frac{w_0^2}{4\alpha_0^2}\right)\\ &\times\left\{\exp\left[-\frac{x^2+\left(y+w_0^2/(2\alpha_0)\right)^2}{w_0^2}\right]+\exp\left[-\frac{x^2+\left(y-w_0^2/(2\alpha_0)\right)^2}{w_0^2}\right]\right\}\cos\left(\frac{x}{\alpha_0}\right).\end{aligned} \tag{7.82}$$

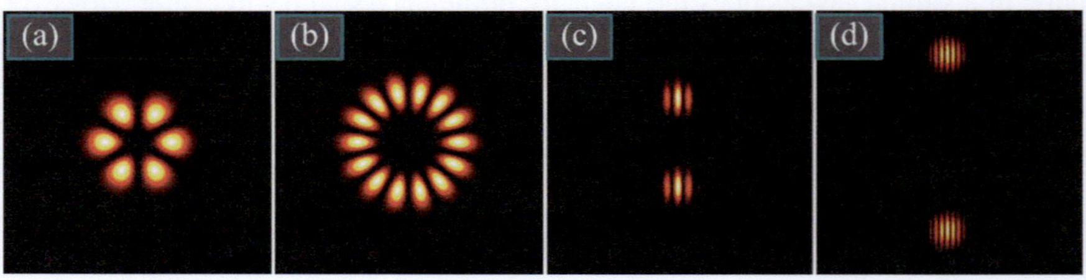

Fig. 7.13 Intensity distribution of the E_x component of two beams with cylindrical polarization (7.81) in the initial plane at $m = 3$ (**a**) and at $m = 7$ (**b**), as well of two beams with an infinite number of polarization singularities (7.50) at a different singularities density: $\alpha_0 = w_0/5$ (**c**) and $\alpha_0 = w_0/10$ (**d**)

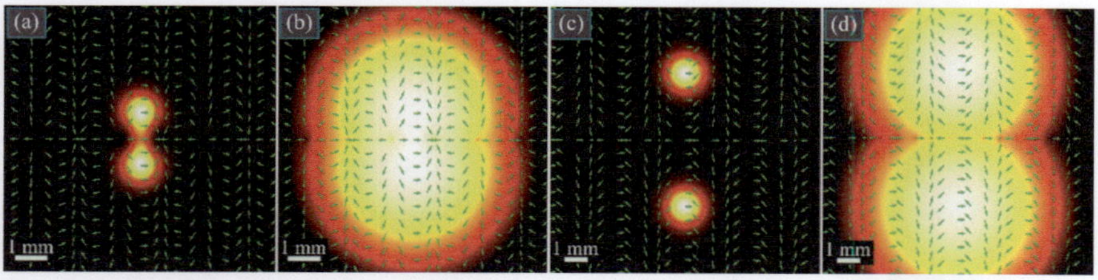

Fig. 7.14 Intensity distributions of two beams (7.50) in the initial plane (**a**, **c**) and the logarithm of this distribution (**b**, **d**), as well as the polarization directions (green arrows) for the following parameters: wavelength $\lambda = 532$ nm, Gaussian beam waist radius $w_0 = 1$ mm, the distance between the polarization singularities $\pi\alpha_0 = \pi w_0/2 \approx 1.57$ mm (**a**, **b**) and $\pi\alpha_0 = \pi w_0/5 \approx 0.63$ mm (**c**, **d**)

Hence, the intensity distribution of the x-component has vertical zero-intensity lines, whose frequency allows determining the density of polarization singularities (Fig. 7.13c,d).

7.2.5 Numerical Simulation

Shown in Fig. 7.14 are intensity distributions of two beams (7.50) (with a different distance between the polarization singularities) in the initial plane in uniform and in logarithmic color scale, as well as polarization directions. The intensity distributions were computed as $I(r, \varphi, 0) = |E_x(r, \varphi, 0)|^2 + |E_y(r, \varphi, 0)|^2$ with using Eq. (7.50), whereas the logarithmic distribution was computed as $\ln(10^{-18} + I(r, \varphi, 0) / \max I(r, \varphi, 0))$, where the constant 10^{-18} was introduced for avoiding the logarithm of zero in points with zero intensity. Polarization directions were computed by the formula $\arg(E_x(r, \varphi, 0) + iE_y(r, \varphi, 0))$.

As seen in Fig. 7.14, the intensity distribution has a shape of two light spots located symmetrically relative to the horizontal coordinate axis, and there are polarization singularities with radial polarization, residing periodically on this axis. Due to the low intensity, they are not visible, but they can be seen on the intensity distribution in logarithmic color scale.

Figure 7.15 illustrates the intensity and SAM density distributions of the vector beam from Fig. 7.14c, d in several transverse planes.

White dots on the SAM density distributions indicate the positions of maxima, obtained by Eq. (7.77). The numbers near the color scales denote the minimal and maximal values. According to Fig. 7.15, both light spots split, each into two spots, one of which shifts to the left and the other shifts to the right. It is also seen that the maximal SAM density decreases upon propagation, which is consistent with Eq. (7.80).

Figure 7.16 depicts the intensity and SAM density distributions of the vector beam from Fig. 7.14a, b in several transverse planes. In contrast to Fig. 7.14c, d, light spots in Fig. 7.14a, b are closer to each other and, thus, they do not split so

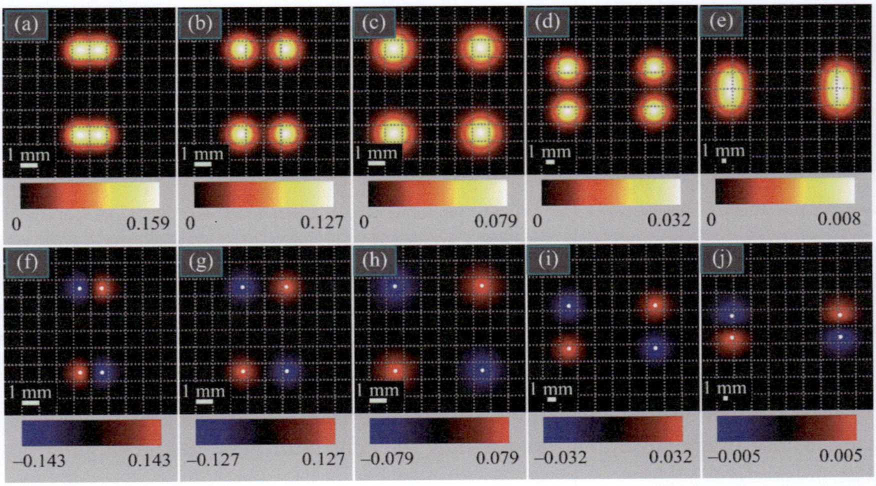

Fig. 7.15 Intensity (**a–e**) and SAM density (**f–j**) distributions of the beam (7.50) from Fig. 7.14**c**, **d** in several transverse planes for the following parameters: wavelength $\lambda = 532$ nm, Gaussian beam waist radius $w_0 = 1$ mm, the distance between the polarization singularities $\pi\alpha_0 = \pi w_0/5 \approx 0.63$ mm, propagation distances from the initial plane $z = z_0/4$ (**a**, **f**), $z = z_0/2$ (**b**, **g**), $z = z_0$ (**c**, **h**), $z = 2z_0$ (**d**, **i**), $z = 5z_0$ (**e**, **j**)

fast upon propagation into spots with left and right circular polarization, compared to Fig. 7.15. Therefore, maximal SAM density in Fig. 7.15 decreases immediately from $z = z_0/4$ till $z = 5z_0$, whereas in Fig. 7.16 it at first increases at distances up to $z = z_0/2$ and then decreases. The dependence of the maximal SAM density on the propagation distance is illustrated in Fig. 7.17.

Figure 7.17 reveals that the maximal SAM density is achieved closer and close to the initial plane, when the distance between the polarization singularities decreases. This effect has a physical explanation since, as seen from Eq. (7.50), decreasing value α_0 leads not only to moving the light spots away from each other, but also to increasing space frequency along the axis y. Therefore, with decreasing α_0, each light spot splits into two spots with opposite circular polarization faster.

In this section, we have constructed a vector Gaussian beam with an infinite number of polarization singularities residing on a straight line [35]. For such a beam, the intensity distribution was derived analytically and it turned out that the polarization singularities appear only in the initial plane and in the far field. We found that the polarization singularities index (Poincaré-Hopf index) is infinite. After propagation from the initial plane to the far field, the azimuthal angle of polarization singularities increases by $\pi/2$, i.e., initial radial polarization is converted into azimuthal and vice versa. We obtained a distribution of the longitudinal component of the spin angular momentum density. Similarly to the intensity distribution, it is independent of the azimuthal angle of polarization singularities. When such a vectorial field propagates in free space, an infinite number of C-points appears, where polarization is circular. The distance to the transverse plane with the maximal spin angular

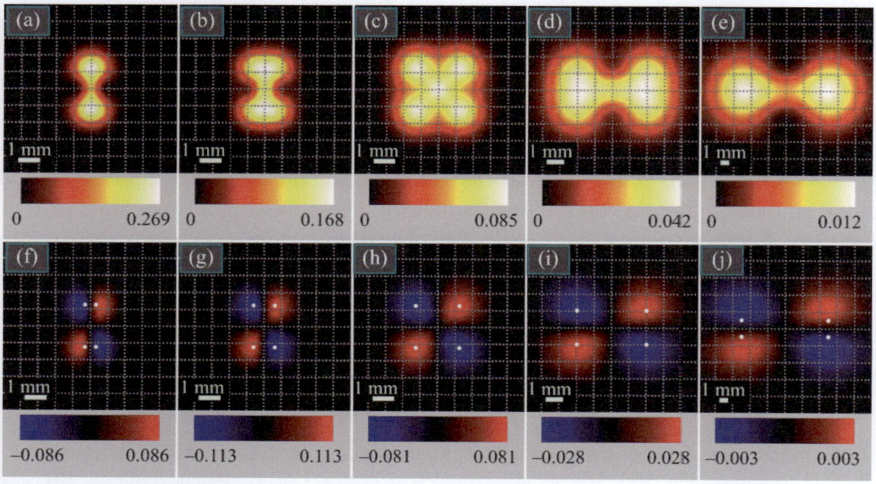

Fig. 7.16 Intensity (**a–e**) and SAM density (**f–j**) distributions of the beam (7.50) from Fig. 7.14**a**, **b** in several transverse planes for the following parameters: wavelength $\lambda = 532$ nm, Gaussian beam waist radius $w_0 = 1$ mm, the distance between the polarization singularities $\pi\alpha_0 = \pi w_0/2 \approx 1.57$ mm, propagation distances from the initial plane $z = z_0/4$ (**a**, **f**), $z = z_0/2$ (**b**, **g**), $z = z_0$ (**c**, **h**), $z = 2z_0$ (**d**, **i**), $z = 5z_0$ (**e**, **j**). White dots in the SAM density distributions denote the positions of maxima computed by Eq. (7.77). The numbers near the color scales denote the minimal and maximal values

Fig. 7.17 Dependence of the maximal intensity and of the maximal SAM density on the propagation distance for the distance between the polarization singularities equal to $\pi\alpha_0 = \pi w_0/2 \approx 1.57$ mm (**a**), $\pi\alpha_0 = \pi w_0/5 \approx 0.63$ mm (**b**), $\pi\alpha_0 = \pi w_0/10 \approx 0.31$ mm (**c**)

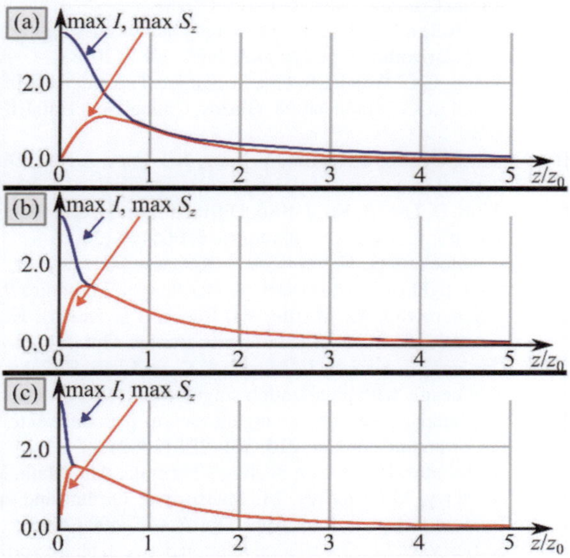

momentum density decreases with decreasing the distance between the polarization singularities in the initial plane. Generation of alternating areas with left and right circular polarization, despite linear polarization in the initial plane, manifests about the optical spin Hall effect. Application areas of the results obtained are designing micromachines for optical driving microscopic objects. The SAM causes particles to rotate around their centers of mass and engineering the SAM density distribution of the studied light field can allow simultaneous manipulating an ensemble of four particles. Another application is optical information transmission where the density of polarization singularities can be used for encoding the data.

References

1. G. Indebetouw, Optical Vortices and Their Propagation. J. Mod. Optic. **40**, 73–87 (1993)
2. E.G. Abramochkin, V.G. Volostnikov, Spiral-type beams. Opt. Commun.Commun. **102**, 336 (1993)
3. Q. Wang, C.H. Tu, Y.N. Li, H.T. Wang, Polarization singularities: progress, fundamental physics, and prospects. APL Photonics **6**, 040901 (2021)
4. Q. Zhan, Cylindrical vector beams: from mathematical concepts to applications. Adv. Opt. Photon. **1**, 1–57 (2009)
5. S.C. Tidwell, D.H. Ford, W.D. Kimura, Generating radially polarized beams interferometrically. Appl. Opt. **29**, 2234–2239 (1990)
6. A.A. Kovalev, V.V. Kotlyar, Gaussian beams with multiple polarization singularities. Opt. Commun.Commun. **423**, 111–120 (2018)
7. A.A. Kovalev, V.V. Kotlyar, Tailoring polarization singularities in a Gaussian beam with locally linear polarization. Opt. Lett. **43**, 3084–3087 (2018)
8. H. Wang, C.C. Wojcik, S. Fan, Topological spin defects of light. Optica **9**, 1417–1423 (2022)
9. A. Kavokin, G. Malpuech, M. Glazov, Optical Spin Hall Effect. Phys. Rev. Lett. **95**(13), 136601 (2005)
10. M. Kim, D. Lee, T.H. Kim, Y. Yang, H.J. Park, J. Rho, Observation of enhanced optical spin Hall effect in a vertical hyperbolic metamaterial. ACS Photonics **6**, 2530–2536 (2019)
11. M. Kim, D. Lee, B. Ko, J. Rho, Diffraction-induced enhancement of optical spin Hall effect in a dielectric grating. APL Photonics **5**, 066106 (2020)
12. S.S. Stafeev, A.G. Nalimov, A.A. Kovalev, V.D. Zaitsev, V.V. Kotlyar, Circular polarization near the tight focus of linearly polarized light. Photonics **9**, 196 (2022)
13. M. Dienerowitz, M. Mazilu, P.J. Reece, T.F. Krauss, K. Dholakia, Optical vortex trap for resonant confinement of metal nanoparticles. Opt. Express **16**, 4991–4999 (2008)
14. F. Cardano, E. Karimi, L. Marrucci, C. de Lisio, E. Santamato, Generation and dynamics of optical beams with polarization singularities. Opt. Express **21**, 8815–8820 (2013)
15. M.R. Dennis, Polarization singularities in paraxial vector fields: morphology and statistics. Opt. Commun.Commun. **213**, 201–221 (2002)
16. H.A. Robbins, Remark on Stirling's Formula. Am. Math. Mon. **62**, 26–29 (1955)
17. M.V. Berry, M.R. Jeffrey, M. Mansuripur, Orbital and spin angular momentum in conical diffraction. J. Opt. A Pure Appl. Opt. **7**, 685–690 (2005)
18. M.V. Berry, W. Liu, No general relation between phase vortices and orbital angular momentum. J. Phys. A: Math. Theor. **55**, 374001 (2022)
19. A.A. Kovalev, V.V. Kotlyar, S.S. Stafeev, Spin hall effect in the paraxial light beams with multiple polarization singularities. Micromachines **14**, 777 (2023)
20. S. Matta, P. Vayalamkuzhi, N.K. Viswanathan, Study of fractional optical vortex beam in the near-field. Opt. Commun.Commun. **475**, 126268 (2020)

21. V.V. Kotlyar, A.A. Kovalev, S.S. Stafeev, A.G. Nalimov, An asymmetric optical vortex generated by a spiral refractive plate. J. Opt. **15**, 025712 (2013)
22. D.A. Ikonnikov, S.A. Myslivets, V.G. Arkhipkin, A.M. Vyunishev, Near-field evolution of optical vortices and their spatial ordering behind a fork-shaped grating. Photonics **10**, 469 (2023)
23. A.J. Jesus-Silva, E.J.S. Fonseca, J.M. Hickmann, Study of the birth of a vortex at Fraunhofer zone. Opt. Lett. **37**, 4552 (2012)
24. J. Wen, L. Wang, X. Yang, J. Zhang, S. Zhu, Vortex strength and beam propagation factor of fractional vortex beams. Opt. Express **27**, 5893 (2019)
25. S. Liu, S. Chen, S. Wen, H. Luo, Photonics spin hall effect: fundamentals and emergent applications. Opto-Electr. Sci. **1**, 220007 (2022)
26. S. Liu, S. Qi, Y. Li, B. Wei, P. Li, J. Zhao, Controllable oscillated spin Hall effect of Bessel beam realized by liquid crystal Pancharatnam-Berry phase elements. Light. Sci. Appl. **11**, 219 (2022)
27. V.V. Kotlyar, A.A. Kovalev, E.S. Kozlova, A.M. Telegin, Hall effect at the focus of an optical vortex with linear polarization. Micromachines **14**, 788 (2023)
28. A.A. Kovalev, V.V. Kotlyar, Spin hall effect of double-index cylindrical vector beams in a tight focus. Micromachines **14**, 494 (2023)
29. M.R. Dennis, Rows of optical vortices from elliptically perturbing a high-order beam. Opt. Lett. **31**, 1325–1327 (2006)
30. A.A. Kovalev, V.V. Kotlyar, Propagation-invariant laser beams with an array of phase singularities. Phys. Rev. A **103**, 063502 (2021)
31. M.V. Berry, Optical vortices evolving from helicoidal integer and fractional phase steps. J. Opt. A Pure Appl. Opt. **6**, 259–268 (2004)
32. P. Vaity, A. Aadhi, R.P. Singh, Formation of optical vortices through superposition of two Gaussian beams. Appl. Optics **52**, 6652–6656 (2013)
33. I. Freund, Polarization singularity indices in Gaussian laser beams. Opt. Commun.Commun. **201**, 251–270 (2002)
34. V.V. Kotlyar, A.A. Kovalev, S.S. Stafeev, A.G. Nalimov, S. Rasouli, Tightly focusing vector beams containing V-point polarization singularities. Opt. Las. Tech. **145**, 107479 (2022)
35. A.A. Kovalev, V.V. Kotlyar, A.G. Nalimov, Spin hall effect in paraxial vectorial light beams with an infinite number of polarization singularities. Micromachines **14**, 1470 (2023)

Conclusion

In this book, using the example of some laser beams with inhomogeneous polarization, it is shown that in a sharp focus (focus size 1–3 microns for visible light), under certain conditions, subwavelength regions can form, in some of which the light has left elliptical polarization, and in some—right (spin Hall effect). In this case, almost in the same areas at the focus, there is a transverse flux of light energy. In some of them, the energy rotates clockwise, and in some—counterclockwise (orbital Hall effect). Below, for definiteness, we list several specific light fields for which the Hall effect takes place.

With sharp focusing of a Gaussian laser beam with linear polarization in transverse planes, four local regions are formed in which the polarization is elliptical before and after the focus. Moreover, in two diagonal regions, there is left, and in the other two diagonal regions, right elliptical polarization. When passing through the plane of focus, the direction of rotation of the polarization vector in these regions changes to the opposite. At the focus in these regions, the polarization is linear.

New types of beams with high-order circular-azimuth polarization can be obtained by passing light with cylindrical vector polarization through quarter-wave plates. This is a new type of inhomogeneous hybrid polarization that combines the properties of m-order cylindrical polarization and circular polarization. The intensity at the focus for such beams has $2(m + 1)$ local maxima located along a closed contour with zero intensity at the center (on the optical axis). At the focus, there are $4m$ vortices of the transverse energy flow, the centers of which are located between the local intensity maxima. The transverse energy flux changes the direction of rotation $2(2m + 1)$ times when passing in the focus plane around the optical axis, the longitudinal projection of the spin angular momentum at the focus changes sign $4m$ times. In those areas of the focus plane where the transverse energy flux rotates counterclockwise, the longitudinal projection of the spin angular momentum is positive and the polarization vector also rotates counterclockwise. And vice versa, where the energy flow rotates clockwise, there the polarization vector rotates clockwise and the longitudinal projection of the spin angular momentum is negative.

V. V. Kotlyar et al., *Optical Hall Effect in the Sharp Focus of Laser Light*,
https://doi.org/10.1007/978-3-031-64683-6

With sharp focusing of vector cylindrical laser beams of fractional order, in which the linear polarization vector has a different inclination to the horizontal axis in the initial plane of each point, local regions are formed in the focus plane in which the transverse energy flux rotates and the polarization is circular (elliptical). Such regions are formed by pairs with mutually opposite directions of transverse energy rotation. In the areas in focus where the energy rotates, at each point the polarization vector also rotates in the same direction. The separation of regions in focus with different transverse energy flux is a new demonstration of the optical Hall effect.

The spin Hall effect at the focus can be used to capture dielectric microparticles, the size of which is comparable to the size of subwavelength regions in the focus with different directions of rotation of the polarization vector, and rotate these particles around their center of mass in different directions.

MIX
Papier aus verantwortungsvollen Quellen
Paper from responsible sources
FSC® C105338

If you have any concerns about our products,
you can contact us on
ProductSafety@springernature.com

In case Publisher is established outside the EU,
the EU authorized representative is:
Springer Nature Customer Service Center GmbH
Europaplatz 3, 69115 Heidelberg, Germany

Printed by Libri Plureos GmbH
in Hamburg, Germany